E. C. Constable

Metals and Ligand Reactivity

© VCH Verlagsgesellschaft mbH, D-69451 Weinheim (Federal Republic of Germany), 1996

Distribution:
VCH, P. O. Box 10 11 61, D-69451 Weinheim, Federal Republic of Germany
Switzerland: VCH, P. O. Box, CH-4020 Basel, Switzerland
United Kingdom and Ireland: VCH, 8 Wellington Court, Cambridge CB1 1HZ, United Kingdom
USA and Canada: VCH, 220 East 23rd Street, New York, NY 10010-4606, USA
Japan: VCH, Eikow Building, 10-9 Hongo 1-chome, Bunkyo-ku, Tokyo 113, Japan

ISBN 3-527-29277-2 (Softcover)   ISBN 3-527-29278-0 (Hardcover)

Edwin C. Constable

Metals and Ligand Reactivity

An Introduction to the Organic Chemistry of Metal Complexes

New, revised and expanded edition

VCH

Weinheim • New York •
Basel • Cambridge • Tokyo

Prof. Dr. Edwin C. Constable
Institut für Anorganische Chemie
Universität Basel
Spitalstr. 51
CH-4056 Basel
Switzerland
e-mail: CONSTABLE@UBACLU.UNIBAS.CH

This book was carefully produced. Nevertheless, author and publisher do not warrant the information contained therein to be free of errors. Readers are advised to keep in mind that statements, data, illustrations, procedural details or other items may inadvertently be inaccurate.

Published jointly by
VCH Verlagsgesellschaft, Weinheim (Federal Republic of Germany)
VCH Publishers, New York, NY (USA)

The first edition of *Metals and Ligand Reactivity* was published by Ellis Horwood, in 1990.

Cover: The reaction of 1,2-diaminoethane(en) with formaldehyde and ammonia in the presence of Co(II): an example of the introduction of three-dimensional structure into the ligand (see Section 7.2.3 for details).

Editorial Director: Dr. Thomas Mager
Production Manager: Dipl.-Ing. (FH) Hans Jörg Maier

Library of Congress Card No. applied for.

A catalogue record for this book is available from the British Library.

Die Deutsche Bibliothek Cataloguing-in-Publication Data:

Constable, Edwin C.:
Metals and ligand reactivity : an introduction to the organic
chemistry of metal complexes / Edwin C. Constable. – New,
rev. and expanded ed. – Weinheim ; New York ; Basel ;
Cambridge ; Tokyo : VCH, 1995
ISBN 3-527-29277-2 brosch.
ISBN 3-527-29278-0 Gb.

Composition: Mitterweger Werksatz GmbH, D-68723 Plankstadt
Printing: belz-druck gmbh, D-64291 Darmstadt
Cover design: Annemargret Sommer-Meyer/mmad
Printed in the Federal Republic of Germany

To Catherine, Philby and Isis

Preface to the Second Edition

Most of sentiments expressed in the preface to the earlier edition of this book still hold. In the past five years metal-directed chemistry and the use of metal ions to control the stereochemical course of reactions have undergone something of a renaissance. The interface of inorganic and organic chemistry is found in supramolecular chemistry. It is true to say that an understanding of the interactions between metal ions and ligands is at the basis of much contemporary supramolecular and metallosupramolecular chemistry.

The text has been substantially revised, many new examples incorporated and errors corrected. A substantial new chapter dealing with supramolecular chemistry has been incorporated. Once again, a deliberate decision was made to try to limit references to the secondary rather than the primary literature. Where structural data have been presented, the use of the files of the Cambridge Crystallographic Data Centre and the Brookhaven Protein Structure Data Base are gratefully acknowledged.

I should like to thank all of the colleagues and correspondents over the years who have pointed out errors or suggested ways of improving the treatment, and also the students in Cambridge and Basel who have acted as guinea pigs for the various chapters. Especial thanks must go to Dr Diane Smith and my wife, Dr Catherine Housecroft, for their diligent proof reading and infuriating ability to find errors in 'final' versions of the text. Finally, thanks again to Catherine for her tolerance whenever I went off to 'talk to Mac'.

Basel 1995   Ed Constable

Preface to the First Edition*

Inorganic chemistry underwent a renaissance in the late 1950s and early 1960s with the discovery and exploitation of the rich transition metal organometallic chemistry. The great promise of organometallic chemistry lay in the ability to isolate and then utilise novel organic fragments bonded to a transition metal centre.

This book attempts to describe alternative approaches to ligand reactivity involving 'normal' co-ordination complexes as opposed to organometallic compounds. In part, a justification for this view comes from a study of natural systems. With very few exceptions, organometallic compounds are not involved in biological systems; it is equally true that numerous enzymes bind or require metal ions that are essential for their activity. If enzymes can utilise metal ions to perform complex and demanding organic chemical reactions in aqueous, aerobic conditions at ambient temperature and pressure, it would seem to be worthwhile to ask the question whether this is a better approach to catalysis.

This book does not offer solutions to major industrial problems, it does not discuss bio-inorganic systems in any great detail, it does not do more than offer the promise of a future technology. What this book **does** try to do is present an array of 'organic' reactions promoted, catalysed or initiated by metal ions. I hope that the reader is as fascinated as I am by the many and varied reactions that proceed in the presence of metal ions. If it stimulates any reader to delve deeper into this subject, it has served its purpose.

I have attempted to keep this book as readable as possible for the non-specialist reader. In pursuing this goal I have deliberately not included references to individual reactions discussed - for this I apologise to all my colleagues who might recognise their work without attribution. The interested reader will soon be led to the primary literature by following the suggestions for further reading. I have also made the decision to omit detailed presentation of kinetic and activation parameters, in the hope of concentrating upon synthetic utility. Once again, these parameters are readily extracted from the review literature.

Finally, I should like to thank all of my colleagues for their advice, assistance and encouragement during the period of writing. I must thank Dr Ken Seddon for his seminal role in convincing an organic chemist that co-ordination chemistry was not quite interred, and Professor Lord Lewis for indulging me in my initial studies in the area. Of the many colleagues who have contributed, wittingly or otherwise, to the birth pangs of this book, I should like to particularly thank Professor Lord Lewis, Professor Bob Gillard and Dr Len Lindoy. Special thanks go to Malcolm Gerloch for answering all my naïve questions in a (usually) patient manner, to Philip Wood for proof reading, and to Rachael Jones for cajoling a recalcitrant author in such a lady-like manner! And, of course, thanks go to Catherine for helping this book through its infancy, and for persevering in teaching theoretical chemistry to a very suspicious synthetic chemist.

Cambridge 1990   Ed Constable

* Published, in 1990, by Ellis Horwood

Contents

Abbreviations and Some Common Ligands

Ac	CH_3CO
acac	2,4-pentanedionate, acetylacetonate
Ar	aryl group
ATP	adenosine triphosphate
bpy	2,2'-bipyridine
bpym	2,2'-bipyrimidine
^tBu	*tert*-butyl, 1,1-dimethylethyl
CFSE	crystal field stabilisation energy
4,4'-Cl_2bpy	4,4'-dichloro-2,2'-bipyridine
diamsar	diaminosarcophagine
dichlorsar	dichlorosarcophagine
diNOsar	dinitrosarcophagine
dmf	*N,N*-dimethylformamide, Me_2NCHO
dmso	dimethylsulfoxide
δ	absolute right-handed configuration of a chelate ring
Δ	absolute right-handed configuration of a complex
Δ_{oct}	crystal field splitting in an octahedral complex
Δ_{tet}	crystal field splitting in a tetrahedral complex
e_g	a pair of degenerate orbitals resulting from an octahedral crystal field
E	element, electrophile
en	1,2-diaminoethane, ethylenediamine, $H_2NCH_2CH_2NH_2$
Et	ethyl
[H]	reducing agent
Hacac	2,4-pentanedione, acetylacetone
Himid	imidazole
HOMO	highest occupied molecular orbital
L	ligand
λ	absolute left-handed configuration of a chelate ring
Λ	absolute left-handed configuration of a complex
log	decadic logarithm
LUMO	lowest unoccupied molecular orbital
M	metal
Me	methyl
NADH	nicotinamide adenine dinucleotide
5-nitrophen	5-nitro-1,10-phenanthroline
nmr	nuclear magnetic resonance

Nu, Nu⁻	nucleophile
[O]	oxidising agent
Ph	phenyl
phen	1,10-phenanthroline
Pr	propyl
^{i}Pr	*iso*propyl, 2-propyl
py	pyridine
R, R'	alkyl group
r	radius
sep	sepulchrate
S_N1	monomolecular nucleophilic substitution reaction
S_N2	bimolecular substitution reaction
t_{2g}	three degenerate orbitals resulting from an octahedral crystal field
tpy	2,2':6',2"-terpyridine
Ts	4-methylphenylsulfonyl, tosyl
VSEPR	valence shell electron pair repulsion model
w/v	weight per volume
χ	electronegativity

1 Introduction

1.1 General

The didactic approach to chemistry divides the subject into the three major disciplines of organic chemistry, inorganic chemistry and physical chemistry. As originally defined, the distinctions between organic and inorganic chemistry were clear; 'air, water and minerals … are neither produced, nor continued in being, by means of organs; they are therefore called inorganic bodies, and the branch of chemistry relating to them is termed Inorganic Chemistry'[1]. More recently, inorganic chemistry has been considered in terms of the chemistry of elements other than carbon. It has, however, become increasingly apparent that traditional inter-disciplinary barriers are merely conveniences for the presentation of the subject. Indeed, it might be claimed that the overly rigid adoption of such delineations has been to the detriment of the subject as a whole.

Inorganic chemists are concerned with the interactions of atoms, ions and electrons. Such interactions tend to be proximal, and within the electrostatic or covalent bonding regime. One of the major areas of interest is co-ordination chemistry, in which the interaction of a central atom with surrounding atoms, ions or molecules is studied. This chapter acts as a brief introduction to co-ordination chemistry.

1.1.1 Co-ordination Compounds

The terms co-ordination compound, dative covalent compound, complex, or donor-acceptor compound are used synonymously for those compounds formed by the interaction of a molecule containing an empty orbital with one that possesses a filled orbital. The term complex was used to distinguish materials such as $CoBr_3 \cdot 3NH_3$ from 'simple' salts such as $CoBr_3$. Notice the use of a dot to indicate some kind of a dative covalent interaction. The molecule or ion with the filled orbital is termed the *donor* and that with the empty orbital is called the *acceptor*. Examples are known in which only main group compounds are involved (Fig. 1-1).

However, the most common examples of co-ordination compounds are those in which metals (and particularly transition metals) act as the acceptors. The metal may be in any oxidation state, and the donor-acceptor compounds may be neutral or charged (Fig. 1-2).

---

[1] This definition comes from J.P. Bidlake, *Text-book of Elementary Chemistry for the Use of Schools and Junior Students, (with new notation, etc.)*, 7th ed., London, **1879**.

$$F_3B + NMe_3 \rightleftharpoons F_3B.NMe_3$$

$$PF_5 + F^- \rightleftharpoons [PF_6]^-$$

Figure 1-1. The formation of co-ordination compounds may involve only main group donors and acceptors, as shown in the formation of the adduct between trimethylamine and boron trifluoride, and in the formation of the hexafluorophosphate anion.

$$HgCl_2 + Cl^- \rightleftharpoons [HgCl_3]^-$$

$$FeCl_3 + Cl^- \rightleftharpoons [FeCl_4]^-$$

$$OsF_5 + F^- \rightleftharpoons [OsF_6]^-$$

Figure 1-2. Some examples of transition metal co-ordination compounds. In each case, the metal changes co-ordination number, but not oxidation state.

The donor is called a ligand (Latin *ligare,* to bind). Ligands may be neutral, negatively or (more rarely) positively charged. Complex molecules or ions are conventionally indicated within square brackets. For example, the use of square brackets indicates that the salt $[Mn(H_2O)_6][PF_6]_2$ contains discrete $[Mn(H_2O)_6]^{2+}$ and $[PF_6]^-$ ions, rather than Mn^{2+}, P^{5+} and F^- ions and water molecules. The resultant species are described as *co-ordination compounds* or *complexes* (as opposed to *simple* salts such as NaCl or $CoBr_2$). Co-ordination chemistry is usually regarded as the province of the 'inorganic' chemist. Inasmuch as co-ordination compounds have been investigated by inorganic chemists, the emphasis has been upon the behaviour of the metal atoms or ions. The use of a dot to represent the donor-acceptor interaction has been introduced in the formula $F_3B \cdot NMe_3$. Very often, an arrow is also used to indicate this type of interaction; the direction of the arrow indicates the transfer of electrons from the donor to the acceptor. Thus, the representations $F_3B \cdot NMe_3$ and $F_3B \leftarrow NMe_3$ are equivalent.

The foundations of modern co-ordination chemistry were laid by Alfred Werner and Sophus Jørgensen in an extraordinarily productive period around the beginning of the twentieth century; our present understanding of the stereochemistry of complexes, and of the effects of ligand substitution or changing oxidation state, are to a great extent based upon these studies.[2] However, the emphasis in these, and later, studies was upon the properties of the metal, and this was promulgated with the development of theoretical models for such systems. The understanding of the bonding and structure of co-ordination compounds culminated in the singularly successful ligand field and later molecular orbital theories. These models have proved remarkably durable in the rationalisation and prediction of the properties of metal ions and atoms, in isolation or in chemical compounds. The

---

[2] A good collection of papers by and about the beginnings of co-ordination chemistry is to be found in G.B. Kauffman, *Classics in Co-ordination Chemistry.* Part 1, Dover, New York, **1968**; Part 2, Dover, New York, **1976**; Part 3, Dover, New York, **1978**.

fundamental tenet of such models is that the energy levels associated with a metal atom or ion are a sensitive function of the local chemical environment. There is a very considerable accumulation of experimental data to support this view, and ligand field theory provides the basis for our understanding of contemporary transition-metal chemistry. However, it is a necessary corollary that, if the metal-centred energy levels are perturbed by the ligands, the ligand-centred levels are also perturbed by the proximity of the metal. This aspect of the metal–ligand interaction has not been widely investigated by the inorganic community, and forms the subject matter of this book. The remainder of this chapter will present a very brief overview of the models that have been developed to understand the properties of metal complexes.

1.2 Crystal Field Theory

Crystal field theory was introduced in the late 1920s by Bethe and Van Vleck and, although initially formulated and applied by physicists, incorporates the 'inorganic' paradigm of primary concern with the consequences at the metal. The crystal field theory

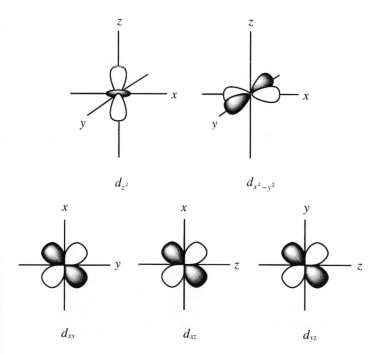

Figure 1-3. Spatial representations of the five real d orbitals indicating their relationship to the principal Cartesian axes.

takes as its remit to explain the behaviour of the metal valence orbitals in the presence of an array of ligands. The principal question addressed was the behaviour of the d orbitals in transition-metal complexes. These orbitals were considered to play a dominant role in determining the properties of the metal ions. Conventional representations of the five (real) d orbitals, based upon hydrogen-like wave-functions, are shown in Fig. 1-3. A brief inspection of Fig. 1-3 reveals that the d orbitals may be considered to be of two types, depending upon their orientation with respect to the principal Cartesian axes.

Two of these orbitals, d_{z^2} and $d_{z^2-y^2}$, possess orbital density aligned along the principal Cartesian axes, whereas the remaining three, d_{xy}, d_{xz} and d_{yz}, are distributed between the principal axes. It is, thus, clear that the various d orbitals might behave differently in non-spherical environments. This is the basis of crystal field theory. The assumption is that the only interaction between the metal and the ligands is electrostatic in origin; no mixing of ligand and metal orbitals is envisaged. The ligands are thus considered to provide a constant electric field about the metal. In an unperturbed metal atom or ion the five d orbitals are degenerate, and such a degeneracy would be expected to persist if the metal were placed in a spherical electric field. However, any real arrangement of ligands about a metal is of lower symmetry than spherical.

The electrostatic repulsions between electrons and a spherical electric field will result in the energy of all of the d electrons being raised in energy with respect to the gas phase atom or ion. The metal in a spherical electric field is our starting point at the left hand side of Figure 1-4. Let us now consider the effect of placing an octahedral array of ligands (or point charges) about that metal. If we place the charges along the Cartesian axes, it is clear that the electrostatic interactions with electrons in the d_{z^2} and $d_{x^2-y^2}$ orbitals (the e_g set) will differ from those with electrons in the d_{xy}, d_{xz} and d_{yz} orbitals (the t_{2g} set).

Electrons in orbitals with large coefficients along the principal axes will experience a greater repulsion from the octahedral array of charges than an electron in an orbital with major coefficients between these axes. These electrostatic interactions will result in the energies of the electrons in the d_{z^2} and $d_{z^2-y^2}$ orbitals being raised with respect to their energies in a spherical electric field. Conversely, those electrons in the d_{xy}, d_{xz} and d_{yz} orbitals will be stabilised with respect to a spherical electric charge distribution. This splitting of the d orbitals into two sets upon the application of a ligand field is indicated in Fig. 1-4.

Each electron placed in one of the t_{2g} orbitals is stabilised by a total of $-\frac{2}{5}\Delta$, whereas those placed in the higher energy e_g orbitals are each destabilised by a total of $\frac{3}{5}\Delta$.

The splitting between the energy levels resulting from an octahedral ligand array is defined as Δ_{oct}. What if we have that other common arrangement of ligands found in metal complexes, a tetrahedron? In a tetrahedral arrangement, the ligands are placed at four of the corners of a cube, and are situated between the Cartesian axes. Although the d_{xy}, d_{xz} and d_{yz} orbitals are not oriented directly towards the corners of the cube, they certainly experience a greater electrostatic interaction with the ligands than the d_{z^2} and $d_{x^2-y^2}$ orbitals (which point towards the centre of a face). This relationship between a tetrahedral ligand arrangement, a cube and the Cartesian axes is indicated in Fig. 1-5.

We use the descriptors t_2 and e for the two groups of orbitals. The subscript g is not used for systems which do not possess an inversion centre. The splitting of the orbitals is thus exactly the reverse of that observed in an octahedral ligand field. Each electron placed in

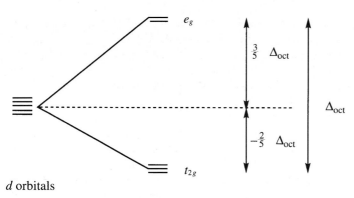

d orbitals

Figure 1-4. The splitting of the *d* orbitals in an octahedral crystal field. The total splitting is given by the quantity Δ_{oct}.

one of the *e* orbitals is stabilised by a total of $-\frac{3}{5}\Delta$, whereas those placed in the higher energy t_2 orbitals are each destabilised by a total of $\frac{2}{5}\Delta$. However, whilst the e_g orbitals are oriented *directly* towards the ligands in an octahedral system, the t_2 orbitals are not pointing *directly* to the corners of a cube in a tetrahedral environment, and so the total destabilisation of the t_2 level in a tetrahedral complex is less than that of the e_g level in an octahedral one. In other words, the splitting, Δ, for a tetrahedral complex is less than that for an octahedral one. Calculations based upon idealised octahedral and tetrahedral geometrics have shown that Δ_{tet} (the splitting of the *d* orbitals in a tetrahedral crystal field) is approximately $\frac{4}{9}\Delta_{oct}$ (Fig. 1-6).

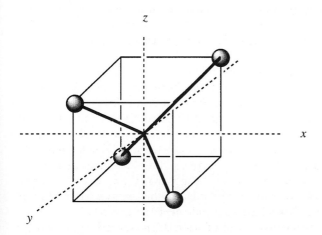

Figure 1-5. The relationship between a tetrahedron and the Cartesian axes within a cube.

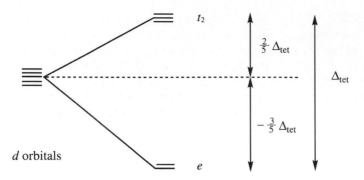

Figure 1-6. The splitting of the d orbitals in a tetrahedral crystal field. The total splitting is given by the quantity Δ_{tet} (which is approximately $\frac{4}{9}\Delta_{\text{oct}}$).

1.2.1 Crystal Field Stabilisation Energies

Let us now consider the consequences of placing electrons into the non-degenerate set of d orbitals resulting from the influence of a non-spherical crystal field. We may calculate the overall stabilisation or destabilisation resulting from the placing of a number of electrons in the orbitals after they have been split by an energy Δ. If we consider a d^1 ion in an octahedral field, we must place the electron in the lowest energy level, which is one of the t_{2g} orbitals. In so doing, we gain a net stabilisation of $-\frac{2}{5}\Delta_{\text{oct}}$ over the same ion in a spherical field of the same magnitude. This quantity of $-\frac{2}{5}\Delta_{\text{oct}}$ is termed the crystal field stabilisation energy (CFSE) for the d^1 ion. We may similarly calculate total CFSE terms for any d^n configuration, simply assigning $-\frac{2}{5}\Delta_{\text{oct}}$ for each electron in a t_{2g} orbital and $+\frac{3}{5}\Delta_{\text{oct}}$ for each electron in an e_g orbital. For example, a d^7 ion with the configuration $t_{2g}^5 e_g^2$ would have a total CFSE of $\{(5 \text{ x} -\frac{2}{5}\Delta_{\text{oct}}) + (2 \text{ x} \frac{3}{5}\Delta_{\text{oct}})\} = -\frac{4}{5}\Delta_{\text{oct}}$. Notice that in this simple discussion we have chosen to ignore the unfavourable energetic terms associated with the electron–electron repulsions experienced between electrons that are placed, spin-paired, in the same orbitals. More sophisticated crystal field treatments attempt to quantify the contribution of this so-called *pairing energy*, P.

We may use exactly similar arguments to obtain total CFSE terms for the various d electron configurations within a tetrahedral crystal field. It is quite possible to construct crystal field splitting diagrams for any of the other geometries commonly adopted in transition-metal complexes, and to calculate the appropriate CFSE terms.

1.2.2 The Spectrochemical Series

The preceding consideration of crystal field theory leads one to ask 'what are the factors which control the magnitude of Δ?'. If we can understand these, we should be able to predict successfully some of the properties of co-ordination compounds. The magnitude of the crystal field experienced will depend both upon the nature of the metal ion and of the co-ordinated ligands. If we concentrate upon the latter factor, what are the properties of the ligands that will influence the magnitude of the crystal field? The energy gap between the t_{2g} and the e_g levels in octahedral complexes is such that it falls within the visible, near ultraviolet or near infrared regions of the electromagnetic spectrum. Thus, an examination of the electronic spectrum of a complex allows us, in principle, to determine directly the magnitude of the energy gap Δ. We should, therefore, be able to rank a series of ligands in the order of the magnitude of Δ which is observed with each. Such an ordering has been made, and is known as the *spectrochemical series* (Fig. 1-7). A very wide range of ligands has been investigated in this context, and some of the data are summarised below, in order of progressively larger crystal field splittings, Δ.

$$I^- < Br^- < SCN^- < Cl^- < S^{2-} < NO_3^- < F^- < OH^- < C_2O_4^{2-} < H_2O$$

$$< NCS^- < py \approx NH_3 < en < bpy \approx NO_2^- \approx phen < PR_3 < CN^- < CO$$

Figure 1-7. The spectrochemical series arranges ligands in the order of their effect upon the magnitude of Δ. Iodide is a weak field ligand and gives small ligand field splittings, whereas carbon monoxide gives a strong field and a large Δ.

This is a generally observed order of orbital splittings, which is more-or-less independent of the metal ion, the oxidation state or the geometry of the complex, although some variations occur. It is also extremely disturbing from a viewpoint of our crystal field model. A ligand to the left of this series is described as a weak field ligand, and one to the right hand end as a strong field ligand. The crystal field model is based upon electrostatic repulsions and, intuitively, one would expect that negatively charged ligands would create a stronger crystal field than neutral ones. The position of the negatively charged halide ions as extremely weak field ligands is, therefore, rather perturbing. Even if we rationalise it in terms of the differing electronegativities of the ligand donor atoms (dubious at best), we are still left with the fact that the hydroxide ion is a weaker field ligand than its parent acid, water, even though the donor atom is the same. Similarly, ammonia is a stronger field ligand than water, although the dipole moment of ammonia (4.90×10^{-30} C m) is less than that of water (6.17×10^{-30} C m). Clearly, something is wrong with the model.

1.2.3 The Nephelauxetic Effect

The spectra of complexes with configurations other than d^1 and d^9 are rather complicated and we cannot always obtain values of Δ directly. Racah has developed methods for the interpretation of the spectra of transition-metal ions in terms of Δ and a number of other parameters. This book is not concerned with the detailed origin and interpretation of these parameters; suffice it to say that one of the parameters, B, is a measure of the electron repulsion terms in the valence shell of the metal ion. It is found that the value of B is a function of the ligands which surround the metal ion, and that it is always lower in a complex than in the free metal ion. In other words, co-ordination of a ligand to a metal ion results in a lowering of electron–electron interactions within the valence shell. Just as we could arrange ligands in the order of the crystal field splittings that they produce to obtain the spectrochemical series, so we can also arrange them in terms of their effect on the Racah *B* parameter. The simplest way to envisage the lessening of the electron repulsion term is to consider the metal-centred orbitals occupied by electrons to have become spatially enlarged in the complex. The sequence so obtained is called the *nephelauxetic series* (from the Greek for 'cloud-expanding'), and is given in Fig. 1-8.

$$F^- < H_2O < NH_3 < en < C_2O_4^{2-} < Cl^- < CN^- < Br^- < N_3^- < I^-$$

Figure 1-8. The nephelauxetic series arranges ligands in the order of their effect upon the Racah B parameter. This equates to the degree of covalency within the metal–ligand bonding. Notice that the order is dramatically different from that of the spectrochemical series in Figure 1-7.

It is clear that this differs from the spectrochemical series in some rather important respects; whereas the larger halide ions are relatively weak field ligands, they exert a powerful nephelauxetic effect. Similarly, cyanide is a very strong field ligand, but it only exerts a moderate nephelauxetic effect. A moment's consideration reveals that the presence of a strong nephelauxetic effect is associated with large, and hence easily polarisable, ligands. This brings us to the crux of the matter: if the metal is sensitive to the polarisability of the ligand, then our simple model that portrays the ligands as simple point charges must be over-naïve. There must be, to some extent, orbital overlap between the ligand and the metal.

1.3 Molecular Orbital and Ligand Field Theories

The above discussion of crystal field theory and its deficiencies has indicated that as a model it underestimates the covalent character of the metal–ligand interaction. In fact, it is quite unreasonable to expect a purely electrostatic approach to bear any resemblance to the reality of the interactions within a complex. As Jørgensen stated, the crystal field model 'combined the rather unusual properties of giving an excellent phenomenological classification of the energy levels of partly filled *d* and *f* shells of transition group complexes and having an absolutely unreasonable physical basis!'.[3] It is remarkable that it does work so well. Whereas the crystal field model of co-ordination compounds is based upon an essentially ionic description of the interactions between metal ions and ligands, the alternative approaches to their description emphasise (some might say over-emphasise) the contribution of covalent interactions to the metal–ligand bonds. It is these widely (and perhaps uncritically) utilised models which we will consider in this section.

---

[3] This interesting comment comes from C.K. Jørgensen, *Modern Aspects of Ligand Field Theory*, Elsevier, New York, **1971.**

1.3.1 Molecular Wave Functions and Valence Bond Theory

The aim of molecular orbital theory is to provide a complete description of the energies of electrons and nuclei in molecules. The principles of the method are simple; a partial differential equation is set up, the solutions to which are the allowed energy levels of the system. However, the practice is rather different, and, just as it is impossible (at present) to obtain *exact* solutions to the wave equations for polyelectronic atoms, so it is not possible to obtain *exact* solutions for molecular species. Accordingly, the application of molecular orbital theory to molecules is in a regime of successive approximations. Numerous rigorous mathematical methods have been utilised in the effort to obtain ever more accurate solutions to the wave equations. This book is not concerned with the details of the methods which have been used, but only with their results.

The valence bond model constructs 'hybrid' orbitals which contain various fractions of the character of the 'pure' component orbitals. These hybrid orbitals are constructed such that they possess the correct spatial characteristics for the formation of bonds. The bonding is treated in terms of localised two-electron two-centre interactions between atoms. As applied to first-row transition metals, the valence bond approach considers that the $4s$, $4p$ and $3d$ orbitals are all available for bonding. To obtain an octahedral complex, two $3d$, the $4s$ and the three $4p$ metal orbitals are 'mixed' to give six spatially-equivalent directed d^2sp^3 hybrid orbitals, which are oriented with electron density along the principal Cartesian axes (Fig. 1-9).

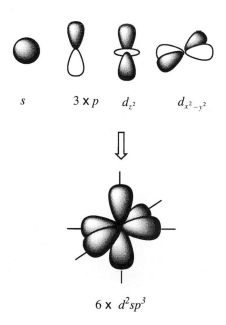

s $3 \times p$ d_{z^2} $d_{x^2-y^2}$

$6 \times d^2sp^3$

Figure 1-9. The construction of six d^2sp^3 hybrid orbitals from six atomic orbitals for use in an octahedral complex.

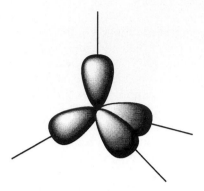

Figure 1-10. The four *sp³* hybrid orbitals used in a tetrahedral complex.

Similarly, the bonding in tetrahedral complexes of first row transition-metal ions is considered in terms of four equivalent sp^3 hybrid orbitals (which are constructed from the $4s$ and $4p$ orbitals of the metal) oriented towards the vertices of a tetrahedron (Fig. 1-10). For a further discussion of the application of the valence bond method to transition-metal complexes, the reader is referred to publications by Pauling.[4] The essential feature is that the bonding consists of localised, two-centre two-electron bonds.

Valence bond theory is somewhat out of favour at present; a number of the spectroscopic and magnetic properties of transition-metal complexes are not simply explained by the model. Similarly, there are a number of compounds (with benzene as an organic archetype) which cannot be adequately portrayed by a single two-centre two-electron bonding representation. Valence bond theory explains these compounds in terms of *resonance* between various forms. This is the origin of the tautomeric forms so frequently encountered in organic chemistry texts. The structures of some common ligands which are represented by a number of resonance forms are shown in Fig. 1-11.

1.3.2 Molecular Orbital Theory

At its simplest, molecular orbital theory considers the symmetry properties of all of the atomic orbitals of all of the component atoms of a molecule. The basis of the calculation is to combine the (approximate) energies and wave-functions of the appropriate atomic orbitals to obtain the best possible approximations for the energies and wave-functions of

---

[4] The best source for an overview of Pauling's ideas and contribution to science is L. Pauling, *The Nature of the Chemical Bond,* 3rd ed., Cornell University Press, Ithaca, **1960**.

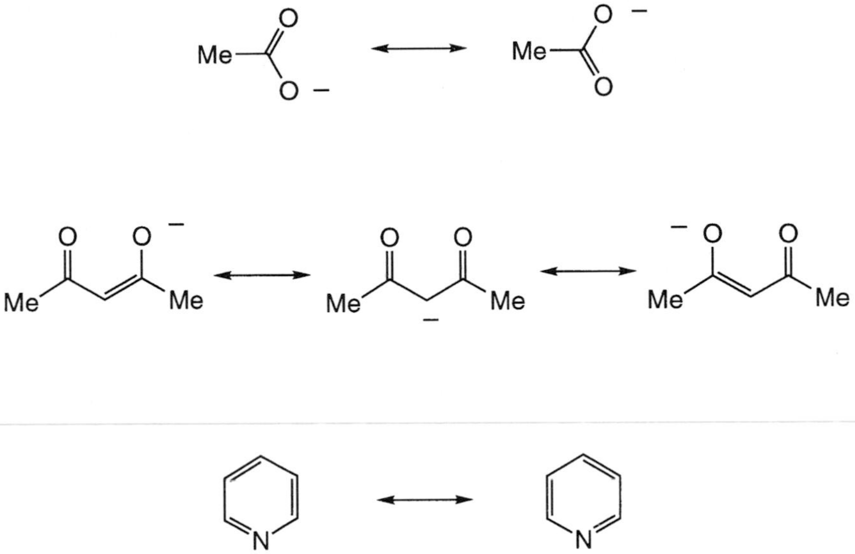

Figure 1-11. The resonance forms of some ligands that cannot be represented by a single valence bond structure (acetate anion, acetylacetonate anion and pyridine).

the molecules. Molecular orbital methods are widely used, and when we apply this model to the first row transition metals we often limit ourselves to a consideration of the valence orbitals, the $4s$, $4p$ and $3d$ levels. More sophisticated methods have been developed, but we will not be explicitly concerned with these.

The sequence of energy levels obtained from a simple molecular orbital analysis of an octahedral complex is presented in Fig. 1-12. The central portion of this diagram, with the t_{2g} and e_g levels, closely resembles that derived from the crystal field model, although some differences are now apparent. The t_{2g} level is now seen to be non-bonding, whilst the antibonding nature of the $e_g{}^*$ levels (with respect to the metal–ligand interaction) is stressed. If the calculations can be performed to a sufficiently high level that the numerical results can be believed, they provide a complete description of the molecule. Such a description does not possess the benefit of the simplicity of the valence bond model.

Representations of the molecular orbitals indicated by the a_{1g} and e_g levels of Fig. 1-12 are shown in Fig. 1-13. They each contain contributions from metal orbitals (the s for the a_{1g}, and the d_{z^2} and the $d_{x^2-y^2}$ for the e_g) and from some or all of the ligands. It is difficult to equate these multi-centred orbitals to the more familiar individual metal–ligand interactions and bonds. The majority of chemists prefer to think in terms of localised bonding, rather than in terms of electrons delocalised over the metal and all six ligands.

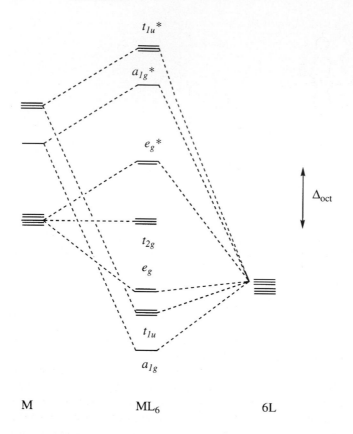

Figure 1-12. Simplified molecular orbital diagram for the formation of an octahedral ML_6 complex in which there are no π-bonding interactions between metal and ligand. The labels on the molecular orbitals refer to their symmetries. Notice the central region may be equated to the crystal field splitting of the d orbitals.

Before leaving this brief introduction to molecular orbital theory, it is worth stressing one point. This model constructs a series of new molecular orbitals by the combination of metal and ligand orbitals, and it is fundamental to the scheme that the ligand energy levels and bonding are, *and must be*, altered upon co-ordination. Whilst the crystal field model probably over-emphasises the ionic contribution to the metal-ligand interaction, the molecular orbital models probably over-emphasise the covalent nature.

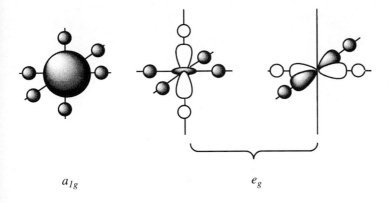

a_{1g} e_g

Figure 1-13. Representations of the a_{1g} and e_g orbitals of Figure 1-12. The ligand donor orbitals are shown as simple s-type orbitals. The shading represents the relative phases of the orbitals. The two molecular orbitals of the e_g level are degenerate.

1.3.3 Ligand Field Theory

One of the criticisms that we levelled at crystal field theory was our inability to explain the observed sequence of ligands in the spectrochemical series. We were confused about the relative positions of water and hydroxide ion, and also about the origin of the very strong crystal fields associated with ligands such as cyanide, 2,2'-bipyridine and carbon monoxide. Does our 'new' molecular orbital theory assist us in the understanding of these observations? We would expect hydroxide to be a stronger σ-donor than water since the electrostatic interaction is greater (although it can be argued that we know nothing *a priori* about the molecular orbitals); similarly, carbon monoxide is not expected to be a particularly strong σ–donor, nor should 2,2'-bipyridine be so much better than ammonia.

We define the spectrochemical series in terms of the magnitude of Δ, the separation between the t_{2g} and $e_g{}^*$ levels. Strong field ligands result in a larger gap between the t_{2g} and $e_g{}^*$ levels. Comparison of Figs. 1-4 and 1-12 shows that this energy gap may be increased in one of two ways; either the level of the $e_g{}^*$ orbitals may be raised, or the energy of the t_{2g} orbitals may be lowered. So far, we have seen the t_{2g} levels as essentially non-bonding, and having little interaction with the ligands. The crystal field model concentrates upon the raising of the $e_g{}^*$ level by the electrostatic interaction(s) between the ligand and the d electrons of the metal. If we are to gain anything more from the molecular orbital model than from the crystal field one, we must explain how lowering of the t_{2g} orbitals might be achieved. In particular, we have to consider how the ligands can interact with those orbitals which are not directed *along* the metal–ligand vector. Consider a ligand which undergoes a σ-bonding interaction with a first row transition-metal ion along one of the principal axes. In the previous discussion of molecular orbital theory, it was clear that the interaction is predominantly with the $4s$ and $4p$ orbitals, or with the e_g set of the $3d$ orbitals. If the ligand donor atom also possesses a p orbital or a π-bonding component orthogonal to the metal–ligand vector, then an interaction between the t_{2g} set and the p orbital may occur (Fig. 1-14). The d orbital may be in phase or out of phase with the ligand

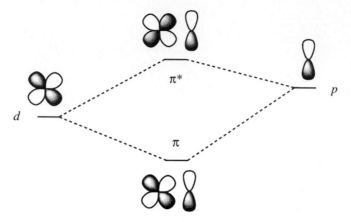

Figure 1-14. Bonding and antibonding combinations of metal t_{2g} orbitals with p orbitals on a ligand giving rise to new molecular orbitals of π- and π^*-symmetry.

p orbital. The interaction, thus, gives rise to a stabilised level of π-symmetry and a destabilised level of π^*-symmetry.

Obviously, the detailed nature of these π-bonding interactions will depend upon the precise occupancy of the metal t_{2g} and ligand p orbitals, and also upon their relative energies and the efficiency of the overlap between them. If the metal t_{2g} levels are filled and the ligand levels are empty, then the interaction will correspond to a net transfer of electron density from the metal to the ligand. Only the π-level will be occupied. In contrast, if the metal t_{2g} levels are vacant and the ligand levels are occupied, then the interaction corresponds to the transfer of electron density from the ligand to the metal. Once again, only the π-bonding level will be occupied. We will see shortly that these two limiting cases allow the ligand to stabilise low and high oxidation states of the metal ion, respectively.

We now begin to understand how the π-interaction can explain the observed spectrochemical series. The 'anomalous' strong field ligands, such as carbon monoxide, pyridine, 2,2'-bipyridine and CN^-, all possess vacant orbitals of π-symmetry of similar energy to the metal valence shell orbitals (Fig. 1-15).

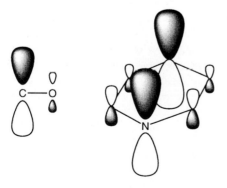

Figure 1-15. The lowest lying π^*-levels of carbon monoxide and pyridine. The size of the orbitals represents their coefficients in the molecular orbital.

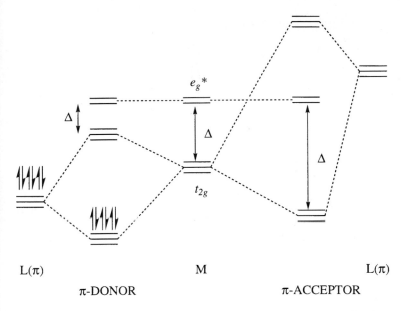

L(π)   M   L(π)

π-DONOR π-ACCEPTOR

Figure 1-16. Correlation diagram showing the ways in which π-bonding interactions with a ligand may vary the magnitude of Δ for an octahedral complex. Vacant, high-lying π-orbitals on the ligand increase the magnitude of Δ, whilst filled, low-lying π-orbitals decrease the splitting.

The orbitals of importance are the empty antibonding π^*-levels of the ligands. The magnitude of Δ increases as a consequence of the interaction of the t_{2g} orbitals of the metal with the π^*-levels of the ligand and their resultant lowering in energy as shown in Fig. 1-16.

Thus, we predict that π-acceptor ligands will be able to stabilise low oxidation states, which possess a high electron density on the metal, by providing a pathway to remove charge from the metal to a more elctronegative centre. Strong field ligands, such as carbon monoxide or 2,2'-bipyridine, are efficient π-acceptors, and act in this manner. In contrast, π-donor ligands are expected to stabilise high oxidation states, which possess al low elctron density at the metal, by providing a source of electrons which may be shared with the metal. Ligands such as fluoride or oxide are efficient π-donors. This combination of molecular orbital considerations with crystal field arguments constitutes ligand field theory as it is commonly accepted. To recap, the inclusion of interactions between the metal and the ligand of π-symmetry enables one to explain the position of ligands in the spectrochemical series, and to make far-reaching predictions regarding the chemical effects of ligands upon the chemistry of complexes.

1.4 The Electroneutrality Principle

We now have two models for the metal–ligand interaction, which place differing importance upon the electrostatic and covalent components. What can we do to determine the relative importance of these two limiting depictions of the metal–ligand bond? This is a problem which was approached by Pauling from a slightly different viewpoint. The valence bond picture of co-ordination compounds results in a large build-up of negative charge upon the metal ion; for example, in the $[Fe(CN)_6]^{4-}$ ion the metal has a formal -4 charge placed upon it. As metals are normally considered to be electropositive this is not a particularly favourable situation. Pauling recognised this and framed the electroneutrality principle, which suggested that the net charge on any metal or ligand would end up close to zero. This can be achieved by a polarisation of the metal–ligand bond resulting from the differences in electronegativity between the metal and the ligand donor atoms, or by back-donation from the metal to the ligand, invoking a π-type interaction. As far as we are concerned, this has important consequences for the chemistry of the metal-bound ligand. The charge on the metal ion will play an important role in determining the polarity of the metal–ligand bond, and the degree of interaction observed with potential π-bonding ligands. Low formal charges upon the metal will result in a large degree of metal-to-ligand charge donation, whereas high oxidation states will favour ligand-to-metal charge donation.

1.5 Reprise

We have now described the theoretical models which have been developed for the metal–ligand interaction, and have seen that they are predominantly concerned with the effect upon the metal ion. Although the majority of the preceding discussion has concerned itself with the effect of co-ordination upon the metal d orbitals in transition-metal complexes, it is apparent that extension to other orbitals in main group metal ions is facile. The special interest of the transition-metal ions is that they possess orbitals of correct symmetry for a π-type interaction with ligand orbitals. This is of particular relevance, since it results in a very major perturbation of the ligand bonding scheme. In the following chapter we will consider the ways in which co-ordination of a ligand to a metal ion is likely to affect the chemistry of the ligand.

1.6 Principles of Reactivity and Mechanism

The fundamental feature exhibited by any chemical reaction is the movement of electrons from one region of space to another. This is seen in the essentially complete transfer of an electron from sodium to chlorine in the formation of ionic sodium chloride from the elements, or in the sharing of an electron between hydrogen and chlorine in the formation of molecular hydrogen chloride. Inasmuch as the movement of electrons characterises a chemical reaction, so anything that alters the distribution, movement or availability of electrons must, in principle, be capable of affecting reactivity. We showed in the first part of this chapter that it was a necessary consequence of the metal-ligand interaction that the electronic arrangement of the ligand must be perturbed, to some extent, upon co-ordination. If changes in the electronic structure may result in changes in the reactivity, an obvious corollary is that the reactivity of co-ordinated ligands may not necessarily parallel that of the free ligands. The ways in which the reactivity may be altered are discussed in the next chapter, and the manifestations of these effects constitute the remainder of this book.

1.6.1 Thermodynamic and Kinetic Constraints Upon Reactivity

One of the most fundamental concepts of chemistry is the distinction between kinetic and thermodynamic factors; nonetheless, such arguments are frequently ignored, or at best only tacitly considered, in wider discussions of reactivity. Chemical thermodynamics is concerned with the energetic relationships between chemical species. The most useful parameter is the Gibbs free energy, $G^{\oplus}$, which, like all thermodynamic terms, is based on an arbitrary scale placing a value of zero upon pure elements in their stable standard states at 298 K and 1 atmosphere pressure. Differences between free energies are denoted by $\Delta G^{\oplus}$, as shown in Eq. (1.1).

$$\Delta G^{\oplus} = G^{\oplus}_{products} - G^{\oplus}_{reactants} \qquad (1.1)$$

Gibbs' free energy terms, like the other thermodynamic parameters commonly encountered ($\Delta H^{\oplus}$, $\Delta S^{\oplus}$, $\Delta V^{\oplus}$) obey simple mathematical rules. The larger the value of $\Delta G^{\oplus}$, the larger the energy gap between the two chemical states. Values of $\Delta G^{\oplus}$ may be either positive or negative; a negative value corresponds to a net release of energy. This is favourable. A negative value of $\Delta G^{\oplus}$ is associated with reactions which require no external input of energy (downhill reactions in the terminology often adopted). The more negative the value of $\Delta G^{\oplus}$, the greater the amount of energy released. The Gibbs free energy is usefully related to the equilibrium constant, K, as in Eq. (1.2)

$$\Delta G^{\oplus} = -RT\ln K \text{ or } K = \exp(-\Delta G^{\oplus}/RT) \qquad (1.2)$$

Does a knowledge of $G^{\oplus}$ values for products and reactants, and hence $\Delta G^{\oplus}$ for the reaction, enable us to make any predictions about the chemical behaviour of reactant systems? To a certain extent, it does. Consider the reaction of metallic sodium with gaseous chlorine (Eq. 1.3), which is strongly exothermic.

$$2Na(s) + Cl_2(g) \rightarrow 2NaCl(s) \qquad (1.3)$$

$$\Delta G^{\oplus}_{reaction} = \Sigma \, G^{\oplus}_{products} - \Sigma \, G^{\oplus}_{reactants} = -384 \text{ kJ mol}^{-1}.$$

We would thus predict that chlorine gas reacts with sodium metal at 298 K to give sodium chloride, and, of course, our prediction is correct. However, it is also clear that the existence of a negative ΔG° is not the sole factor affecting reactivity.

Hydrogen bromide is a well-known chemical species, but ΔG° for the dissociation reaction is -106 kJ mol^{-1}, and yet hydrogen bromide is an isolable compound (Eq. 1.4)!

$$2HBr\ (g) \rightarrow H_2\ (g) + Br_2\ (g) \qquad\qquad (1.4)$$

If we consider organic compounds, nearly all are thermodynamically unstable in air (i.e., possess a negative $\Delta G^\circ_{reaction}$) with respect to conversion to carbon dioxide and water, but the book you are reading has not (I hope!) spontaneously burst into flames. Even more worrying, at least as far as we are concerned, is to consider the odd assortment of organic compounds and inorganic salts that constitutes a living system. There is a massive negative free energy term which favours conversion to carbon dioxide, rust and sea water! Clearly, all that the free energy term tells us is the total energy change involved on going from one state to another. However, although we know the amount of energy released or absorbed during the chemical change, we have no concept of the rate at which that energy change occurs. This book could either burst into flames, or oxidise slowly over a century; the end result is the same, but the immediate temporal effect upon the reader is rather different!

We thus need to introduce the concept of rate of change, which is directly related to the mechanism of reaction. If we return to one of our anthropomorphic models, thermodynamics tells us that Paris is 330 km from London, but it does not tell us how long it takes to travel between the two. This is the role of kinetics, which describes the route taken. In a chemical system, the route which is followed is described by the mechanism. Inherent in the concept of a mechanism is the presence of energy barriers between reactants and products. This is depicted in Figure 1.17. The height of the energy barrier above the reactants is depicted by the term $\Delta G^{\circ\ddagger}$, where the double dagger symbol indicates an activation parameter. Clearly, the overall energy change for the reaction cannot be affected by the magnitude of $\Delta G^{\circ\ddagger}$, and whatever happens along the route from reactants to products, we still start and finish at the same points. However, $\Delta G^{\circ\ddagger}$ is an energy term, and for a molecule to pass over this barrier it must possess energy equal to, or in excess of, $\Delta G^{\circ\ddagger}$. In a very few cases involving small atoms such as hydrogen, quantum mechanical tunnelling processes may allow mechanisms in which it is not necessary to pass over the energy barrier. In most cases, however, energy must be provided to the system by thermal, photochemical or other means such that molecules can pass over the energy barrier.

Although we cannot relate the thermodynamic free energy change for a reaction directly to the rate of the reaction, it is always useful to remember that the thermodynamic stability constant may be related to the rate constants for the forward (k_f) and backward (k_b) reactions by Eq. (1.5).

$$K = k_f / k_b \qquad\qquad (1.5)$$

The more molecules which possess available energy in excess of $\Delta G^{\circ\ddagger}$, the more can pass over this barrier to give products. In other words, the rate of the reaction is proportional to the magnitude of the energy barrier. The larger this barrier, the fewer molecules

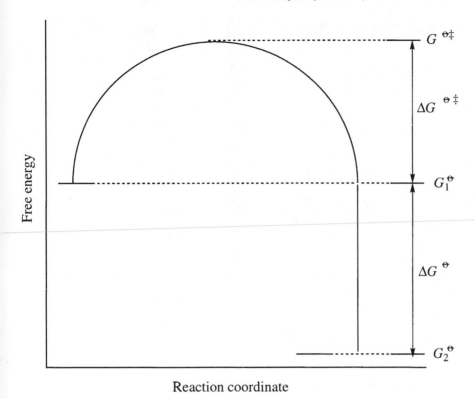

Reaction coordinate

Figure 1-17. The relationships between the various free energy terms involved in a chemical reaction.

that possess enough available energy to cross it, and the slower the rate of reaction. It is because the activation energy is high that this book does not burst into flames, and that living systems are persistent. A description of the way in which atoms, ions and molecules behave as they are transformed from reactants to products is termed the mechanism of the reaction. Why should there be an energy barrier separating the reactants from the products? This arises from our original picture of chemical reactions resulting from the movement of electrons from one equilibrium position to another. Before a chemical reaction can occur, it is necessary to perturb the nuclei and electrons such that they are in a configuration unstable with respect to the products. Since, by definition, any isolable chemical species must be at, or near, the bottom of a local potential energy minimum, it is necessary to provide energy to escape from this potential well before reaction can occur. This may occur by elongation or contraction of internuclear distances, promotion of electrons to higher energy levels, adoption of non-equilibrium conformations or any of a range of other processes. We are now in a position to consider how a metal ion may affect the reactivity of a ligand which is co-ordinated to it. In the light of the previous discussion, we see that any effects may be expressed in one of two ways.

Any effects operative upon the ground state of the products or reactants may be characterised as thermodynamic in origin. In contrast, any effect that is operative upon the transition state or upon the interaction of reactants may be described as kinetic. Although these two effects are readily distinguished in principle, the practice is frequently not so simple! This distinction is clarified slightly when we consider our description of ligand field theory. The thermodynamic effect is seen in the stabilisation or destabilisation of metal orbital energies, i.e., the crystal field (or ligand field) stabilisation energies. These are operative in both product and reactant complexes and will have an effect upon the total energy change associated with the reaction. It is also apparent that ligand field effects may be expressed in the transition state, where they are much more difficult to quantify. The transition state is, by definition, transitory; consequently we do not have the detailed knowledge of the geometry and internuclear distances required for ligand field analyses. Approximations can be made using parameters extrapolated from isolable complexes of unusual stereochemistry or co-ordination number. Algebraic analyses of the crystal fields expected for various geometries have given accurate orbital splitting patterns, but these give no quantitative measure of the absolute magnitude of the crystal field splitting.

There is another, rather more subtle, effect which may be expressed in the reaction rate. The height of the activation barrier is directly related to the availability of a mechanism. A very high barrier reflects a transition state which is unfavourable (for whatever reason). A reaction cannot normally proceed if a relatively low energy mechanism cannot be provided. This is why the book does not inflame. Since a chemical reaction involves the making and breaking of bonds, we would expect the nature and character of the orbitals of the reactants to have a profound effect upon the activation barrier. If the orbitals of the reactants overlap well, then we might expect a low barrier; if they overlap only very poorly, then the barrier is expected to be high. This leads us to anticipate a dependence of the reaction rate on the symmetry properties of the orbitals available. In favourable cases, the majority of the other contributions to the energy changes occurring on passing through the transition state cancel out, leaving the *symmetry* terms as the dominant factor. This is exactly the situation in which *frontier orbital theory* is applicable. This is a model in which the major factors affecting reactivity are considered to be the relative energies and symmetries of the highest occupied and lowest unoccupied molecular orbitals. It is clear that ligand field theory may be regarded as a form of frontier orbital theory – it has as its basis the assumption that the bulk of the properties of the transition metals may be explained from a consideration of the symmetry of the valence orbitals.

1.7 Summary

In this chapter we have briefly discussed the models which have been developed for the description of co-ordination compounds, and have stressed the consequences that the metal–ligand interaction may have for the ligand. In the remainder of this book we will consider the detailed chemical consequences of the metal–ligand interaction. The suggestions for further reading given below will allow the interested reader to delve deeper into co-ordination chemistry should he or she so wish.

Suggestions for further reading

1. F. Basolo, R.C. Johnson, *Co-ordination Chemistry*, Science Reviews, **1986**.
2. M. Gerloch, E.C. Constable, *Transition-Metal Chemistry*, VCH, Weinheim, **1994**.
3. R.L. DeKock, H.B. Gray, *Chemical Structure and Bonding*, Benjamin/Cummings, Menlo Park, **1980**.
4. J.E. Huheey, E.A. Keiter, R.L. Keiter, *Inorganic Chemistry*, 4th ed., Harper Collins, New York, **1993**.
5. W.W. Porterfield, *Inorganic Chemistry*, Addison-Wesley, Reading (Mass.), **1984**.

These basic texts describing co-ordination chemistry provide simple and easy-to-read introductions to this area of chemistry.

2 The Metal-Ligand Interaction

2.1 General

The previous chapter consisted of a résumé of contemporary models for the bonding in transition-metal (and other) co-ordination compounds. It was stressed throughout the account that the emphasis of most such approaches was upon the properties of the metal ion. In this chapter we consider the ways in which a metal ion might affect the properties of a ligand to which it is co-ordinated. Whereas the effect of the ligands upon a metal ion is relatively well understood, and moderately easily quantified, the converse effect is not so. There are a number of reasons for this. Firstly, a metal ion is a mononuclear centre, and, to a certain extent, the orbitals have many of the properties of hydrogen-like atomic orbitals. In contrast, ligands are frequently polynuclear molecular species, in which the orbitals of the donor atoms are also involved in intra-ligand covalent bonding. Secondly, the ligand orbitals may be more or less polarisable than metal orbitals. Thus, whereas metal d orbitals remain more-or-less recognisable as such in co-ordination compounds (as seen from d-d electronic spectra), ligand donor orbitals become rather more perturbed. Thus, it is less meaningful to talk of an electron in the p orbital of a nitrogen atom in the complex cation $[Ni(en)_3]^{2+}$ than it is to discuss an electron in the nickel $3d$ orbitals of the same species.

In particular, the change from lone pair to bonding pair, consequent upon the formation of the co-ordinate bond, has effects upon the other groups which might be bonded to the ligand donor atom(s). Accordingly, we do not have such readily-accessible semi-quantitative methods as crystal field or ligand field theories available for investigating the effect of co-ordination to a metal upon the ligand. Molecular orbital calculations will tell us about the electronic effects at the ligands, but need to be carried out at a sophisticated level for the results to be quantitatively useful. Once again, the molecular orbital description of a complex in terms of multicentre orbitals is alien to our experience; it is a particular problem if we are trying to compare a free ligand with a co-ordinated one. In general, we will only refer to the results of molecular orbital calculations where they provide some useful addition to, or explanation of, experimental data.

In the absence of general quantitative methods, we need to resort to qualitative descriptions and models to describe the changes which result from co-ordination of a ligand to a metal. Let us now consider the detailed manner in which a metal may alter the properties of a co-ordinated ligand.

- *Conformational Changes*

The simplest change that can occur is in the equilibrium conformation of the co-ordinated ligand with respect to the equilibrium conformation of the free ligand. This is a natural consequence of the involvement of a ligand lone pair in bonding to a metal. Conversion of a lone pair to a bonding pair changes the non-bonded interactions in the molecule, and we expect to observe concomitant changes in bond lengths, bond angles and molecular geometry.

- *Polarisation Changes*

If the metal ion to which a ligand is co-ordinated is in a non-zero oxidation state, it will exert an electrostatic effect upon the bonding electrons of the ligand. This will result in the induction of a net permanent dipole in the ligand, with any associated chemical and physical effects. Even zero-oxidation state metal centres may induce a polarisation in the ligand through electronegativity or induced dipole-dipole effects.

- *π-Bonding Changes*

The introduction of π-bonding interactions between the metal and the ligand results in a metal-to-ligand or ligand-to-metal transfer of electron density. This occurs in accord with the electroneutrality principle, and in many cases, it opposes the polarisation effects of the metal ion. Furthermore, the effect will be expressed in orbitals possessing rather specific symmetry properties which may play important roles in the reactivity of the ligands.

These effects, and the combinations of them, are the origins of the modification of ligand reactivity which is observed in co-ordination compounds. An understanding of these effects allows a rationalisation of the chemistry occurring in bio-inorganic and organometallic systems, and gives an insight into the biogenetic processes leading to the evolution of particular metalloenzymes. The simplistic scheme described above proves to be remarkably versatile in describing an enormous range of ligand reactions.

Before we consider the ways in which these effects may be expressed, it is worth mentioning that a number of other classification schemes have been developed to explain the reactivity of co-ordinated ligands. These schemes are most often based upon the observed reactivity of the co-ordinated ligand, and as such are rather more complex than the simple scheme presented above, which is derived from a consideration of the origins of the interaction. For further details of these, the reader is referred to the suggestions for further reading at the end of this chapter.

2.2 Conformational Changes

What are the ways in which co-ordination to a metal ion may affect the conformation of a ligand, and what are the consequences of these changes? The simplest effect is expressed in the ground state equilibrium geometry of the co-ordinated ligand. The bond angles and bond lengths in co-ordinated ligands may differ from those in the free ligand; indeed, it would, perhaps, be more remarkable if they were unchanged. Changes of this type may be detected by a gamut of spectroscopic and structural techniques, and *may* be reflected in the reactivity of the ligand. If the ligand is a polydentate species which binds to the metal through two or more sites, then the entire ligand conformation may be different from that of the free ligand. The adoption of the geometry required for all the donor atoms to be in the proximity of the metal may result in very major changes from the equilibrium solution geometry of the free ligand. Once again, this may be reflected in the ground state spectroscopic properties, or in the reactivity of the ligand. Finally, the chemical properties of the ligand may be rather markedly altered by the "masking" of the lone pair. This might be seen as a reduction in the nucleophilicity or basicity of the ligand, or in the enhancement of some other, previously unobserved, property.

2.2.1 Steric Consequences of Metal–Ligand Bonding

By definition, co-ordination of a ligand to a metal centre involves the donation of electrons from the ligand to the metal. This is the fundamental feature of the donor–acceptor model of co-ordination. The number of electrons donated by the ligand may vary, as may their ultimate origin in bonding, non-bonding or antibonding levels of the ligand. This feature of co-ordination has a number of consequences for the ligand, based upon the redistribution of electrons at the donor atom.

To exemplify this, let us consider a ligand which possesses a non-bonding lone pair of electrons at the donor atom. Upon co-ordination to a metal, this lone pair is converted into a bonding pair, and the total electron density at the ligand donor atom is reduced. This is illustrated in Fig. 2-1. This sharing of electrons has a number of consequences.

How might this interaction affect the geometry of the ligand? If the X and Y groups directly attached to the ligand donor atom (L) are bulky and the other ligands attached to the metal are bulky, steric repulsions will tend to reduce the XLY bond angle in the complex. In contrast, the valence shell electron pair repulsion model (VSEPR) is based upon the assumption that the electrostatic repulsions between electrons in a lone pair and those in a bond are greater than those experienced between electrons in two similarly placed bonds. Applied to the system depicted in Fig. 2-1, in which a lone pair in the free ligand is converted to a bond pair in the complex, the VSEPR model predicts that co-ordination of the ligand to a metal should result in a decrease in the non-bonded interactions between the XL and YL bond pairs and the 'lone pair' on L, with con-

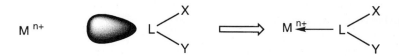

Figure 2-1. Schematic representation of the formation of a co-ordination compound, emphasising the conversion of the ligand lone pair to a bond pair.

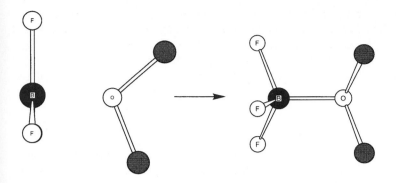

Figure 2-2. Changes in geometry occurring upon the formation of an adduct between boron trifluoride and dimethyl ether.

comitant increases in the XLY bond angle. Finally, if the metal possesses a charge, bond length changes might be expected in the ligand. These will be discussed in more detail in Section 2.3.

What effects (if any) are observed in real donor-acceptor systems? The bulk of the evidence suggests that the net effect upon the geometry of the ligand is relatively minor, although rather more dramatic changes occur at the acceptor molecule. Thus, upon forming the adduct $Me_2O \cdot BF_3$, the BF_3 changes from C_3 symmetry to distorted tetrahedral (C_{3v}) symmetry, whereas the geometry of the dimethyl ether is almost unchanged (Fig. 2-2). This pattern of changes is commonly observed in both main-group and transition-metal donor-acceptor compounds.

Detailed studies of the geometry of water molecules in $[M(H_2O)_6]^{n+}$ cations have shown that significant and systematic variations in H-O bond lengths and in H-O-H bond angles exist. The arrangement of the two hydrogen atoms and the metal centre about an individual oxygen atom may be either trigonal planar or trigonal pyramidal. In co-ordinated trigonal planar water molecules, the H-O-H bond angles increase (in the range $106 - 114°$) as the M-O distance shortens. This angle is also dependent to some extent upon the oxidation state of the metal centre. Similar observations have been made for co-ordinated ammine ligands.

2.2.2 Geometrical Consequences of Chelation

The effects considered above may be expressed rather more dramatically upon moving from a monodentate to a polydentate ligand. If a polydentate ligand binds to a single metal atom by two or more donor atoms, the resultant complex is known as a chelate (Greek *chelos,* claw). Such chelated complexes are kinetically and thermodynamically more stable, with respect to ligand displacement, than comparable compounds with an equal number of equivalent monodentate ligands. What are the steric consequences of chelation?

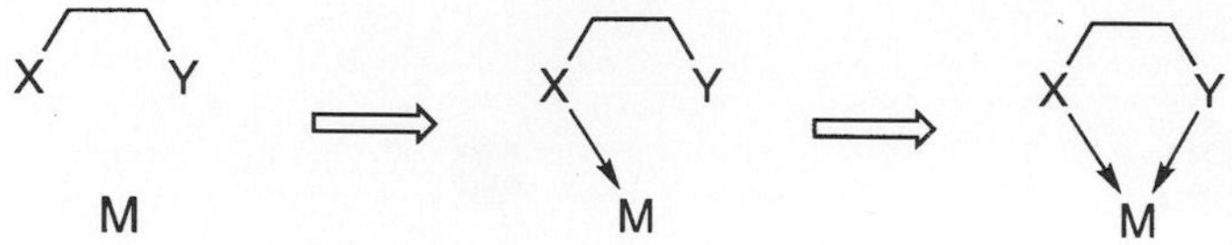

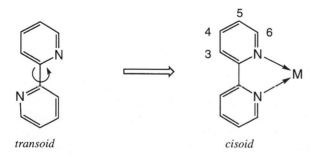

Figure 2-3. The sequential formation of metal–ligand bonds in a chelated complex.

If we consider the co-ordination of a didentate ligand to a metal ion we may anticipate a number of kinetic and thermodynamic effects. The equilibrium solution conformation of the ligand may differ from that which must be adopted in the cyclic chelated complex. If both of the metal–ligand bonds were to be formed simultaneously, it would require a pre-equilibrium in which the ligand adopted the 'unnatural' and higher energy conformation before co-ordination could occur. In practice, co-ordination follows an Eigen–Wilkins pathway in which metal–ligand bond formation is sequential (Fig. 2-3).

Behaviour of this type is typified by the co-ordination of a ligand such as 2,2'-bipyridine. The equilibrium solution conformation of the ligand is *trans*-coplanar, although there is free rotation about the interannular C–C bond, whereas in its metal complexes it usually behaves as a chelating didentate species, and, thus, adopts the *cis*-coplanar conformation (Fig. 2-4).

We may use spectroscopic methods to detect these conformational changes. The ^{1}H nmr spectra of CD_3COCD_3 solutions of 2,2'-bipyridine and the ruthenium(II) complex $[Ru(bpy)_3][BF_4]_2$ are shown in Fig. 2-5. In the free ligand, the lowest field resonance is that due to H_6, whilst in the complex the resonance corresponding to H_6 has moved upfield whereas that of H_3 has moved downfield. The downfield shift for H_3 is a consequence of the steric interactions between the H_3 groups in the *cisoid* conformation of the complex. The upfield shift for H_6 is caused by the arrangement of this proton directly above the π-cloud of another bpy ligand in the complex.

There is also a more subtle effect associated with chelation, which follows as a consequence of ring formation. It is well-known from organic chemistry that carbocyclic and

transoid   *cisoid*

Figure 2-4. *Transoid* to *cisoid* conformational change upon co-ordination of 2,2'-bipyridine.

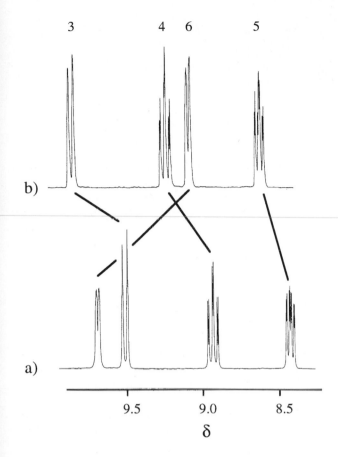

Figure 2-5. ^{1}H nmr spectra of CD_3COCD_3 solutions of a) bpy and b) $[Ru(bpy)_3][BF_4]_2$ showing the downfield shifting of H_3.

heterocyclic rings possess favoured conformations. A chelate ring system is merely a spe-cial class of heterocycle in which one of the heteroatoms is a metal, and similar confor-mational effects are observed in 1,2-diaminoethane (en) chelate complexes to those in other heterocycles. The rings are non-planar, adopting an envelope conformation remi-niscent of cyclopentane. The two conformations of the ring are related as mirror images, as seen in a Newman projection (Fig. 2-6). These projections also emphasise the imposi-tion of a tetrahedral geometry at nitrogen, in contrast to the facile pyramidal inversion observed in free amines. Co-ordination has created new, potentially chiral, centres at nitrogen. It is evident that the presence of a single chelate ring creates rather more struc-tural possibilities than might initially be thought. In practice, the barrier to inversion is low if only one chelated en ligand is present, and it is not usually possible to resolve the two enantiomeric forms of the complex. However, this is not the case if more than one such chelate ring is present in the complex.

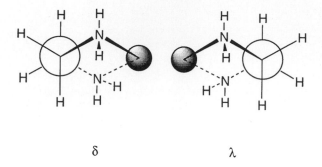

δ λ

Figure 2-6. The two enantiomeric forms of the chelate ring present in complexes of 1,2–diaminoet-hane. The labels δ and λ describe the absolute configuration of the chelate ring.

If we consider the geometry of the $[M(en)_3]^{n+}$ complex ion, we have further possibili-ties to consider. Whereas an octahedral complex with six identical ligands can only exist in one form, one with three didentate chelating ligands is chiral and can exist as two (non-superimposable) enantiomers (Fig. 2-7). The incorporation of polydentate ligands into a co-ordination compound may well lead to a rather considerable increase in the complexi-ty of the system, with regard both to the stereochemical properties and any related che-mical reactivity.

Of course, the physical properties of the two enantiomers of a complex containing three chelating didentate ligands will be identical, and interactions with achiral reagents will also

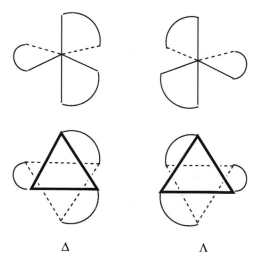

Δ Λ

Figure 2-7. Two representations of the two enantiomeric forms of a complex containing three che-lating didentate ligands. The lower view emphasises the relationship between the donor atoms and the triangular faces of an octahedron. The labels Δ and Λ describe the absolute configuration of the complex.

be identical. However, the interactions of the two enantiomers with chiral reagents or media will differ, and the diastereomeric intermediates or transition states obtained from the Δ or Λ forms will differ. In consequence, rates of reactions of the two enantiomers may differ.

In the case of a planar chelating ligand such as bpy, we only have to consider the two enantiomeric Δ and Λ forms of an $[M(bpy)_3]^{n+}$ cation. This is not the case when we come to consider cations such as $[M(en)_3]^{n+}$. Initially, we must consider the overall chirality of the $[M(en)_3]^{n+}$ cation, which can be either Δ or Λ. Each of the individual chelated en rings may possess a δ or λ absolute configuration. There are, thus, a total of eight different forms of the cation, as indicated in Table 2-1.

Table 2-1. The eight forms of the $[M(en)_3]^{n+}$ cation showing the possible conformations of the tris(chelate) (Δ or Λ) and the individual chelate rings (δ or λ). The entries in the left hand column are related to those in the right as mirror images.

Δ δδδ	Λ λλλ
Δ δδλ	Λ λλδ
Δ δλλ	Λ λδδ
Δ λλλ	Λ δδδ

However, forms such as Λ λλλ and Λ λλδ, which differ in the arrangement of only one chelate ring, are related as diastereomers, and are expected to possess different conformational energies and associated chemistry. The origin of the differing chemical properties of the various diastereomers is most easily seen by considering a square planar complex which has only two en ligands. This could exist as λλ, δδ or λδ diastereomers, with the λλ and δδ forms related as an enantiomeric pair. The λλ and λδ forms are shown in Fig. 2-8.

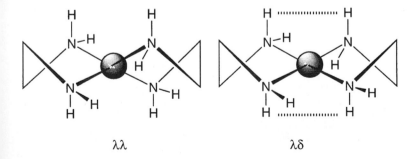

λλ   λδ

Figure 2-8. The λλ and λδ forms of a square planar $[M(en)_2]^{n+}$ complex. The steric interactions between the NH groups indicated in the λδ form destabilise it with respect to the λλ or (enantiomeric) δδ forms.

These steric interactions become more pronounced when we consider the introduction of an additional chelate ring in those complexes containing three didentate en ligands. The $\Delta \lambda\lambda\lambda$ form of an $[M(en)_3]^{n+}$ cation is estimated to be 7–8 kJ mol^{-1} more stable than the $\Delta \delta\delta\delta$ diastereomer. This becomes particularly important when we consider kinetically inert complex cations, such as $[Co(en)_3]^{3+}$, where there is a significant barrier to the inter-conversion of the diastereomers. In practice, the conformation of the chelate rings in $[Co(en)_3]^{3+}$ salts depends upon the nature of both the anions and any additional solvent molecules in the lattice which can form hydrogen bonds with the en ligands. We will return to this topic in Chapter 7, where we discuss some reactions of $[Co(en)_3]^{3+}$ salts in which an extraordinary steric control is exerted.

2.2.3 Chemical Consequences of Lone Pair Non-Availability

In the previous section we considered the formation of the metal–ligand bond in terms of the geometrical changes at the donor atom. A rather more fundamental change is asso-ciated with the 'sharing' of the electrons between the donor atom and the metal. The result is that the lone pair becomes 'less' available to the donor atom; thus any properties of the ligand associated with the presence of the lone pair will be modified. The simplest way in which this might be expressed is in the nucleophilicity of the donor atom. As an example, the nucleophilicity of amines is associated with the possession of a lone pair of electrons on the nitrogen atom; upon co-ordination, these electrons are involved in the metal-nitro-gen bond, and one would expect the nucleophilic properties to be reduced, or to disappear entirely. Even donor atoms bearing two or more lone pairs will be affected, since the posi-tive charge on the metal will be transmitted to the donor atom and result in a contraction of the non-bonded lone pair (Fig. 2-9).

However, the changes indicated in Fig. 2-9 are not meant to imply that an ammonia molecule co-ordinated to a metal will be non-nucleophilic. The reaction of a co-ordinated

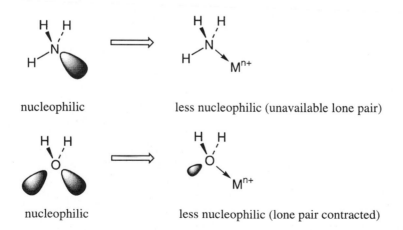

Figure 2-9. Lone pair changes upon co-ordination to a metal ion.

$$[M \leftarrow NH_3]^{n+} \rightleftharpoons M^{n+} + NH_3$$

$$M^{n+} + NH_3 + E^+ \rightleftharpoons M^{n+} + [E-NH_3]^+$$

Figure 2-10. A possible mechanism for the reaction of an ammine complex with electrophiles involves dissociation of the ligand from the metal. This process is likely to be of importance with relatively labile metals.

$$[M \leftarrow NH_3]^{n+} \rightleftharpoons M^{n+} + [E-NH_3]^+$$
$$\llcorner_{E^+}$$

Figure 2-11. A second mechanism involves a concerted process in which the new N-E bond is formed as the M-N bond is broken.

ammonia (or amine) ligand with an electrophile could involve the initial dissociation of the ligand (Fig. 2-10) or alternatively by a concerted process in which the new E-N bond is formed concurrently with the breaking of the M-N bond (Fig. 2-11). Both of these processes are expected to be less rapid than the corresponding reactions of the free ammonia molecule with electrophiles.

2.3 Polarisation of the Ligand

The very act of forming a complex results in the build-up of charge upon the donor. It is instructive to consider a valence bond representation of the formation of a metal complex. Two resonance forms are relevant. In the first we have the metal (atom or ion) showing no covalent interaction with the ligand (which bears a lone pair of electrons). In the second, the electrons are shared equally between the metal and the ligand in a two-electron two-centre bond. The overall effect is the loss of electron density by the ligand and the gain of electron density by the metal. If the two electrons were shared equally between the metal and the ligand donor atom, this would result in the metal acquiring a formal charge of -1 and the ligand donor atom a charge of +1 (Fig. 2-12). The 'resonance' means, of course, that the 'real' distribution of electrons between the metal and the ligand is not equal.

The transfer of charge density from metal ion to ligand, or from ligand to metal ion, represents one of the most obvious ways in which the electronic properties of the ligand may be altered upon co-ordination. In addition to the simple charge separation within the metal–ligand bond described above, there is a further factor to consider. The majority of co-ordination compounds with which we are concerned possess metal ions in 'normal' (+2 or +3) or 'high' (> +3) formal oxidation states; the positive charge associated with the metal centre may further *polarise* the ligand. Such positively charged metal ions are expected to induce substantial positive charges on the ligand donor atom. This may emphasise or oppose any existing polarisation within the ligand. These electrostatic inter-

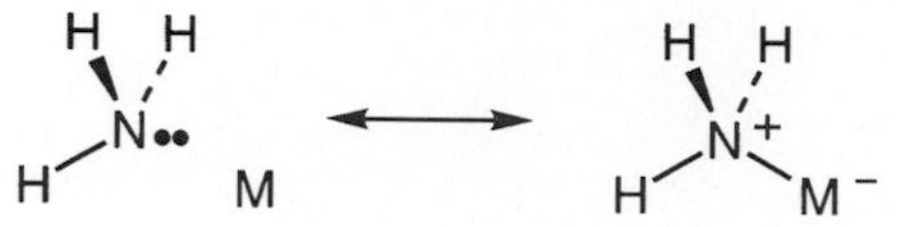

Figure 2-12. Valence bond representation of the formation of a complex. The left hand form is the 'ionic' representation with no covalent interaction between metal and ligand. The right-hand form shows the charge distribution which results from equal sharing of the lone pair.

actions may be transmitted through space or through the ligand σ- and, to a lesser extent, π-bonding framework.

This naive picture of the metal-ligand interaction suggests that no additional polarisation of the ligand beyond that involved in forming the donor–acceptor bond is to be expected in complexes of formally zero-valent metal ions. Whilst this would be correct if we only considered Coulombic interactions of formally charged species, it is obviously an over-simplification once we remember that the electronegativities of the metal and the ligand donor atoms may differ. If this is the case, then the metal-ligand bond will be polar, regardless of the formal charges on the two atoms. The perturbation of charge density in the co-ordinated ligands should be reflected in the physical and chemical properties of the complex.

2.3.1 The Nature of Ionic Interactions

A charged species is expected to propagate its charge through the region of space close to itself. The simplest model of the metal-ligand interaction places the ligand at a distance r from a point charge of magnitude ze in a medium with permittivity ε_0. This is an inversion of the concern of the classical crystal field model with the electrostatic effects of the ligand upon the metal. The Coulomb potential relates the experienced potential to the distance between the point charges, as depicted in Eq. (2.1).

$$d(r) = (ze\,/4\pi\varepsilon_0)(1/r) \qquad (2.1)$$

The force experienced between two charges z_1e and z_2e is given by Eq. (2.2).

$$F = (z_1z_2e^{\,2}/\,4\pi\varepsilon_0)(1/r^2) \qquad (2.2)$$

If we consider two point charges of equal but opposite unitary charge at a separation of 2.5×10^{-10} m, the attractive force is equal to 3.7×10^{-9} N.

Naturally, this potential will be perturbed in any real system, in which the nuclei, the core electrons and the bonding electrons all have associated magnetic and electric fields contributing to the total electronic environment, and in which the permittivity differs somewhat from that of free space. Nevertheless, we still expect any charge located at the metal nucleus to be experienced at relatively great distances. The effect of the charge is to polarise the electrons towards or away from the metal. This will be particularly important when electrons involved in ligand bonding are so polarised, i.e., with ligand atom valence electrons.

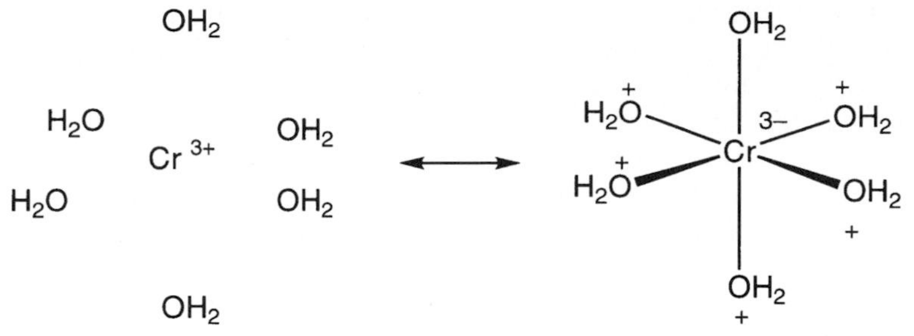

Figure 2-13. Valence bond representations of the $[Cr(H_2O)_6]^{3+}$ ion showing the charge distributions.

This is, of course, an aspect of the question that Pauling addressed and that led to the definition of the electroneutrality principle. If we consider the interaction of six water molecules with a Cr^{3+} ion, we may write two limiting valence bond structures. One of these forms localises the positive charge on the metal centre and depicts a Cr^{3+} ion surrounded by six water molecules. The covalent representation places a single positive charge on each of the water oxygen atoms and a -3 charge on the chromium centre (Fig. 2-13).

The 'truth' lies somewhere between these two extremes, and the assessment of covalent and ionic contributions to bonding has attracted considerable attention. Pauling devised the electroneutrality principle and suggested that the relative importance of ionic and covalent components was such that the overall charge on any one atom did not exceed ± 1. In the context of the $[Cr(H_2O)_6]^{3+}$ ion, this leads to a description in which the metal centre is neutral and each oxygen bears a charge of $+\frac{1}{2}$. The Cr-O bonds may be described as 50 % covalent.

Various semi-empirical relationships have been developed to describe the ionic potential actually observed by the valence electrons in an A–B bond. An example of such a relationship is shown in Eq. (2.3).

$$C(AB) = b\{\frac{z_1 e^2}{r_1} - \frac{z_2 e^2}{r_2}\}\exp\{-k_s \frac{(r_1 + r_2)}{2}\} \tag{2.3}$$

The exponential term is a Thomas–Fermi screening factor which accounts for the screening by the core electrons. Direct measurement of the 'ionic character' of a bond is a complex operation. In principle, a number of techniques such as X-ray or neutron diffraction, nmr, photoelectron or Mössbauer spectroscopy provide information about electron distribution and charge density; in practice the results are usually far from unambiguous.

However, whatever the precise electron distribution in the metal–ligand bond, a charged metal ion *must* induce polarisation changes within the co-ordinated ligand. This has a number of possible consequences for the ligand. If any of the ligand bonds possess, or indeed the entire ligand possesses, a permanent polarisation towards the metal, then this will be amplified; the magnitude of this effect depends upon the distance of the relevant group or bond from the metal, the charge on the metal, the electronegativities of the metal and the ligand, and the nature of the bonding in the intermediate region of space. If the ligand does not possess such a permanent dipole, then a purely Coulombic effect will induce such polarisation. Once again, the magnitude of the effect will depend upon the precise nature of the bonding and the distance from the metal ion.

The final possibility is that the bond or bonds of interest are polarised in the opposite sense to that induced by the metal. In a case such as this, the ligand may well be de-activated towards reactions it undergoes in the free state, but may now exhibit some new type of reactivity.

If we take a positively charged metal ion, the primary effect will be to polarise the ligand electrons towards the metal. If we consider any bond in the ligand, the electron density will be enhanced in the direction of the metal ion when compared with that of the free ligand. Taken to the extreme, electronegativity differences in ligand atoms may result in an overall dipole in the opposite sense. (A typical example would be in a ligand such as 2-chloroaniline co-ordinated to a metal through the nitrogen atom (**2.1**); the carbon–chlorine bond will still be polarised in the sense C→Cl, but the magnitude of this dipole will be lower in the complex than in the free ligand.)

2.1

2.3.2 Enhancement of Ligand Polarisation – Intrinsic Effects

As established in the previous sections, co-ordination of a ligand to a positively charged metal centre will result in the development of positive charge on the ligand donor atom. If any group attached to the donor atom is capable of leaving as a cation, then this leaving group ability will be enhanced by co-ordination. The simplest example of a cationic leaving group is the proton; the pK_a of protic acids, as defined in Eqs (2.4) and (2.5), will be lowered upon co-ordination to a positively charged or electropositive metal centre.

$$HL = L^- + H^+ \tag{2.4}$$

$$pK_a = -\log K_a = -\log\{[H^+][L^-]/[HL]\} \tag{2.5}$$

This phenomenon has been known since the investigations of Werner and is the basis for the formation of olated and oxolated complexes. The formation of 'basic' salts and the 'acidity' of aqueous iron(III) solutions provide further examples (Fig. 2-14).

In all the cases above, the ligand is water and we are looking at the effect of co-ordination upon the first and second ionisation processes (Fig. 2-15).

In the case of tripositive ions, this may be a significant effect and the pK_a values of $[Al(H_2O)_6]^{3+}$ and $[Fe(H_2O)_6]^{3+}$ are 4.97 and 2.20, respectively. The biological, geological and technological importance of this phenomenon cannot be over-emphasised; it is the

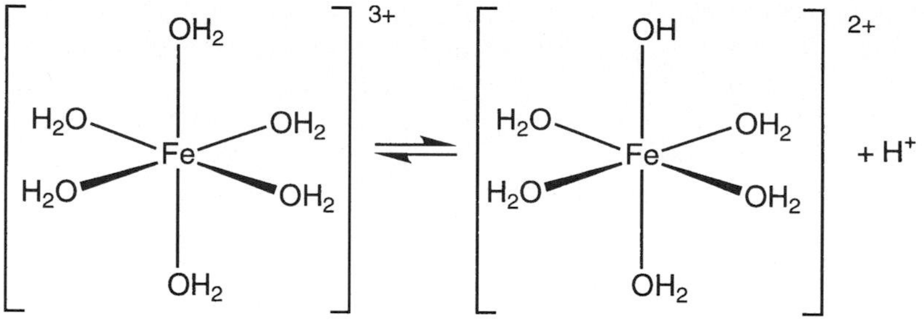

Figure 2-14. Deprotonation of a water molecule co-ordinated to an iron(III) centre.

$$2H_2O \rightleftharpoons OH^- + H_3O^+$$

$$H_2O + OH^- \rightleftharpoons O^{2-} + H_3O^+$$

Figure 2-15. Co-ordinated water ligands show deprotonation processes corresponding to the first and second ionisation steps.

basis of the mode of action of the zinc metalloenzyme carbonic anhydrase, the deposition of metals as oxide minerals in the biosphere and the 'rusting' of ferrous materials. The effect is not unique to water, with alcohols, carboxylic acids and amines all exhibiting similar equilibria. If we consider the case of co-ordinated ammonia, mechanistic studies have long indicated that the reactions of certain ammine complexes with hydroxide are rapid, and obey bimolecular rate expressions, apparently consistent with associative pathways (Fig. 2-16).

Initial explanations in terms of an associative S_N2-type reaction proved untenable, and the reaction is now thought to involve deprotonation of a co-ordinated amine (Fig. 2-17). This is the S_N1cb or Dcb mechanism. The key step is the formation of the amide intermediate, $[Co(NH_3)_4(NH_2)Cl]^+$, which undergoes halide loss to generate the reactive five-co-ordinate intermediate $[Co(NH_3)_4(NH_2)]^{2+}$ (**2.2**) (Fig. 2-18).

Effects such as these may be transmitted a considerable distance through the molecule. Thus, the cation derived by the protonation of pyrazine (**2.3**) has a pK_a of 0.6, and hence it is a relatively strong acid (Fig. 2-19).

$$[Co(NH_3)_5Cl]^{2+} + OH^- \rightleftharpoons [Co(NH_3)_5(OH)]_2^+$$

$$\text{rate} = k[Co(NH_3)_5Cl_2^+][OH^-]$$

Figure 2-16. The reactions of certain cobalt(III) amine complexes with base obey second order kinetics. The kinetically inert cobalt(III) ion is unlikely to undergo rapid associative processes, and another mechanism must be found.

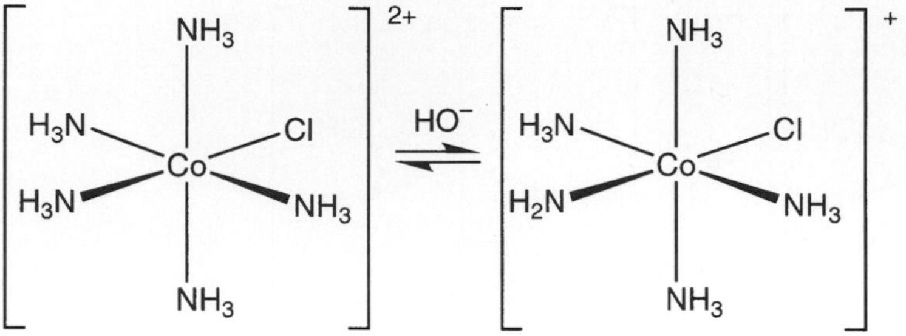

Figure 2-17. The first step in the S_N1cb mechanism for a kinetically inert cobalt(III) complex containing ammonia ligands involves a deprotonation of the co-ordinated NH_3 to generate an amido complex.

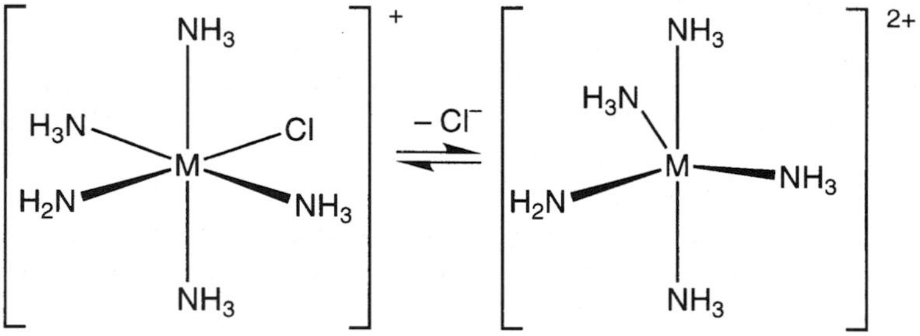

2.2

Figure 2-18. The second step of the S_N1cb mechanism involves the loss of chloride and the formation of a five-co-ordinate species.

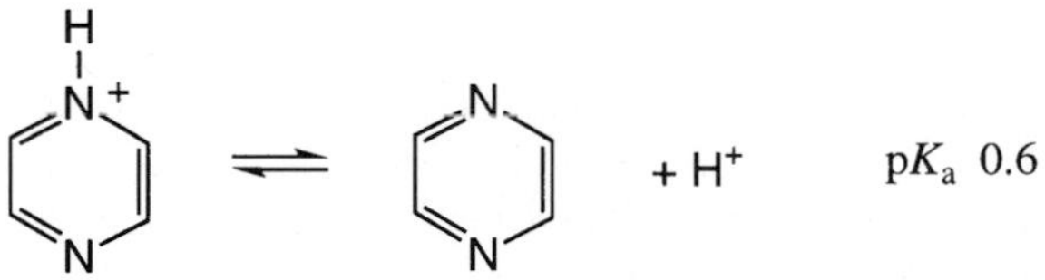

$$+ H^+ \qquad pK_a \ 0.6$$

2.3

Figure 2-19. The deprotonation reaction of the pyrazinium cation.

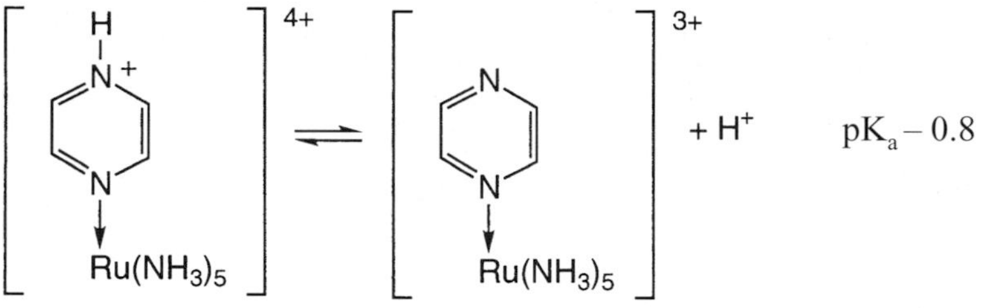

$$\left[\begin{array}{c} \overset{H}{\underset{|}{N^+}} \\ \diagdown \\ N \\ | \\ Ru(NH_3)_5 \end{array}\right]^{4+} \rightleftharpoons \left[\begin{array}{c} N \\ \diagdown \\ N \\ | \\ Ru(NH_3)_5 \end{array}\right]^{3+} + H^+ \qquad pK_a - 0.8$$

2.4

Figure 2-20. The co-ordination of the pyrazinium cation to a ruthenium(III) centre makes it a stronger acid. This is due to the electron-withdrawing metal centre destabilising the cation.

The pyrazinium cation still possesses a lone pair of electrons on the non-protonated nitrogen, and may act as a ligand to a metal centre (Fig. 2-20). Upon co-ordination to a ruthenium(III) centre the acidity is considerably enhanced, and the pK_a of complex **2.4** is now − 0.8.

This is mainly on account of the electron-withdrawing ruthenium(III) centre destabilising the protonated nitrogen centre. This is exactly parallel to the effect of placing more conventional electron-withdrawing substituents on a carboxylic acid (*c.f.* the pK_a of acetic *versus* that of trifluoroacetic acid). We will observe in a later section that this effect may be modified by π-interactions with the ligand.

Although the most common positively-charged leaving group is the proton, a number of others may be envisaged. One of the simplest such groups is an alkyl cation; however, such cations are high energy species, and would normally represent relatively unfavourable pathways. Attack by a nucleophile upon such a group is closely related to the formal loss of a cation and the two processes may be compared to the limiting S_N2 and S_N1 mechanisms in organic chemistry. In fact, such a process represents the stabilisation of a negatively charged leaving group by co-ordination to a metal ion, and will be considered in more detail in the next section. A type of reaction which corresponds to the formal loss of a methyl cation is seen in the Arbusov-type dealkylation of a trimethylphosphite complex (Fig. 2-21). The metal ion stabilises the *P*-bonded phosphonate ester (**2.5**).

The polarisation of the ligand resulting from co-ordination to a metal can be reflected in the ground state properties of the ligand, as discussed above. How may changes in the ligand polarisation be detected by other molecules in the immediate environment? Consider a ligand co-ordinated to a positively charged metal ion. We expect to see the build-up of positive charge on the ligand. This will result in an enhanced electrostatic interaction between the ligand and any negatively charged or dipolar molecules in the environment. In other words, any reactions involving nucleophilic attack at the ligand are likely to be enhanced.

We could imagine two types of nucleophilic attack upon a ligand. The first involves the formation of a new bond to the incoming nucleophile and the associated loss of some other leaving group. This corresponds to nucleophilic displacement. The second type of reaction does not involve the loss of any other species and corresponds to an addition reaction. There are a number of ways in which the polarisation of the ligand can enhance a

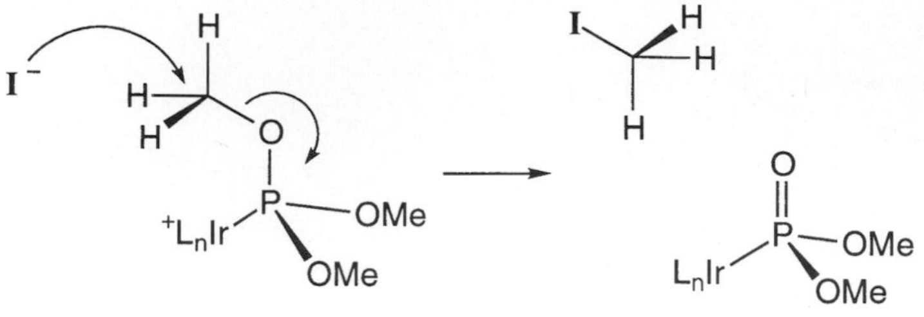

2.5

Figure 2-21. The dealkylation of an iridium phosphite complex.

nucleophilic displacement reaction. In the simplest case, the leaving group shows no interaction with the metal ion. This is exemplified in the hydrolysis of amino acid esters, which is enhanced dramatically in the presence of a metal ion. The oxygen atom of the carbonyl group is bonded to the metal, which enhances the polarisation of the O=C bond, thus increasing its susceptibility towards attack by water or hydroxide (Fig. 2-22). Many other nucleophiles react with co-ordinated amino acid derivatives in an analogous manner.

A slightly different effect is observed if the leaving group is co-ordinated to the metal. In this case, the metal may serve a dual role in activating the electrophilic site and in increasing the leaving group ability of the outgoing ion. This is seen in the mercury-mediated transesterificaton reactions of thioesters. A common synthetic application is seen in Fig. 2-23, where mercury(II) is used to remove a protecting group. The interaction of the mercury(II) with the sulfur atom enhances the leaving ability of the SR group, and increases the rate by many orders of magnitude.

The well-known action of silver(I) salts on nucleophilic substitution in alkyl halides is another commonplace example of this effect. The silver ion interacts with the halide, thus weakening the carbon–halogen bond and enhancing the leaving ability of the halide (Fig. 2-24).

The polarisation of a ligand may also exert a deactivating effect upon the reactions of the ligand. Co-ordination to a positively charged metal is expected to deactivate any reactions with electrophiles (in the absence of competing conjugate base formation at the ligand). These reactions may be transmitted a considerable distance through the ligand. Thus, the cobalt(III) complex $[Co(H_2NCH_2CH_2NHCH_2CH_2OH)_3]^{3+}$ does not react with acylating agents at either oxygen or nitrogen, even under forcing conditions.

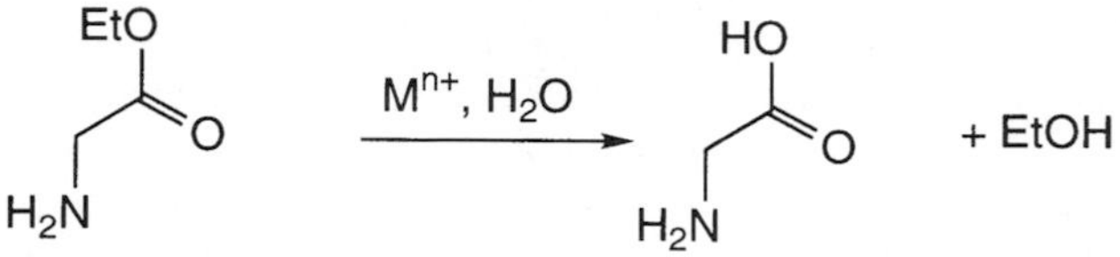

Figure 2-22. The enhanced hydrolysis of an amino acid ester in the presence of a metal ion.

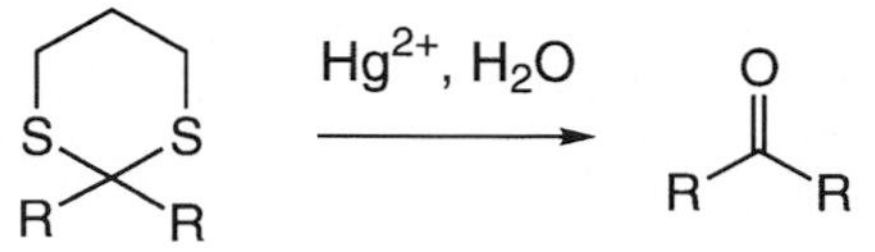

Figure 2-23. Mercury(II)-assisted deprotection of a dithioketal.

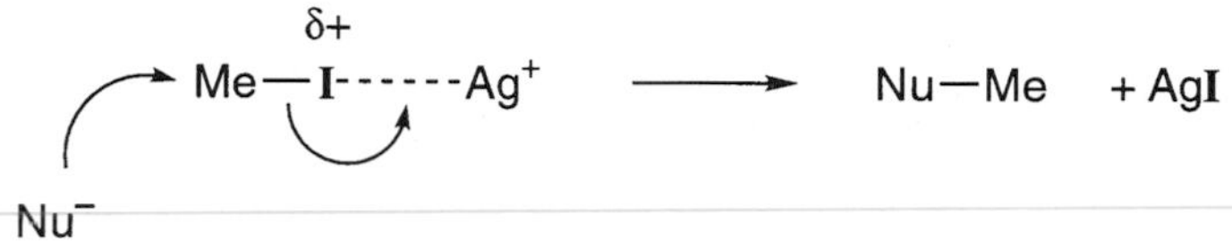

Figure 2-24. Silver-assisted reaction of an alkyl halide with a nucleophile.

2.4 π-Bonding Changes

We have now seen that co-ordination to a metal ion may affect the ligand in a number of ways. The conformation of the ligand may change, or the electron density may change as a result of through-space electrostatic interactions or through-bond interactions. In parti- cular, we have considered the propagation of electronic effects through the σ-bonding fra- mework of the molecule. However, as we saw in Chapter 1, there is another type of bon- ding which may be perturbed by the metal ion, i.e., any π-bonding system which may be present. A π-bond differs from a σ-bond in that the electron density is not located along the internuclear axis. There are a number of π-bonding interactions which we will encoun- ter in co-ordination complexes. As far as the metal ion is concerned, the *d* and *p* orbitals are suitable for π-bonding; these may be either full or empty (Fig. 2-25).

If the ligand possesses a π-bonding system, then its existing π- or π*-orbitals may be of the correct symmetry for overlap with metal *p* or *d* orbitals. Furthermore, any other

M-*d*   L-π*   M-*p*   L-π*

Figure 2-25. The interaction of metal *d* or metal *p* orbitals with the π*-levels of a ligand such as car- bon monoxide or an imine.

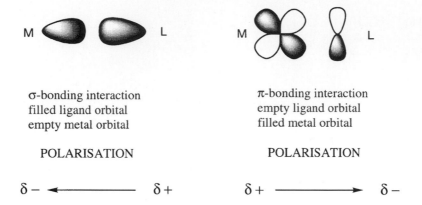

σ-bonding interaction
filled ligand orbital
empty metal orbital

π-bonding interaction
empty ligand orbital
filled metal orbital

POLARISATION

POLARISATION

$\delta-$ ◄─────── $\delta+$ $\delta+$ ───────► $\delta-$

Figure 2-26. Competing σ- and π-effects in the metal ligand interaction. The overall electron distribution will represent the balance of these two components.

orbitals on the ligand derived from suitable p or d orbitals may also enter into π-bonding interactions with metal ions if the energies are similar. The relative energies and occupancy of metal and ligand orbitals will determine the consequences of the π-bonding interactions. This interaction may result in the occupancy of previously unoccupied ligand bonding, non-bonding or antibonding levels or of metal levels.

The chemical consequences of these interactions are less readily defined than those resulting from conformational or polarisation changes. In particular, the π-bonding changes frequently oppose those resulting from the σ-polarisation. How can this occur? Let us consider a ligand possessing a vacant π^*-orbital co-ordinated to a positively charged metal ion. The metal ion will polarise the ligand and induce a positive charge on the donor atom and a smaller charge on the other ligand atoms. If the metal possesses filled d or p orbitals of suitable symmetry and energy for overlap with the π^*-level, then electron density may be transmitted from the metal to the ligand, in opposition to the polarisation in the σ-framework. Similarly, high oxidation state metal ions will be stabilised by π-bonding between vacant metal orbitals and filled π-, non-bonding or π^*-levels of the ligand (Fig. 2-26). Interactions involving the π-orbitals also differ from those only involving ligand polarisation in another important manner; the symmetry of the orbitals involved has crucial effects upon the reactivity of the molecular system.

2.4.1 Symmetry and Reactivity

Our understanding of organic chemistry has improved dramatically over the past few decades with the development of molecular orbital theory. In particular, the contributions of Woodward and Hoffmann have emphasised the importance of orbital symmetry in dictating the course of organic reactions. More recently, the symmetry properties of the orbitals close to the potential energy surface, the so-called frontier orbitals, have been shown to be of paramount importance. The basics of frontier orbital theory are

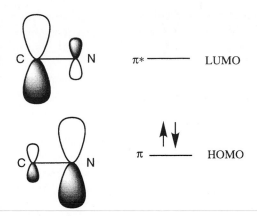

π* ———— LUMO

π ——↑↓—— HOMO

Figure 2-27. The π-molecular orbitals of an imine. The HOMO is occupied by a pair of electrons and the LUMO is vacant.

admirably described elsewhere; suffice it to say that the interactions of the highest occupied molecular orbitals (henceforth, HOMO) and the lowest unoccupied molecular orbitals (LUMO) frequently dictate the favoured reaction pathways.

Let us consider a real case – the hydrolysis of an imine, $R_2C=NR'$. The π-molecular orbitals of an imine are shown in Fig. 2-27.

The π-orbital is the HOMO and the π* the LUMO. Notice that the coefficients of the orbitals are unequal, since nitrogen is more electronegative than carbon, and that the magnitude of the coefficients alternates from HOMO to LUMO. We may now imagine a water molecule approaching the imine. On the basis of orbital symmetry rules, the important interactions could be the LUMO of the water with the HOMO of the imine, or the HOMO of the water with the LUMO of the imine. This selectivity is on the basis of better matching of orbital energies. It is commonly found that the important interaction is that of the HOMO of the nucleophile with the LUMO of the electrophile (Fig. 2-28). The

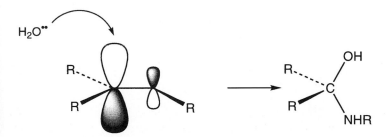

Figure 2-28. The interaction of the lone pair of a water molecule with the π* LUMO of an imine. The product of the reaction is an aminol, which might then react further to yield a carbonyl compound, $R_2C=O$, and an amine.

HOMO of the water molecule will approximate to one of the sp^3 lone-pairs on the oxygen atom. This will interact with the LUMO of the imine in such a way that the overlap of orbitals is maximised, i.e., at the carbon atom, which has the larger coefficient, rather than at the nitrogen atom.

We may now consider the effect of bonding this imine to a metal ion possessing a filled d orbital of correct symmetry for overlap with the π^*-level. Back-donation may occur, and the net effect is to transfer electron density from the metal to the ligand π^*-level. The molecular orbital description of this system is similar to that for π-allyl. The three molecular orbitals resulting from the overlap of the three atomic orbitals are sketched in Fig. 2-29 (relative coefficients of the orbitals have not been indicated). This is an orbital correlation diagram showing how the character of the new molecular orbitals (for the co-ordinated imine in the centre of the diagram) relates to those of the imine π-levels (on the left-hand side) and the metal d orbital (on the right).

If the metal possesses two electrons in the appropriate d orbital, the molecular orbitals Ψ_1 and Ψ_2 are filled. The important orbital is Ψ_2, which is derived (in part) from the old π^*-level of the ligand. Placing electron density within this orbital results in a build-up of electron density in the π-symmetry orbitals on the carbon atom of the imine. This will result in a repulsion being experienced by any incoming nucleophile, and a deactivation of the imine towards nucleophilic attack.

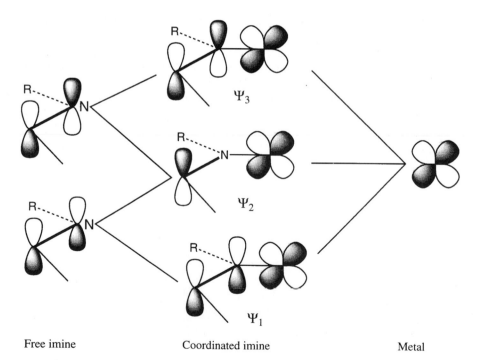

Free imine   Coordinated imine   Metal

Figure 2-29. Molecular orbital diagram showing the interaction of a metal d orbital with the π-molecular orbitals of an imine. If the metal possesses electrons in the relevant d orbital, there will be electron density in Ψ_2.

It is often found that imines are stabilised towards hydrolysis by co-ordination of the nitrogen to a π-bonding transition metal. Of course, in the absence of significant π-bonding interactions, ligand polarisation is expected to have the opposite effect and activate the imine towards nucleophilic attack.

There is copious experimental evidence to support the importance of back-donation in controlling the ground state geometry and reactivity of co-ordinated ligands. The best documented examples are those observed with the carbon-donor ligand carbon monoxide, commonly encountered as a ligand in organometallic complexes. In these compounds we may distinguish the combination of σ-donor and π-acceptor components of the metal–ligand interaction. Let us consider the infrared stretching mode associated with the C-O bond (Table 2-2).

Table 2-2. Infrared $\nu(CO)$ frequencies (cm^{-1}) for some carbonyl complexes

CO	Ni(CO)$_4$	[Co(CO)$_4$]$^-$	[Fe(CO)$_4$]$^{2-}$
2155	2046	1890	1730

All of these metal complexes are isoelectronic and isostructural with each other. Upon co-ordination of the carbon monoxide to the nickel(0) centre, the metal ion acquires negative charge. The carbon monoxide ligand possesses empty π^*-orbitals, and back-donation from the metal to the ligand reduces the electron density on the metal.

The bond order of the free ligand is 3. Occupancy of the π^*-level results in a lowering of the bond order and a corresponding lowering of the C-O stretching frequency. If we move from the nickel(0) complex to the anionic complex formed with cobalt(–1), the build-up of electron density on the metal ion is likely to be even less favourable. The result is that the back-donation is more efficient and the stretching frequency is lowered even more. Similarly, the stretching frequency lowers sequentially as we move to a metal with a formal oxidation state of –2.

A more subtle effect of π-bonding is seen in ruthenium complexes of pyrazine. We saw in Fig. 2-20 that co-ordination of a pyrazinium cation to a ruthenium(III) centre resulted in an increase in its acidity. However, co-ordination to a ruthenium(II) centre results in a *decrease* in acidity. The lower oxidation state ruthenium(II) centre is stabilised by back-

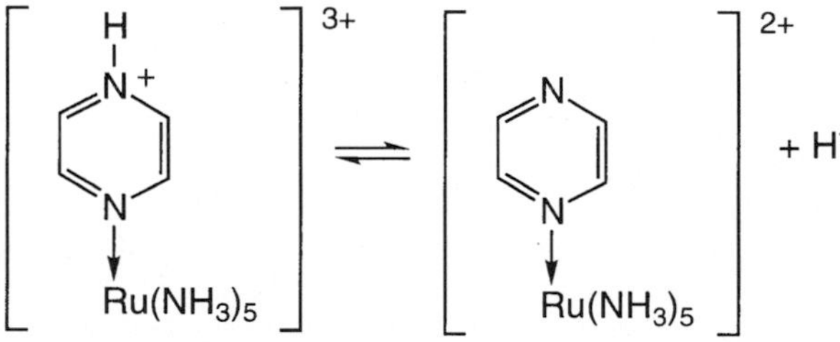

2.6

Figure 2-30. Competing polarisation and back-bonding effects control the acidity of a pyrazinium cation co-ordinated to ruthenium(II)

donation. This results in transfer of electron density from the metal to the ligand, and a consequent stabilisation of the positive charge on the protonated ligand, **2.6**. We see in Fig. 2-30 the true competition between the polarisation and π-bonding effects. Even though the ruthenium(II) complex is positively charged, the π-bonding interaction is sufficient to reverse the expected increased pK_a resulting from polarisation.

2.5 Conclusions

In this chapter we have seen the three ways in which co-ordination to a metal ion may modify the properties of a ligand. In any real metal-ligand system, the observed reactivity is expected to be a subtle blend of all three. In the remainder of this book we will consider the ways in which these various effects work in real complexes.

Suggestions for further reading

1. J.P. Collman, in *Transition-Metal Chemistry*, Vol. 2, (ed. R.L. Carlin), Dekker, London, **1960**.
   – A rather old, but nonetheless interesting account which shows how organometallic chemistry developed from co-ordination chemistry.
2. *Reactions of Co-ordinated Ligands.* Adv. Chem. Ser. No 37, (ed. D.H. Busch), American Chemical Society, Washington D.C., **1963**.
   – A volume devoted to pioneering studies in ligand reactivity and organometallic chemistry.
3. *Werner Centennial*, Adv. Chem. Ser. No 62, (ed. R.F. Gould), American Chemical Society, Washington D.C., **1967**.
   – There are a number of relevant chapters in this volume.
4. M.M. Jones, *Ligand Reactivity and Catalysis*, Academic Press, New York, **1968**.
   – A long forgotten book which laid the grounds for the study of the organic chemistry of metal complexes. If you can find this book, it is worth reading in its entirety.
5. J.P. Candlin, K.A. Taylor, D.T. Thompson, *Reactions of Transition-Metal Complexes*, Elsevier, Amsterdam, **1968**.
   – This volume contains a lot of interesting material regarding early studies.
6. R.P. Hanzlik, *Inorganic Aspects of Biological and Organic Chemistry*, Academic Press, New York, **1976**.
   – Although this is written from the point of view of the bioinorganic chemist, it is an extremely valuable, and easy to read, text.
7. *Co-ordination Chemistry*, Vol. 2, Adv. Chem. Ser. No 174, (ed. A.E. Martell), American Chemical Society, Washington D.C., **1978**.
   – Another multi-author volume containing much material of interest.

8. R.P. Houghton, *Metal Complexes in Organic Chemistry*, Cambridge University Press, Cambridge, **1979.**
    – A text book written very much from the point of view of the organic and organo-metallic chemist. Somewhat dated in its presentation, but worth looking at.
9. P.A. Challoner, *Handbook of Co-ordination Catalysis in Organic Chemistry*, Butterworths, London, **1986**.
    – Predominantly organometallic in emphasis, but with a number of relevant sections.
10. P.S. Braterman, *Reactions of Co-ordinated Ligands*, *Vol.* 2, Plenum, New York, **1989**.
    – A little like the curate's egg.
11. *Comprehensive Co-ordination Chemistry*, (eds. G. Wilkinson, R.D. Gillard, J.A. McCleverty), Pergamon, Oxford, **1987**.
    — There are a number of excellent chapters related to the reactivity of co-ordinated ligands in this multi-volume work.

There is also a continuing series of reviews concerning itself with the mechanistic aspects of the reactivity of co-ordinated ligands in *Mechanisms of Inorganic and Organometallic Reactions*, Plenum.

3 Reactions of Co-ordinated Carbonyl Compounds with Nucleophiles

3.1 General

A very large number of reactions of carbonyl compounds has been shown to be sensitive to the presence of metal ions. Carbonyl compounds are of paramount importance in the formation of C-C bonds in organic chemistry and metal-ion control of their reactivity offers a way to promote, inhibit or control such reactions. The co-ordination of the oxygen atom of a carbonyl compound to a metal centre is likely to modify the reactivity of that group. These reactions are of such importance and illustrate the principles of ligand reactivity so well that this chapter is devoted to them. Reactions in which a non-co-ordinated carbonyl group is attacked by a nucleophile that is co-ordinated to a metal are considered in more detail in Chapter 5. This distinction between attack by a free or by a co-ordinated nucleophile is not as clear-cut as might initially be thought. Both the metal and the carbonyl carbon atom may possess electrophilic character, and a nucleophile might initially attack at *either* the carbon of the carbonyl *or* at the metal centre. Products formally derived by attack at the carbonyl group may arise *either* by direct attack at that carbon atom *or* by attack at the metal followed by attack at carbon by the *co-ordinated* nucleophile. There has been (and still is) a healthy debate over the relative importance of these two competing mechanisms in biological and biomimetic systems. These two mechanisms may well both be operative in some cases. The principal reactions of carbonyl compounds are of two types, which we shall now consider.

The first class is that in which the nucleophile reacts with the electrophilic carbon atom of a carbonyl group to generate a new tetrahedral centre (Fig. 3-1).

The fate of this tetrahedral intermediate is dependent upon the nature of the substituents R and X and upon the incoming nucleophile. If X^- is a better leaving group than Nu^-, the overall reaction is a nucleophilic substitution of X by Nu, with the resultant loss of X^- (Fig. 3-2). This is the mechanism which is typically adopted in the alkaline hydrolysis of esters.

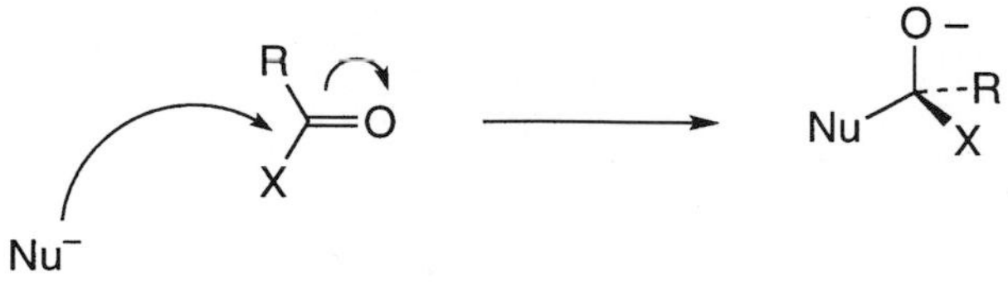

Figure 3-1. The formation of a tetrahedral intermediate in the reaction of a nucleophile with a carbonyl compound.

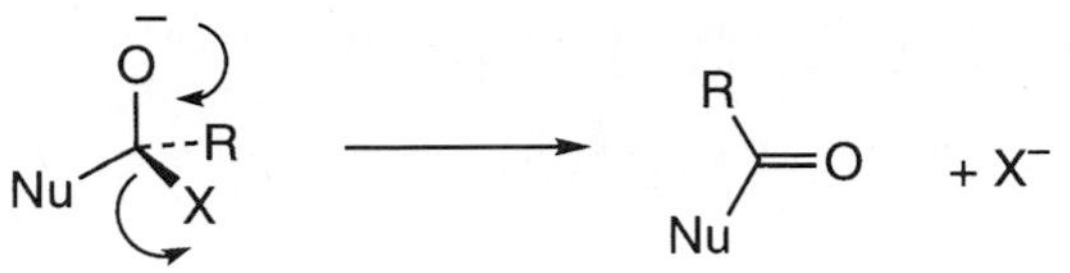

Figure 3-2. Collapse of the tetrahedral intermediate with loss of X⁻ and the formation of a new carbonyl compound.

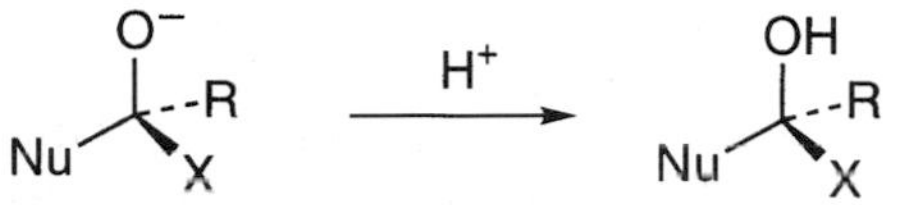

Figure 3-3. Trapping of the tetrahedral intermediate with a proton to generate an alcohol.

In contrast, if X is not a better leaving group than Nu⁻, then either Nu⁻ is lost (with no overall resulting reaction) or a substituted alcohol may be formed by reaction of the tetra-hedral anion with a proton (Fig. 3-3). In many cases the alcohol may then undergo further reactions.

The second type of reaction which is commonly associated with carbonyl compounds involves the generation of a nucleophilic enol or enolate ion. Although the conversion of a ketone to the tautomeric enol does not necessarily involve any other species, the gene-ration of an enolate requires a base (Fig. 3-4). In this latter reaction, the putative nucleo-phile may act as a general base.

The carbon atom of the enol or enolate is nucleophilic and may undergo facile reac-tions with electrophiles. In other cases, the electrophile can react with the oxygen atom (Fig. 3-5). The relative importance of these two modes of attack depends upon the nature of the electrophile and, to a lesser extent, upon the specific enolate.

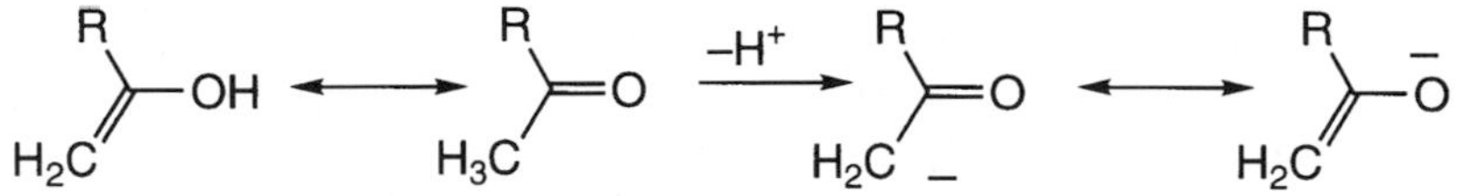

Figure 3-4. Generation of enol and enolate anions from the keto form of a carbonyl compound.

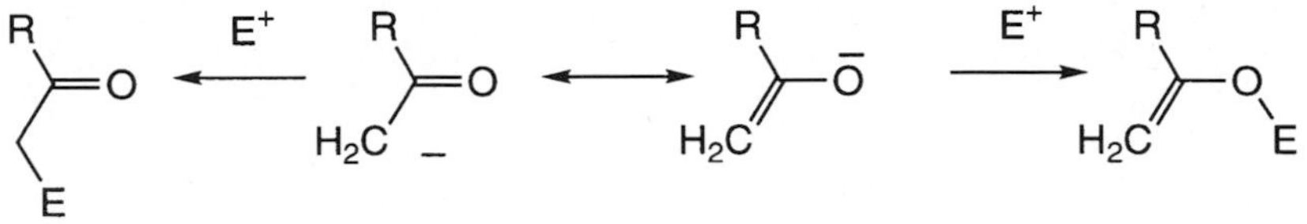

Figure 3-5. Reactions of an enolate anion with an electrophile (E⁺) at carbon or oxygen.

Both of these fundamental types of reactivity of a carbonyl compound may be modified by co-ordination to metal ions. The polarisation effect means that the electrophilic character of a carbonyl carbon atom may be increased by co-ordination of the oxygen to a metal ion, and so any reactions in which the rate-determining step is nucleophilic attack upon this site might be accelerated. Similarly, the presence of an electropositive metal co-ordinated to the oxygen atom might favour enolisation and accelerate reactions in which enolisation or reaction of an enolate is rate-determining. The pK_a of the enol will also be sensitive to co-ordination of the oxygen, and generation of the enolate may be facilitated.

Numerous examples of metal-ion modification of the reactions of carbonyl compounds are known, and some will be presented in this chapter.

3.2 Nucleophilic Attack on Co-ordinatcd Estcrs and Amides

One of the simplest reactions we might consider involves nucleophilic attack on the carbonyl group of an ester. However, esters are, in general, only very weakly co-ordinating ligands, and metal-directed reactions are best observed with derivatives containing donor atoms other than oxygen. The complexes of such ligands are stabilised by the chelate effect. The best examples of these are the amino acid compounds considered in the next section. However, examples of nucleophilic attack upon esters containing other good donor atoms are known, and these *are* accelerated by metal ions – reactions include amination, hydrolysis, transesterification and reduction.

The majority of these reactions involve attack by a co-ordinated nucleophile and are considered in more detail in Chapter 5. Reactions of amides involving attack by extra-nuclear nucleophiles are better established. For example, the hydrolysis of *N,N*-dimethyl-formamide (Me$_2$NCHO, dmf) in the complex [Co(NH$_3$)$_5$(dmf)]$^{3+}$ is ten thousand times faster than that of the free ligand. The dmf is co-ordinated to the metal through the oxygen atom of the amide group.

One of the problems associated with the study of reactions of co-ordinated ligands is in establishing that the products that are observed do indeed arise from reactions of the co-ordinated, rather than the free, species. This may be established by kinetic investigations if they demonstrate that the rates of formation of the products are dramatically faster than those in the corresponding reactions of the free ligands. If the reactions are slower than those of the free ligands, it is difficult to distinguish between a rate-determining step involving attack upon the co-ordinated ligand or loss of the ligand from the metal centre followed by rapid reaction of the free ligand. Two tricks are used to allow the study of these reactions to be performed more readily. Both increase the stability of the complex such that dissociation of the ligand and reaction of the free species are disfavoured. The first involves the use of poly-dentate ligands, where the chelate effect stabilises the complexes to ligand dissociation. This is a thermodynamic stabilisation of the co-ordination compound. The second approach involves the use of kinetically inert metal centres. It is found that transition-metal ions possessing d^3 or low-spin d^6 configurations are stabilised with respect to ligand exchange reactions. This is a kinetic stabilisation effect reflecting the high CFSE terms associated with these configurations – the activation energies required for the formation of either five co-ordinate dissociative or seven co-ordinate associative intermediates or transition states are particularly high. Complexes of chromium(III) (d^3), cobalt(III) (d^6), iridium(III) (d^6) and rhodium(III) (d^6) are of particular use in this context. One or other of the effects may be sufficient to unambiguously observe a particular type of reactivity, as we saw above by the use of a cobalt(III) dmf complex. Many of the best examples of reactions of co-ordinated ligands occur with kinetically inert complexes containing chelating ligands. In these cases, the new organic molecules often remain co-ordinated to the metal centre.

3.2.1 Attack by Co-ordinated Nucleophile

In the above discussion we have considered the ability of the metal ion to polarise the carbonyl group and increase the electrophilic character of the carbon atom. We mentioned, however, that in some reactions, a competing metal-mediated mechanism may occur which involves attack by a metal-bonded hydroxide or other nucleophile. The products of these reactions are often identical to those which arise by the attack of an external nucleophile upon the co-ordinated carbonyl compound, and a number of subtle investigations involving the use of chelating ligands and kinetically inert metal centres have been performed. These reactions are considered in detail later. In Fig. 3-6 competing mechanisms involving attack by internal (pathway 1) and external (pathway 2) hydroxide nucleophile for the formation of the complex [Co(en)$_2$(H$_2$NCH$_2$CO$_2$)]$^{2+}$ are shown.

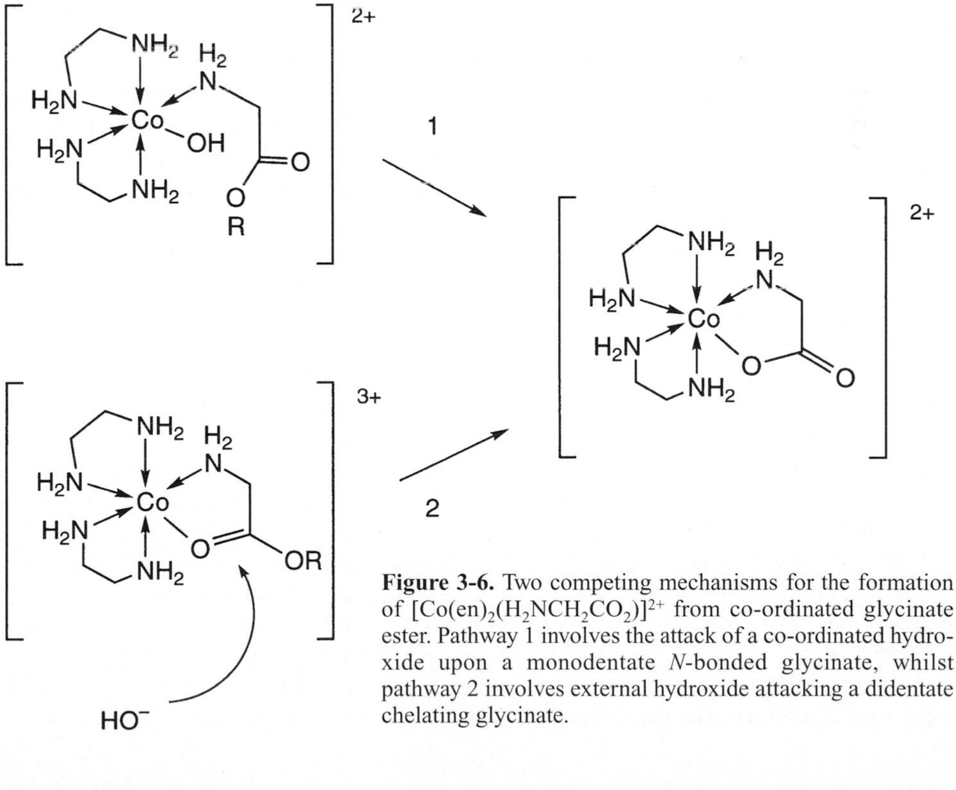

Figure 3-6. Two competing mechanisms for the formation of [Co(en)$_2$(H$_2$NCH$_2$CO$_2$)]$^{2+}$ from co-ordinated glycinate ester. Pathway 1 involves the attack of a co-ordinated hydroxide upon a monodentate *N*-bonded glycinate, whilst pathway 2 involves external hydroxide attacking a didentate chelating glycinate.

3.3 Hydrolysis of Amino Acid Esters and Amides

The metal-accelerated hydrolysis of amino acid esters or amides comprises one of the best investigated types of metal-mediated reaction (Fig. 3-7). One of the reasons for this is the involvement of chelating ligands, which allows chemical characterisation of inter-mediates and products in favourable cases, and allows detailed mechanistic studies to be made. The reactions have obvious biological relevance and may provide good working models for the role of metals in metalloproteins.

The rates of hydrolysis of amino acid esters or amides are often accelerated a million times or so by the addition of simple metal salts. Salts of nickel(II), copper(II), zinc(II) and cobalt(III) have proved to be particularly effective for this. The last ion is non-labile and reac-tions are sufficiently slow to allow both detailed mechanistic studies and the isolation of intermediates, whereas in the case of the other ions ligand exchange processes are suffi-ciently rapid that numerous solution species are often present. Over the past thirty years the interactions of metal ions with amino acid derivatives have been investigated intensively, and the interested reader is referred to the suggestions for further reading at the end of the book for more comprehensive treatments of this interesting and important area.

In principle, a number of possible mechanisms may be considered involving monoden-tate or chelated amino acid derivatives with either intra- or intermolecular attack by water or hydroxide ion. Let us start by considering the amino acid derivative acting as a mon-odentate nitrogen donor. The complex cation $[Co(en)_2(H_2NCH_2CO_2R)Cl]^{2+}$ (**3.1**) is an example of a compound containing such a ligand. In this compound there is no direct interaction of the metal ion with the oxygen atom of the carbonyl group, nor is there any direct interaction between the metal ion and the nucleophile. Because cobalt(III) is a kine-tically inert centre there is no readily accessible pathway for the displacement of chloride by either water or by the oxygen of the ester under neutral or acidic conditions. Of cour-se, in alkaline conditions the S_N1cb (p. 109) mechanism becomes a possibility, and we shall consider the implications of this in a later chapter. In **3.1** the only effect of the metal is a long-range polarisation of the carbonyl group which slightly increases the δ+ charge on the carbon of the carbonyl group.

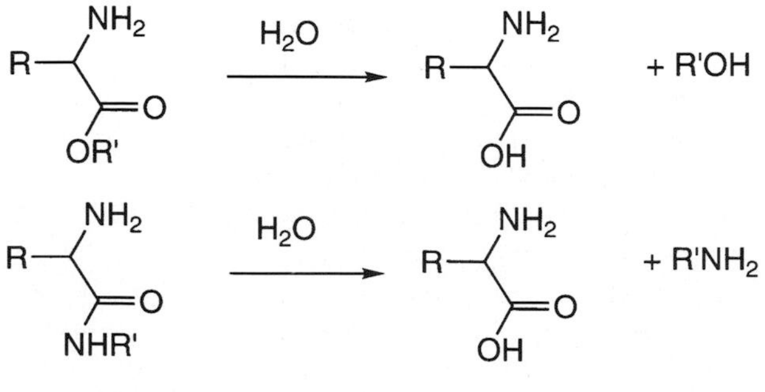

Figure 3-7. Hydrolysis of amino acid esters and amides.

What effect does co-ordination of the amino acid derivative **3.1** have upon the rate of hydrolysis? The rates of the hydrolysis reactions depicted in Fig. 3-8 are only slightly more rapid than those of the free amino acid esters, and, in general, the rates of reactions involving monodentate *N*-bonded ligands very closely resemble those for acid-catalysed hydration. This monodentate bonding mode is only exhibited with non-labile ions such as cobalt(III) or chromium(III) and is relatively rare even then.

With labile metal ions much greater rate enhancements are observed. This is illustrated in the data for the hydrolysis of methyl glycinate, $H_2NCH_2CO_2Me$, presented in Table 3-1. The uncatalysed rate refers to hydrolysis in water at pH 7. A modest enhancement of about twenty times is observed in the presence of protic acids (which protonate the terminal nitrogen and slightly polarise the carbonyl group), but a very dramatic rate enhancement of about 10,000 times is seen when copper(II) salts are added to the solution. This suggests that the copper(II) ion is acting in a rather different way to the proton and also to the cobalt(III) ion in **3.1**. This is consistent with the formation of *NO*-donor chelated complexes and that it is these compounds that undergo the rapid hydrolysis. This is fully compatible with the rapid formation of chelated complexes at the labile metal centre.

Table 3-1. Rate constants for the hydrolysis of methyl glycinate, $H_2NCH_2CO_2Me$, in neutral water, in dilute acid and in the presence of copper(II) salts.

	rate/ $M^{-1}s^{-1}$
k_{uncat}	1.28
k_{H^+}	28.3
$k_{Cu^{2+}}$	7.6×10^4

Rate enhancements of $10^4 - 10^6$ are typically associated with the formation of chelated complexes in which the carbonyl oxygen atom is also co-ordinated to the metal (**3.2**). This results in a considerably greater polarisation of the C–O bond.

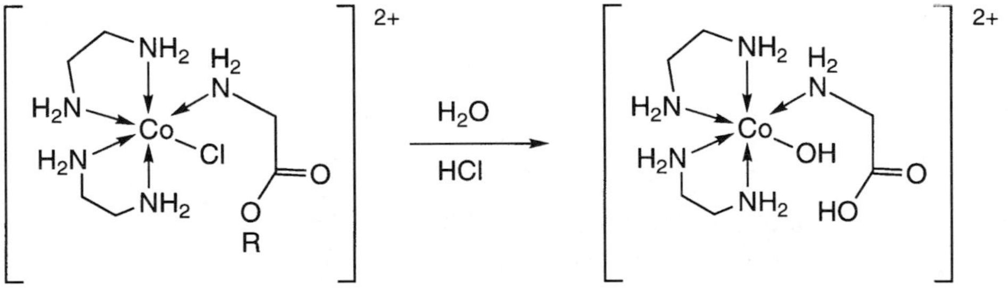

3.1

Figure 3-8. The hydrolysis of a kinetically inert complex containing a monodentate amino acid ester co-ordinated through nitrogen. The only effect of the metal is a long-range polarisation which slightly increases the electrophilic character of the carbonyl carbon atom.

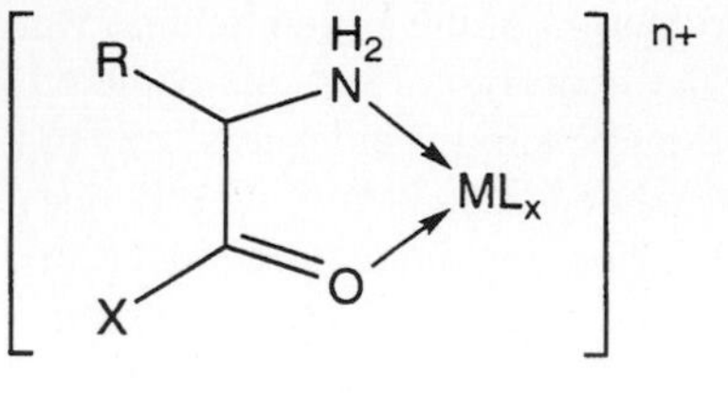

3.2

Although very dramatic rate enhancements have been observed with labile metal ions such as copper(II) and nickel(II), most studies have involved kinetically inert d^6 cobalt(III) complexes. In general, copper(II) complexes have been found to be the most effective catalysts for these reactions.

In the case of inert cobalt(III) complexes it is possible to isolate the chelated products of the reaction. Let us return to the hydrolysis of the complex cations [Co(en)$_2$(H$_2$NCH$_2$CO$_2$R)Cl]$^{2+}$ (**3.1**), which contain a monodentate *N*-bonded amino acid ester, that we encountered in Fig. 3-8. The chelate effect would be expected to favour the conversion of this to the chelated didentate *N,O*-bonded ligand. However, the cobalt(III) centre is kinetically inert and the chloride ligand is non-labile. When silver(I)

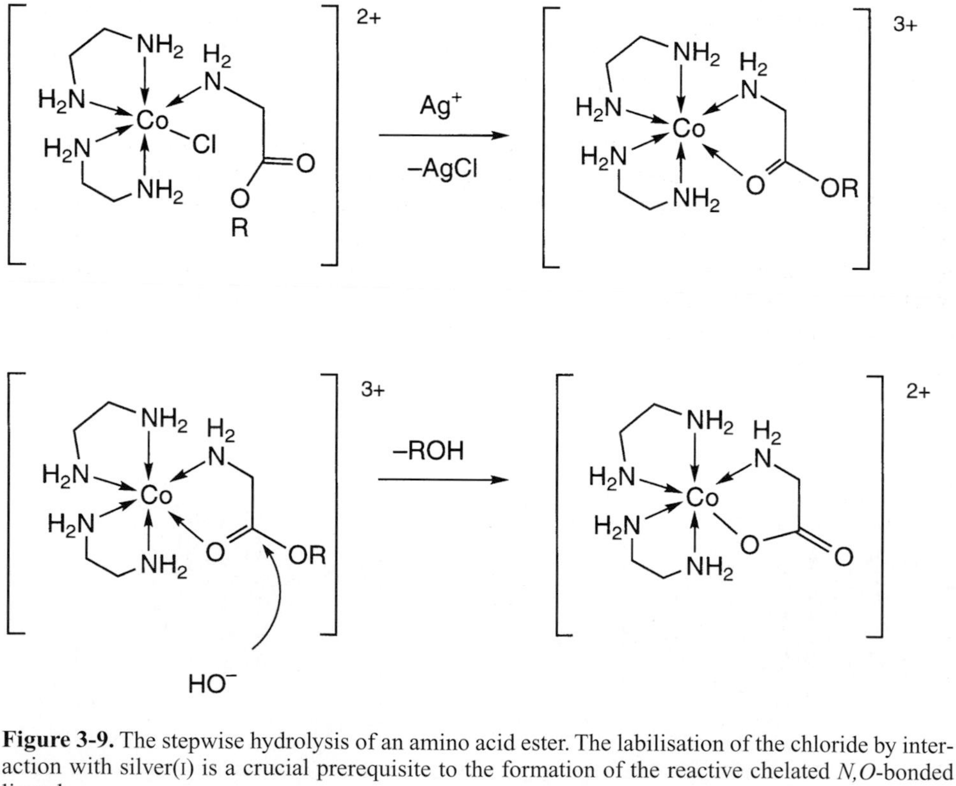

Figure 3-9. The stepwise hydrolysis of an amino acid ester. The labilisation of the chloride by interaction with silver(I) is a crucial prerequisite to the formation of the reactive chelated *N,O*-bonded ligand.

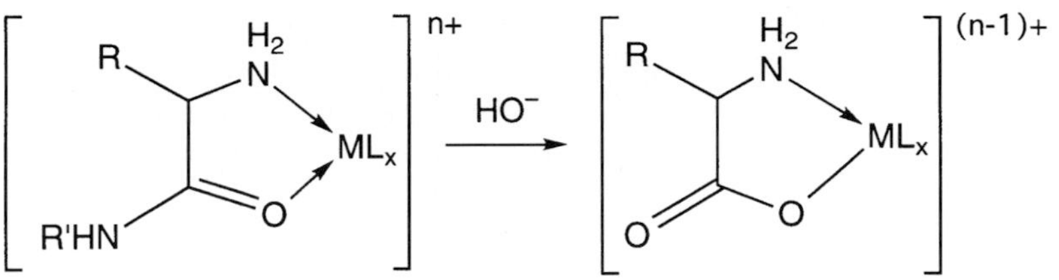

Figure 3-10. Hydrolysis of an amino acid amide (peptide).

salts are added to the solution, there is an interaction between the silver(I) ion and the chloride ligand which weakens the Co-Cl bond in the same way that we saw silver weakening the carbon-halogen bond in Fig. 2-16. The effect is to labilise the chloride ion, an effect which receives a thermodynamic boost from the insolubility of silver(I) chloride in common solvents. This is a general method for the formation of solvent complexes of cobalt(III) from the corresponding halo complexes. However, in this case, the oxygen atom of the amino acid ester replaces the chloride ion to generate the cation **3.3**, which contains a chelated N,O-bonded ligand. The cobalt centre now strongly polarises the carbonyl group, and external hydroxide can attack the carbon of the carbonyl group to generate a tetrahedral intermediate, which collapses with loss of alcohol, ROH, to give $[Co(en)_2(H_2NCH_2CO_2)]^{2+}$ containing the amino acid anion (Fig. 3-9). Similar mechanisms occur in the metal-mediated hydrolysis of co-ordinated amino acid amides (Fig 3.10).

The ability of a metal ion to increase the rate of hydrolysis of a peptide has enormous implications in biology, and many studies have centred upon the interactions and reactions of metal complexes with proteins. However, hydrolysis is not the only reaction of this type which may be activated by chelation to a metal ion, and chelated esters are prone to attack by any reasonably strong nucleophile. For example, amides are readily prepared upon reaction of a co-ordinated amino acid ester with a nucleophilic amine (Fig. 3-11). In this case, the product is usually, but not always, the neutral chelated amide rather than a deprotonated species.

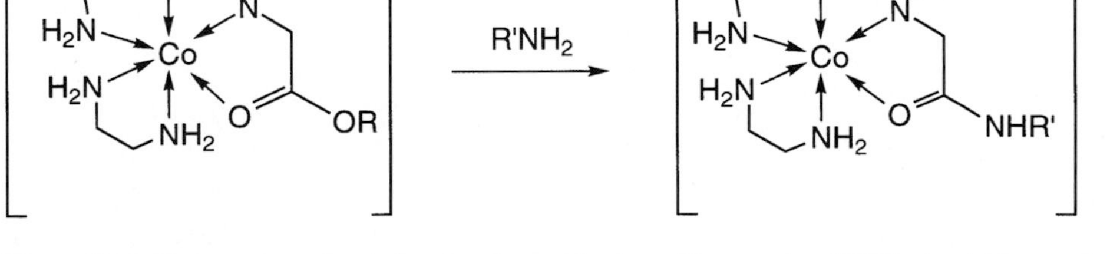

Figure 3-11. The reaction of a chelated amino acid ester with an amine, R'NH$_2$, to yield an amino acid amide complex.

This tendency to react with a range of nucleophiles is reflected in the general rate equation for reactions of this type, as seen for the hydrolysis of $[Co(en)_2(H_2NCH_2CO_2{}^iPr)]^{3+}$. Typically, a three-term rate equation is obtained. This is indicative of a process in which *at least* three parallel reaction pathways are being followed. In the case of the hydrolysis of $[Co(en)_2(H_2NCH_2CO_2{}^iPr)]^{3+}$, the k_1 term refers to attack of the chelated ester by water, the k_2 term to attack by hydroxide and the k_B term to general base attack by any other nucleophile which is present in solution. The rate is defined in terms of the loss of the starting complex cation, rather than the formation of any specific product of the reaction.

$$rate = -d[Co(en)_2(H_2NCH_2CO_2{}^iPr)]/dt$$

$$rate = \{k_1 + k_2[HO^-] + k_B[B]\}[Co(en)_2(H_2NCH_2CO_2{}^iPr)]$$

There are a number of useful synthetic applications of these reactions of chelated amino acid esters (Fig. 3-12). For example, if the attacking nucleophile is not a simple amine, but is another amino acid ester or an *O*-protected amino acid, then peptide or polypeptide esters are formed in excellent yields. This may be developed into a general methodology for the metal-directed assembly (and, in the reverse reaction, the hydrolysis) of polypeptides.

When we consider the reverse reactions, the metal-directed hydrolysis of amides, the polarising metal ion may also play a second, and often undesirable, role. In addition to polarising the carbonyl group and activating the carbon atom to nucleophilic attack, the metal may also polarise an amide N–H bond. If we consider the amino acid amide **3.4**, the polarisation may be transmitted through the ligand framework to the amide N–H bond. This polarisation may be sufficient to lower the pK_a so as to allow deprotonation under the desired reaction conditions (Fig. 3-13).

As indicated in Fig. 3-13, this is very often associated with a change in the co-ordination mode of the chelate from *N,O*- to *N,N'*-bonded. The deprotonation and formation of the amido anion results in an *increase* in electron density at the carbonyl carbon atom, and the complex is *less* prone to attack by a nucleophile. Note also that even if attack by a nucleophile did occur, breaking of the C-N bond would generate a monodentate *N*-bonded amino acid derivative. If the reaction is performed at a non-labile metal centre,

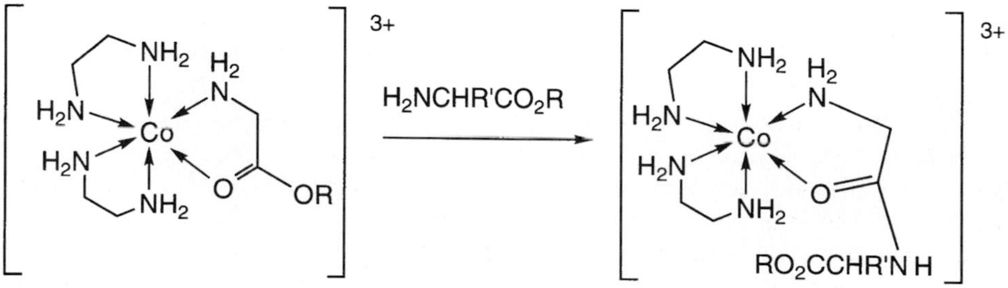

Figure 3-12. The reaction of a chelated amino acid ester with another amino acid ester to give a metal complex of a dipeptide.

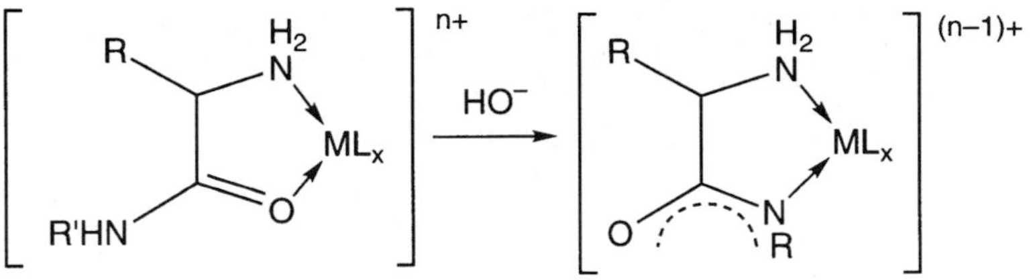

3.4

Figure 3-13. Polarisation of a chelated amino acid amide leading to deprotonation of the amido group. Notice the change in bonding mode from *N,O* to *N,N'* associated with this process.

the consequence would be a monodentate amino acid derivative and an amino compound co-ordinated to the metal. The deprotonation of co-ordinated amide ligands provides a way to selectively deactivate specific sites, and has been utilised as a synthetic procedure in peptide chemistry.

A very specific example of the hydrolysis of an amide is seen in the metal-promoted hydrolysis of urea, $CO(NH_2)_2$ as depicted in Eq. (3.1).

$$CO(NH_2)_2 + H_2O \rightarrow CO_2 + 2NH_3 \qquad (3.1)$$

In the complex $[Rh(NH_3)_5(NH_2CONH_2)]^{3+}$, the urea is *N*-bonded to the metal and hydrolysis to $[Rh(NH_3)_6]^{3+}$ is ten thousand times faster than the hydrolysis of urea itself. The mechanism is not known with certainty, but it has been proposed that the first step in the reaction involves ammonia loss to generate a co-ordinated isocyanate, which is then attacked by the water nucleophile. The intermediate $[Rh(NH_3)_5(NCO)]^{2+}$ complex has been isolated from this reaction; the ammonia loss is probably promoted by polarisation of the co-ordinated urea, but the precise details of this step are somewhat obscure. Note that the use of the non-labile rhodium(III) centre results in the urea nitrogen atom remaining attached to the metal throughout this sequence of reactions. Urea may also co-ordinate to a metal through the oxygen atom, and when complexes of non-labile metal ions containing *O*-bonded urea are hydrolysed, the oxygen of the urea is eventually found in a co-ordinated water molecule. These reactions are of relevance to the biological hydrolysis of urea by the enzyme urease, which contains two nickel ions at the active site. A number of mechanisms have been proposed for the interaction of the substrate urea with the metal centres in the enzyme.

Another 'special' example of the metal-promoted hydrolysis of an amide is seen with the lactam rings of cephalosporins or penicillins. The hydrolysis of penicillin, **3.5**, is accelerated 100 million times in the presence of copper(II) salts. Unfortunately, the precise mechanism of the reaction, whether it involves intra- or intermolecular attack by hydroxide or water, or even the site of co-ordination, is not known with any certainty.

A further example is seen in the rapid hydrolysis of benzylpenicillin (**3.6**) by copper(II) which is thought to proceed through an intermediate of type **3.7**. Notice that in this case the activation of the carbonyl group to nucleophilic attack seems to be through

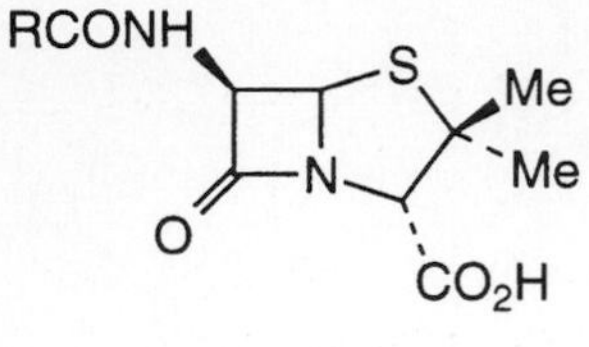

3.5

a longer range polarisation resulting from the co-ordination of the nitrogen rather than a direct polarisation of the C=O bond as a result of co-ordination of the oxygen atom (Fig. 3-14).

In conclusion, the hydrolytic and other reactions of co-ordinated amino acid derivatives with nucleophiles may proceed by two major routes. The first involves a moderate acceleration by general acid catalysis of a monodentate *N*-bonded ligand, whilst the second may involve very dramatic rate increases (by a factor of a million or so) associated with didentate chelating *N,O*-bonded ligands. There is little evidence for the widespread involvement of co-ordinated nucleophiles attacking the carbonyl in amino acid derivatives, although some special, and well characterised, examples with cobalt(III) complexes are considered in the next chapter.

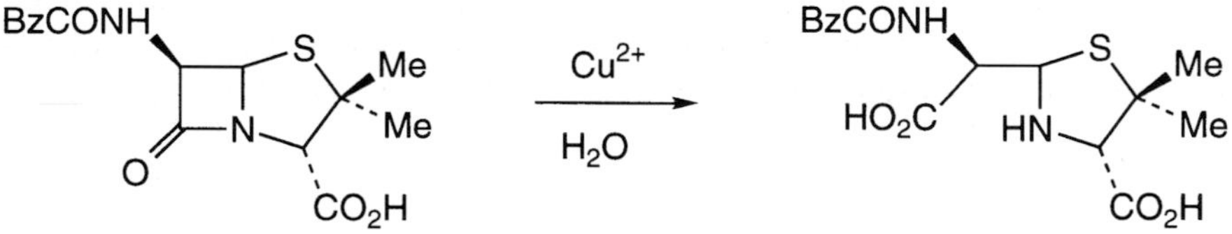

3.6

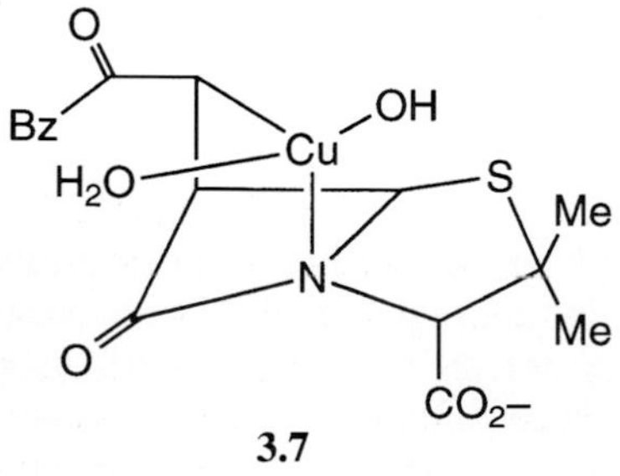

3.7

Figure 3-14. The copper(II)-directed hydrolysis of benzylpenicillin (**3.6**) showing the postulated intermediate, **3.7**.

3.4 Other Reactions of Nucleophiles and Carbonyl Compounds

We showed in Figs. 3-2 and 3-3 that the tetrahedral intermediate which is initially formed from the reaction of a nucleophile with a carbonyl compound may further react in a number of different ways. In this section, we will consider some reactions which proceed along the pathway indicated in Fig. 3-3. The hydration of ketones is a reaction analogous to the hydrolysis of an ester, with the first step of the reaction involving nucleophilic attack of water on the carbonyl group. The tetrahedral intermediate is trapped by reaction with a proton to yield the hydrated form of the ketone, the *geminal* diol (Fig. 3-15). Similar reactions occur with alcohols as nucleophiles to yield, initially, hemiacetals.

In practice, there is often an equilibrium between hydrated forms and ketones in an aqueous medium, although the equilibrium usually lies far over towards the ketone form. If however, one (or both) of the R groups is strongly electron-withdrawing (such as CF_3 or CCl_3), the polarisation of the C-C σ-bond connecting the R group to the carbonyl is sufficient to increase the electrophilic nature of the carbonyl carbon and increase the equilibrium percentage of the hydrated form. In some cases, these polarisations can be so great that the ketone (or aldehyde) is present almost entirely in the hydrated form. A typical example is seen when water is added to the liquid compound, CCl_3CHO; an immediate reaction occurs to yield the crystalline hydrated form $CCl_3CH(OH)_2$ ('chloral hydrate').

This suggests that in metal-mediated hydrations of ketones it might not be necessary for the carbonyl oxygen atom to be co-ordinated to the metal centre – the induced polarisation from another more remote co-ordination site might be sufficient. There are many examples in which a ketone group in a polydentate ligand is attacked by water or an alcohol to give products analogous to those of Fig. 3-15. The reactions of bis(2-pyridyl)keto-

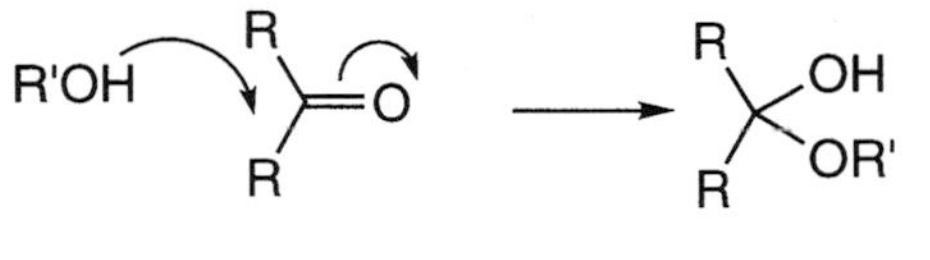

Figure 3-15. The formation of a hydrate (R' = H) or a hemiacetal (R' = alkyl or aryl) in the reaction of a ketone with water or an alcohol.

ne, **3.8**, with metal ions have been particularly well studied over the years. A range of nucleophiles, including water and alcohols, have been shown to add to the carbonyl group upon co-ordination of the pyridine rings to the metal. Note that this is a relatively long range effect – there is often no direct interaction of the carbonyl oxygen with the metal centre. In some cases, an alkoxy product is stabilised by co-ordination to a metal. The reaction of **3.8** with water in the presence of copper(II) salts gives copper(II) complexes containing the hydrated form of the ligand; in general, these complexes contain *N,N'*-bonded ligands, and there is no direct interaction between the metal and the oxygen in the product (Fig. 3-16).

A slightly different pattern of reactivity is seen when **3.8** interacts with antimony(III) fluoride in methanol. In this case, the product contains a ligand derived from the hemiacetal, but the hydroxy group is deprotonated and co-ordinated to the metal centre, to give an *N,N',O*-bonded anionic ligand (Fig. 3-17). It is not known whether the co-ordination of the oxygen to the metal centre occurs *after* the hydration reaction (in which case we are

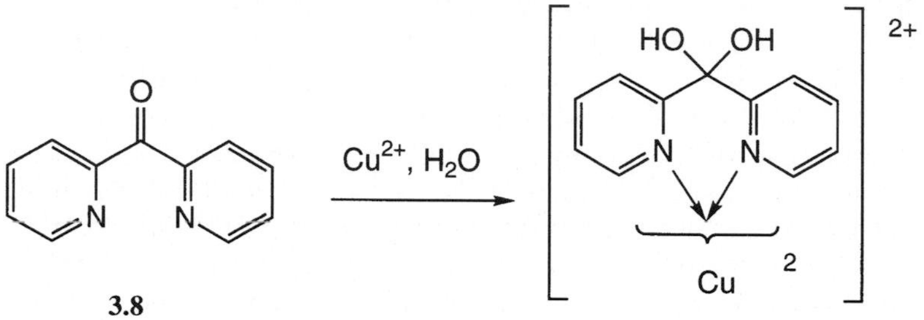

Figure 3-16. The hydration of bis(2-pyridyl)ketone upon reaction with aqueous copper(II) salts.

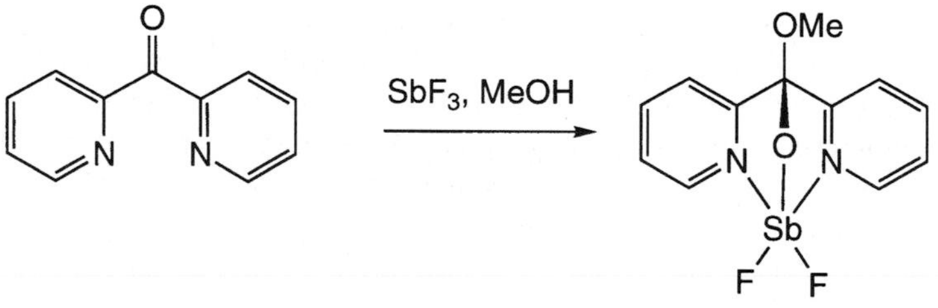

Figure 3-17. Methanolysis of **3.8** in the presence of antimony(III) fluoride.

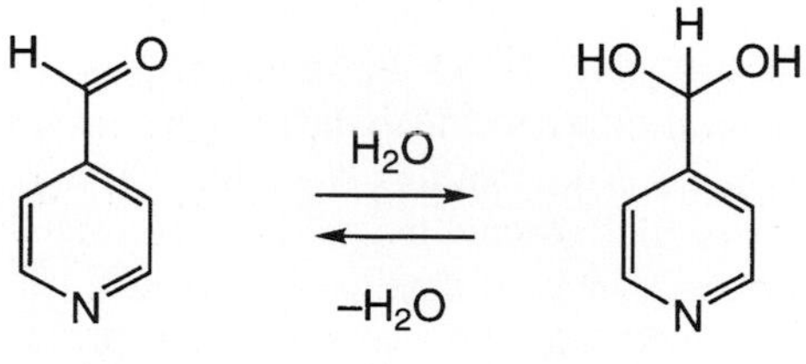

Figure 3-18. The hydration of 4-pyridinecarbaldehyde.

seeing a polarisation effect of the metal upon the hydroxy group leading to its deprotonation) or *prior* to the reaction, in which case we are seeing a direct activation of the carbonyl towards nucleophilic attack.

The effect of charge on the cation is important, and is clearly seen in the hydration of
4-pyridinecarbaldehyde, **3.9**. In aqueous solution the free ligand is only partially hydrated, and at equilibrium the ratio of the carbonyl form to the hydrate is about 55:45 (Fig. 3-
18). Upon co-ordination of the nitrogen atom of the pyridine to ruthenium(III) in the complex [Ru(NH$_3$)$_5$(**3.7**)]$^{3+}$, the ligand is over 90 % hydrated at equilibrium.

However, the role of the metal is not quite as simple as it might seem. Thus, the co-ordination of the 4-pyridinecarbaldehyde to ruthenium(II) in the complex cation
[Ru(NH$_3$)$_5$(**3.9**)]$^{2+}$ results in a favouring of the carbonyl form rather than the hydrate form,
and at equilibrium only about 10 % of the hydrate is present. At first sight this is rather
surprising considering the formal positive charge on the metal centre. The amount of the
hydrated form present at equilibrium is expected to be less with the ruthenium(II) than the
ruthenium(III) complex, as indeed it is, but even the ruthenium(II) complex is expected to
have more of the hydrate present than the free ligand. However, the previous assumptions
only rely upon the polarisation changes associated with co-ordination to the metal centre.
We can explain this observation in terms of a π-bonding effect and the back-donation from
the ruthenium centre into the π*-levels of the ligand. The lower oxidation state rutheni-
um(II) centre has a greater electron density at the metal and will be more effective in back-
donation than the less electron rich ruthenium(III) centre. The extended conjugation between the aromatic ring and the carbonyl group in the aldehyde form makes this a more
effective π-acceptor than the hydrated form. In consequence, back-donation from the
metal to the carbonyl form of the ligand is more effective with the ruthenium(II) complex.
This results in an additional stabilisation of the carbonyl form, and the hydrate form is
destabilised in the ruthenium(II) complex. There may also be a kinetic barrier towards
attack of the nucleophile on the ruthenium(II) complex, in which the back-donation reduces the positive charge on the carbonyl carbon atom.

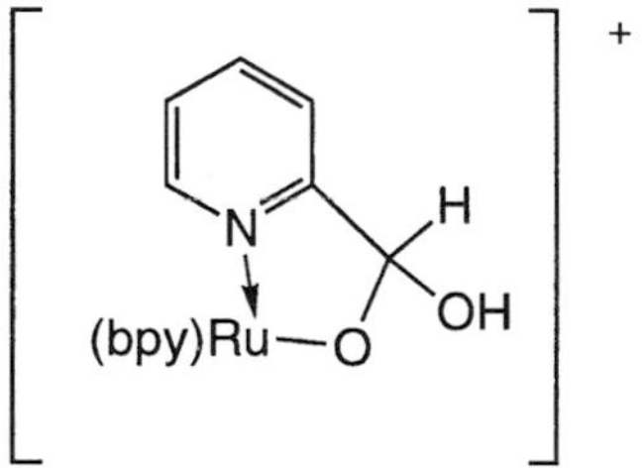

3.10

A direct metal involvement in the hydration of an aldehyde is seen in the ruthenium(II)
complex of the hydrated form of 2-pyridinecarbaldehyde (**3.10**).

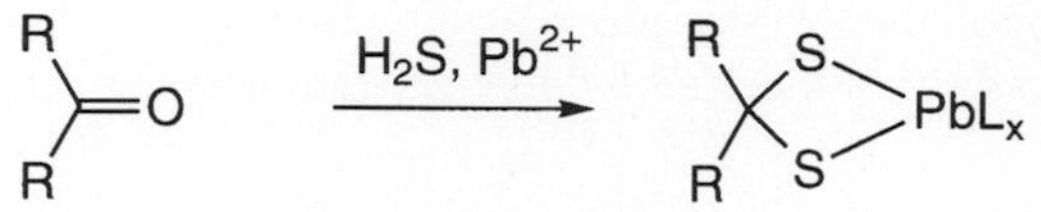

Figure 3-19. The reaction of a ketone with hydrogen sulfide in the presence of lead(II) to give a chelation stabilised *geminal* dithiolate. The other ligands co-ordinated to the lead centre are indicated by L_x.

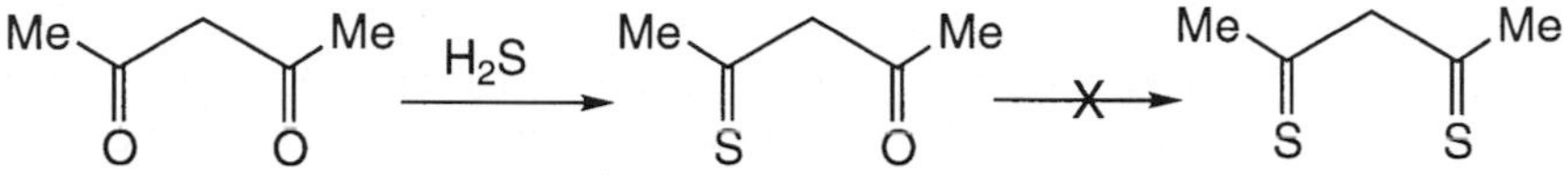

3.11

Figure 3-20. The reaction of a 1,3-diketone with hydrogen sulfide only gives a monothio derivative, **3.11**, rather than a 1,3-dithioketone.

Similar reactions are observed when other nucleophiles interact with carbonyl compounds. For example, the reactions of ketones with hydrogen sulfide are metal-ion dependent. In the same way that a *geminal* diol may be formed from the reaction of a carbonyl compound with water, so *geminal* $R_2C(SH)(OH)$ and $R_2C(SH)_2$ compounds may result from the reaction with hydrogen sulfide. The *geminal* dithiol is usually unstable with respect to loss of hydrogen sulfide and formation of the thiocarbonyl compound or hydrolysis. However, it may be stabilised by co-ordination of the deprotonated dithiolate form to a soft metal ion such as lead(II). In this case, the more important role of the metal ion is likely to be in stabilising the product of the reaction, rather than in the polarising of the carbonyl group (Fig. 3-19).

Similar reactions may occur with 1,3-diketones and 1,3-diketonates, and these reactions have been exploited by co-ordination chemists to prepare novel sulfur containing chelating ligands. The metal-directed reactions are synthetically useful since it is not possible to prepare 1,3-dithioketones by direct reactions of free 1,3-diketones with hydrogen

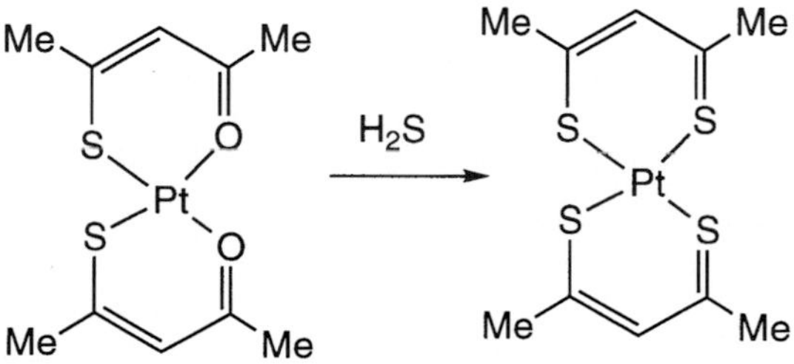

Figure 3-21. The metal directed reaction of a monothio compound with hydrogen sulfide to yield a chelated 1,3-dithioketonate.

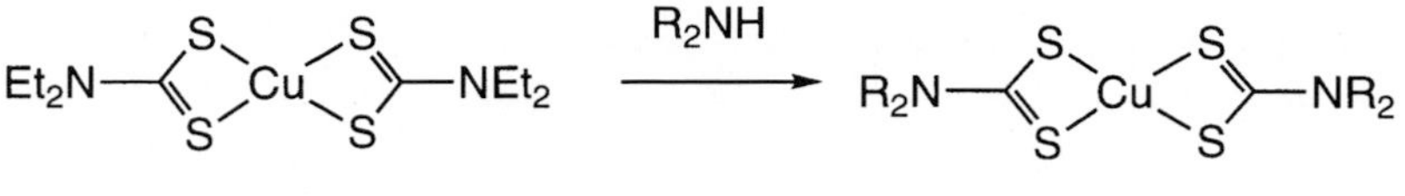

$$Et_2N-\underset{S}{\overset{S}{\diagdown}}Cu\underset{S}{\overset{S}{\diagup}}-NEt_2 \quad\xrightarrow{\quad R_2NH\quad}\quad R_2N-\underset{S}{\overset{S}{\diagdown}}Cu\underset{S}{\overset{S}{\diagup}}-NR_2$$

Figure 3-22. Transamination of bis(*N,N*-diethyldithiocarbamato)copper(II) with a secondary amine.

sulfide. The reaction only proceeds as far as the monothio derivative, **3.11** (Fig. 3-20). The dithio derivatives are expected to be excellent ligands for soft metal centres such as platinum(II) or palladium(II).

However, the dithioketones may be readily prepared by reaction of the metal complexes of the monothio derivatives, **3.11**, with hydrogen sulfide (Fig. 3-21). In this case, the metal probably plays a dual role, both in polarising the ligand and stabilising the product. Clearly, the use of a soft metal ion such as lead(II) or platinum(II) is to be preferred in a reaction of this type.

Although transesterification and transamidation reactions of simple carbonyl ligands are usually associated with the involvement of a co-ordinated nucleophile, there are some well-documented processes which unambiguously involve attack by an external nucleophile upon a co-ordinated electrophile. Dithiocarbamate complexes contain a chelated *S,S'*-bonded dithioamide ligand and undergo facile transamination reactions upon treatment of the copper complexes with amines (Fig. 3-22).

'Unusual' nucleophiles may also attack co-ordinated carbonyl groups. The biological interconversion of carbonyls and alcohols is usually achieved by hydride transfer catalysed by a metalloenzyme (see Chapter 10). The biological transfer agent is a dihydropyridine (NADH), which transfers a hydride to the carbonyl acceptor and forms a pyridinium salt (NAD⁺) (Fig. 3-23).

The transfer of the hydride to the carbonyl is likely to be promoted by increased polarisation of the carbonyl group by co-ordination to a metal. Clearly the best results will be obtained with a π-neutral Lewis acid in which the polarisation of the ligand is not countered by any back-donation from the metal to the ligand. Transition-metal ions which possess *no* electrons in *d* orbitals, or those in which the t_{2g} orbitals are not at a suitable energy for efficient overlap with the ligand π^*-levels, are likely to be the most efficient Lewis acid catalysts. A d^{10} configuration, as found in zinc(II), proves to confer excellent Lewis acid properties, and it is indeed zinc that is found in liver alcohol dehydrogenase, one of the enzymes which achieves this transformation in higher animals. A series of model compounds have been developed to improve our understanding of hydride transfer reactions of

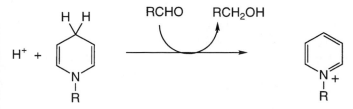

Figure 3-23. Representation of the hydride transfer reaction involved in the biological reduction of aldehydes to alcohols.

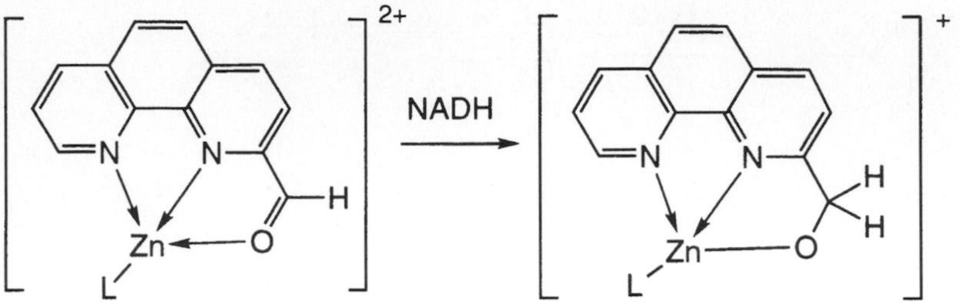

3.12

Figure 3-24. A zinc(II) complex which acts as a functional model for the hydride transfer reaction which occurs at the active site of the enzyme liver alcohol dehydrogenase.

this type. The complex **3.12** has a carbonyl-zinc bond which is stabilised by the chelation. The co-ordination of the 1,10-phenanthroline-2-carbaldehyde to zinc(II) dramatically activates it towards the transfer of hydride from a dihydropyridine, and detailed mechanistic studies have been made on this, and related, reaction systems (Fig. 3-24).

3.5 Conclusions

In this chapter we have seen that many of the fundamental 'organic' reactions of carbonyl compounds may be controlled by co-ordination to a metal ion. Reactions involving the attack of nucleophiles upon the electrophilic carbon atom of a carbonyl group are enhanced by co-ordination of the oxygen atom to a metal ion. This enhancement is particularly strongly expressed in the reactions of chelating or potentially chelating ligands.

In the next chapter we will observe further reactions of co-ordinated carbonyl compounds in which the nucleophile is also co-ordinated to the metal centre. These intramolecular reactions are sufficiently different in character to merit separate treatment.

Suggestions for further reading

1. R.W. Hay in *Comprehensive Co-ordination Chemistry, Vol. 6*, (eds. G. Wilkinson, R.D. Gillard, J.A. McCleverty), Pergamon, Oxford, **1987**.
   – An excellent account of the role of metal ions in the solvolysis of amides and related compounds.
2. R.W. Hay in *Reactions of Co-ordinated Ligands. Vol. 2*, (ed. P.S. Braterman), Plenum, New York, **1989**.
   – Very similar to the article above.

4 Other Reactions of Co-ordinated Ligands with Nucleophiles

4.1 General

In Chapter 2 we described the ways in which the co-ordination of a ligand to a positively charged or electropositive metal centre might enhance the reactions of nucleophiles with that ligand. In the previous chapter we saw specifically how the reactions of carbonyl compounds with nucleophiles may be dramatically accelerated on co-ordination of the carbonyl ligand to a metal ion. In this chapter, we will consider related reactions of a series of other electrophilic or potentially electrophilic co-ordinated ligands with nucleophiles. Once again, many of the examples involve the use of polydentate ligands and complexes that are either kinetically or thermodynamically stabilised to allow the observation of intermediate species or detailed mechanistic studies.

The reaction of a co-ordinated ligand with a nucleophile may be one of two types, which differ both in mechanistic detail and in synthetic consequence. The first type is that which we have already considered in some detail for carbonyl groups in Chapter 3. In these reactions, the nucleophile is added to an *unsaturated* carbon centre. This results in a change of hybridisation at carbon (Fig. 4-1). Specifically we have only so far discuss-

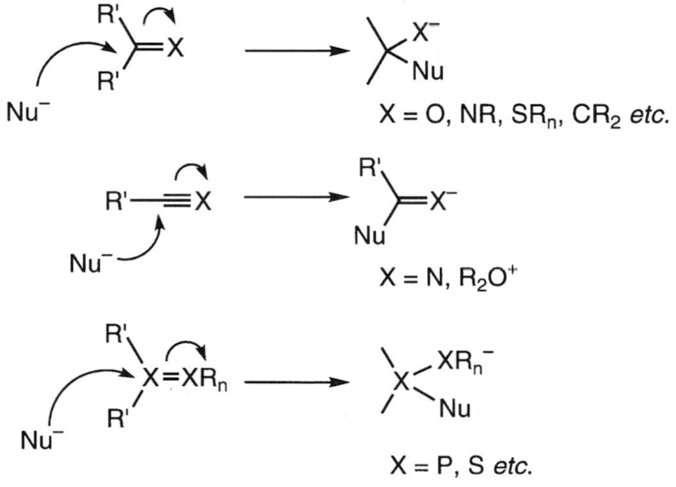

Figure 4-1. The addition of nucleophiles to a variety of unsaturated compounds showing the formation of an intermediate species with an increased co-ordination number.

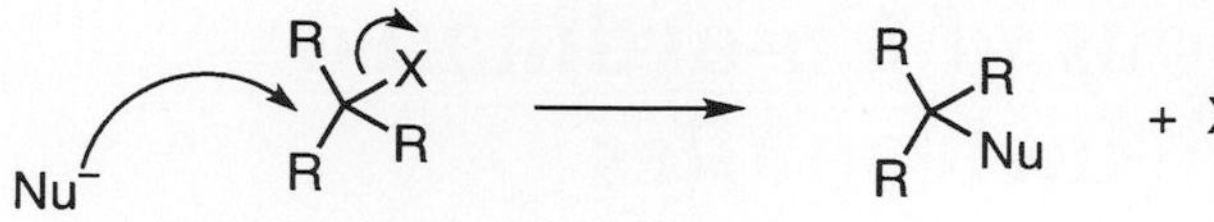

Figure 4-2. A classical S_N2 substitution process.

ed reactions of sp^2 hybridised carbon atoms in C=O compounds, but exactly analogous reactions occur at other sp^2 hybridised carbon centres ($R_2C=X$) and at sp hybridised centres. Similarly, reactions are not limited to the nucleophile attacking an unsaturated carbon centre, and many examples are known in which the unsaturated electrophilic centre is
an element other than carbon. The fate of the new compound with an increased co-ordination number at carbon (or at another element, X) may vary; it might be an isolable species, or might undergo further reaction to regenerate an unsaturated species.

The second type of reaction involves the attack of the nucleophile upon a *saturated*
centre, with concomitant loss of a suitable leaving group. This second type of reaction corresponds to a classical S_N2 process (Fig. 4-2).

Although the loss of some anionic or other good leaving group is the usual consequence of such reactions at carbon centres, this is not the case when the electrophilic centre is an element such as silicon, sulfur, phosphorus or a transition metal with available *d*
orbitals in the valence shell. In these cases, a simple addition reaction may occur to give
a product of increased co-ordination number. This may involve simple reactions such as
the addition of an anion to a neutral molecule or more complex processes (Fig. 4-3).

A metal ion might play a number of roles in modifying the behaviour of co-ordinated
electrophiles. Co-ordination of a donor atom X to the metal polarises the C–X bond and
increases the electrophilic character of the carbon. In the case of an unsaturated centre this
polarisation may be countered or even completely reversed by back-donation from filled
(*d*) orbitals on the metal into vacant π^*-(or similar) orbitals of the ligand. The X^- group
generated after nucleophilic attack on the carbon atom may also be stabilised by co-ordi-

$$PF_5 + F^- \rightleftharpoons [PF_6]^-$$

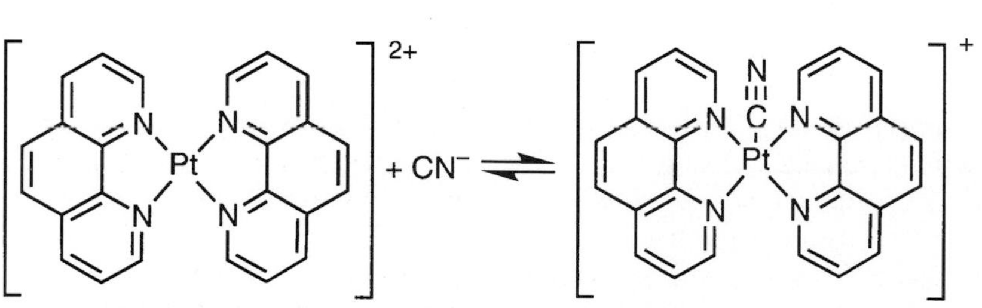

Figure 4-3. Addition of a nucleophile may lead to the formation of stable species with an increased
co-ordination number. This may involve neutral acceptors or charged metal complexes.

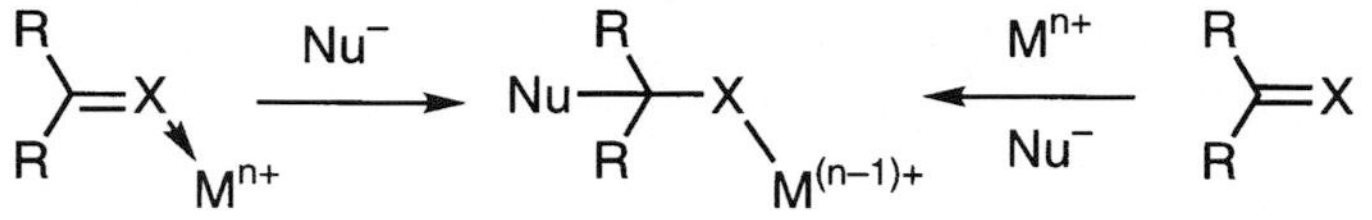

Figure 4-4. Metal-ion control in the reaction of a nucleophile with an unsaturated ligand.

nation to the metal centre. With attack upon an unsaturated centre, the intermediate that is formed may experience a variety of fates. The metal ion might polarise the ligand *and* stabilise the intermediate, or it might only stabilise the intermediate (Fig. 4-4). The former case will be typically characterised by a change in the *rate* of reaction with the nucleophile, the latter by changes in the overall position of equilibrium and product distribution.

Alternatively, the metal may act by assisting the departure of a leaving group from the intermediates generated in Fig. 4-4, as is shown in Fig. 4-5.

We shall start our study by considering reactions involving attack of nucleophiles upon unsaturated centres within ligands co-ordinated to a metal.

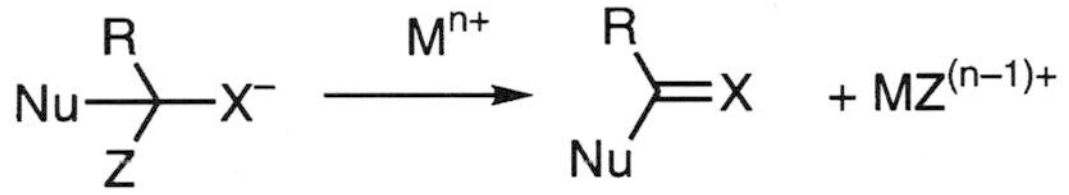

Figure 4-5. Metal-assisted loss of a leaving group.

4.2 Reactions of Nucleophiles with sp Hybridised Carbon Centres

In this book we are not concerned with 'organometallic' reactivity of compounds containing M–C bonds, and so to all intents and purposes this section is about the reactions of nitriles or cyanides with nucleophiles. Reactions of co-ordinated alkynes are adequately discussed elsewhere.

4.2.1 Nucleophilic Attack on Co-ordinated Nitriles

Many of the reactions which have been reported for carbonyl compounds (hydration, alcoholysis, aminolysis, *etc.*) are also shown by co-ordinated nitrile ligands. The reaction of a nucleophile with a nitrile is a two-stage process, which may involve one or two molecules of the incoming nucleophile (Fig. 4-6).

In the case of water as a nucleophile, the initially produced hydroxyimine may tautomerise to an amide, which in turn generates a carboxylic acid upon further hydrolysis. Hydrolysis of a nitrile is, of course, one of the standard classical methods for the synthesis of carboxylic acids (Fig. 4-7).

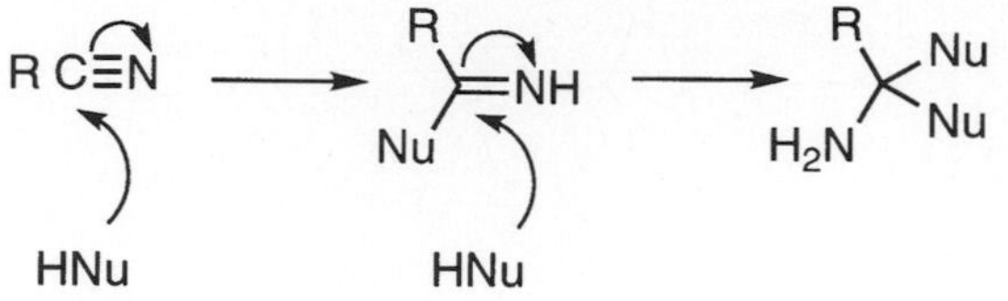

Figure 4-6. Nucleophilic attack upon a nitrile.

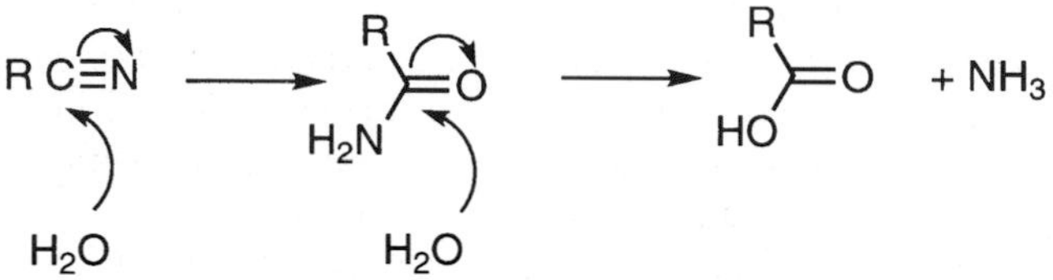

Figure 4-7. The hydrolysis of a nitrile to generate a carboxylic acid.

The metal-promoted reactions with nucleophiles generally yield metal complexes of the intermediate imino species and it is quite unusual to observe the carboxylate products derived from the attack of a second equivalent of nucleophile. A wide variety of products have been observed. The ligands may be monodentate *O*- or *N*-bonded species, or exhibit a didentate *N,O*-mode. The ligands may be in the amide or the hydroxyimine tautomeric forms, and may be neutral or deprotonated (Fig. 4-8).

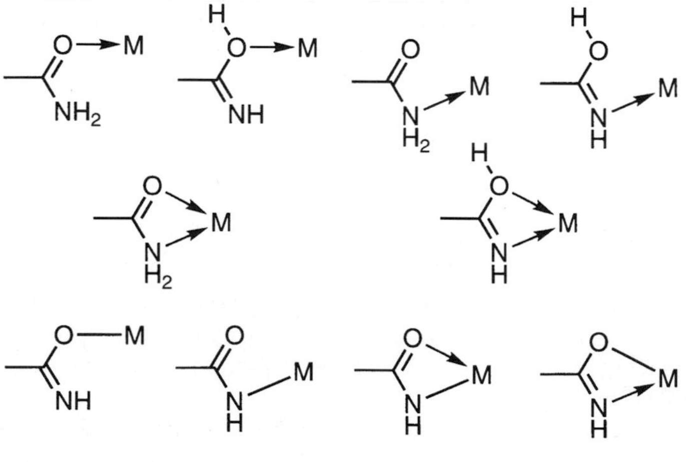

Figure 4-8. Some possible amide or amido bonding modes.

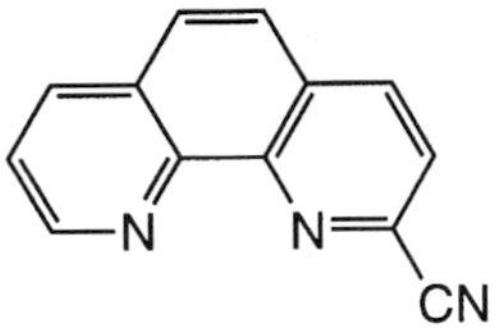

4.1

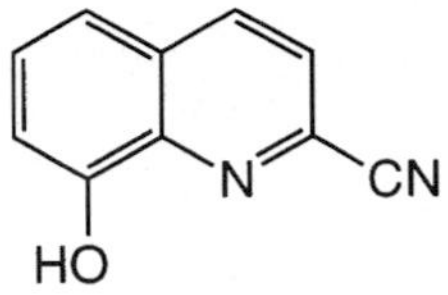

4.2

The hydrolysis of nitriles has been investigated in some detail. Numerous metal salts have been shown to be effective in promoting the hydrolysis of nitriles to amides. The attack has been shown in some cases to be by co-ordinated hydroxide ion, whereas in other cases it is by an external nucleophile. In this section we are only concerned with the latter case. We will return to the involvement of metal-co-ordinated nucleophiles in Chapter 5. The polarisation of the ligand is one of the principal features in these reactions. Thus, the rate of hydrolysis of the nitrile ligands in ruthenium(II) complexes such as $[Ru(NH_3)_5(N{\equiv}CR)]^{2+}$ is a million times slower than that in the analogous ruthenium(III) species $[Ru(NH_3)_5(N{\equiv}CR)]^{3+}$. This is as expected for a reaction in which the principle acceleration is due to ligand polarisation. Apparently, back-bonding effects are not very pronounced with the nitrile ligands. In the case of $[Co(NH_3)_5(N{\equiv}CR)]^{3+}$ complexes, the rate of hydrolysis is so dramatically enhanced that conversion to co-ordinated amide is essentially instantaneous on reaction with hydroxide. The initial products of these reactions are deprotonated N-bonded amides (Fig. 4-9), although subsequent processes may result in protonation, ligand loss or rearrangement to other bonding modes. Once again, notice the use of kinetically inert metal centres to allow the isolation of metal-bound intermediates.

Particularly well-studied examples of the hydration of nitriles are seen with the chelating ligands 2-cyano-1,10-phenanthroline (**4.1**) or 2-cyano-8-hydroxyquinoline (**4.2**). The products of the reactions are the appropriate amides, and the rate of hydration is enhanced by up to ten million times on co-ordination to nickel(II), copper(II) or cobalt(III).

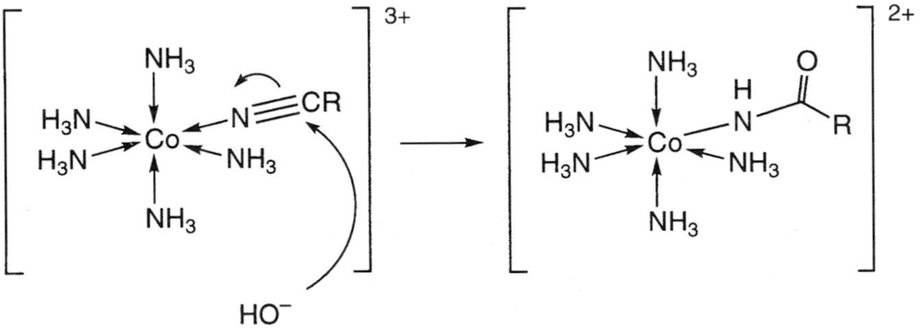

Figure 4-9. Conversion of a co-ordinated nitrile to a monodentate N-bonded amido species at a kinetically inert cobalt(III) centre.

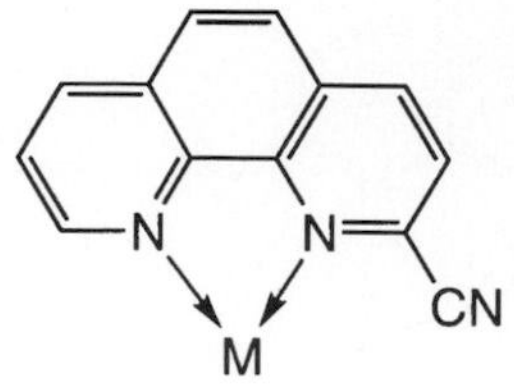

4.3

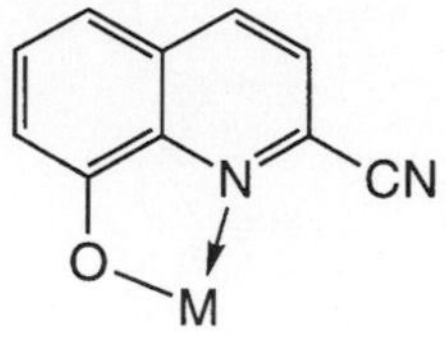

4.4

At first sight these reactions are simple examples of metal-activated nucleophilic attack upon the nitrile carbon atom. However, the geometry of the co-ordinated chelating ligand is such that the nitrile nitrogen atom is not co-ordinated to the metal ion (**4.3** and **4.4**)! It was initially thought that this provided evidence for a mechanism involving intra-molecular attack by co-ordinated water or hydroxide (Fig. 4-10). However, detailed mechanistic studies of the pH dependence of the reaction have demonstrated that the attack is by external *non-co-ordinated* water (or hydroxide) (Fig. 4-11).

The metal ion is thought to act in two ways. First, a long range polarisation through the ligand σ-bonding framework may increase the electrophilic character of the nitrile carbon, and, second, stabilisation of the intermediate anionic species by co-ordination to the charged centre. This latter process is thought to be more important, and is reflected in a very favourable activation entropy term for the reaction. Circumstantial support comes from the observation that the hydrolysis of 2-cyanopyridine, but not that of 3-cyano- or 4-cyanopyridine, is enhanced by metal ions. Only in the case of the 2-cyanopyridine is it possible for the intermediate to be stabilised by co-ordination to the metal.

A special example of the hydrolysis of a nitrile is seen in the copper-promoted addition of water (or an alcohol) to 3-cyano-2-azamaleodinitrile (**4.5**). It is of interest that the nitrile is specifically attacked in preference to the imine (Fig. 4-12).

The addition of nucleophiles to co-ordinated nitriles is not limited to water. We have already alluded to the use of alcohols. The reaction with an alcohol differs in one major way from that with water: the intermediate cannot tautomerise and an amidate ester is obtained (Fig. 4-13).

The alcoholysis parallels the hydrolysis in that the metal may act either by polarising the ligand or by stabilising the intermediate. The latter is seen in the very rapid ethanol-

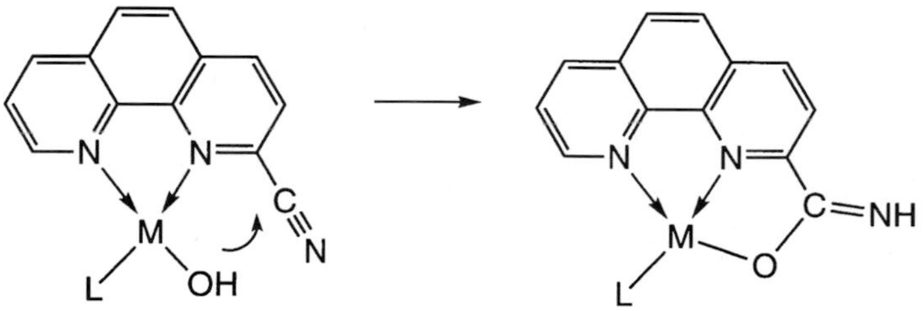

Figure 4-10. The initially proposed, but incorrect, mechanism for the nickel(II)-assisted hydrolysis of **4.1** involving nucleophilic attack by a co-ordinated hydroxide.

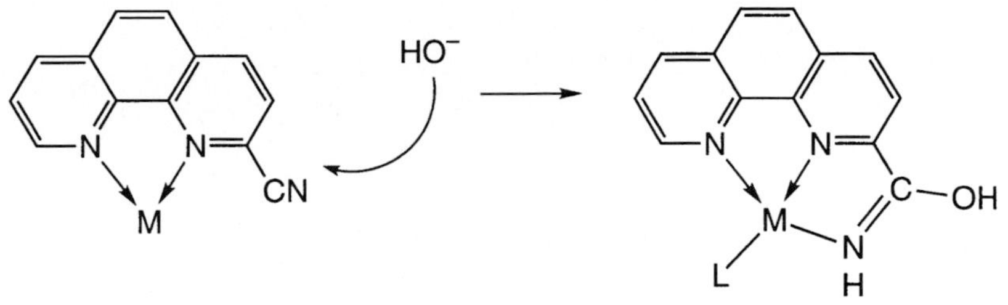

Figure 4-11. The attack of an external nucleophile upon a complex containing the nitrile ligand **4.1**.

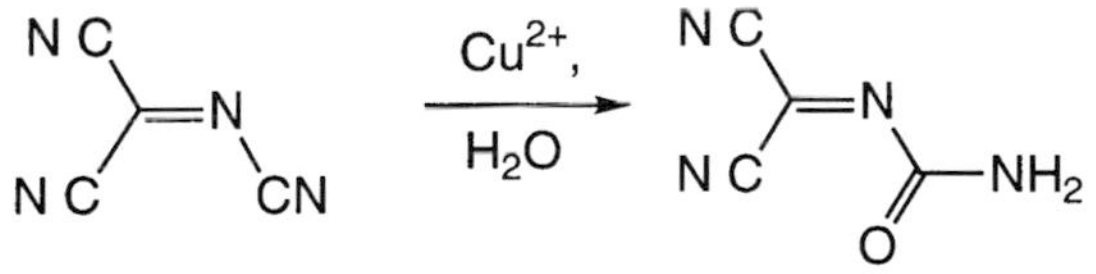

4.5

Figure 4-12. The hydrolysis of 3-cyano-2-azamaleodinitrile in the presence of copper(II) salts.

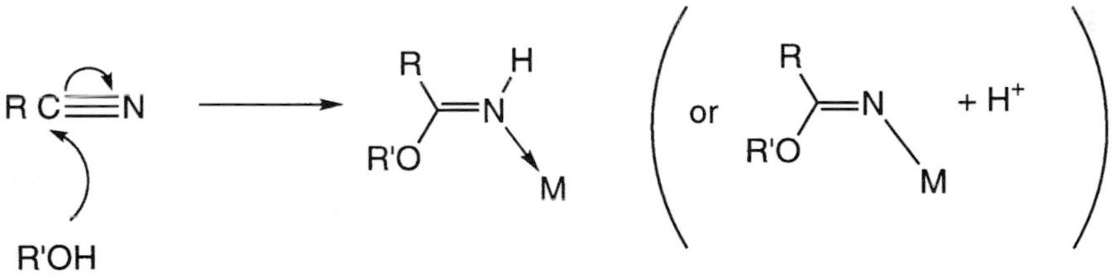

Figure 4-13. The alcoholysis of a co-ordinated nitrile.

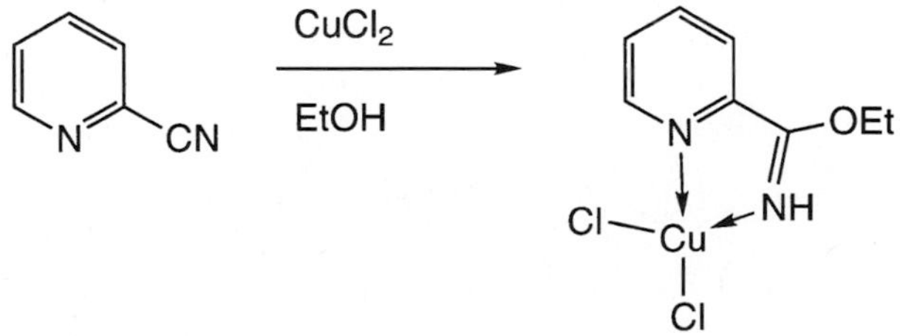

Figure 4-14. The copper-promoted ethanolysis of 2-cyanopyridine.

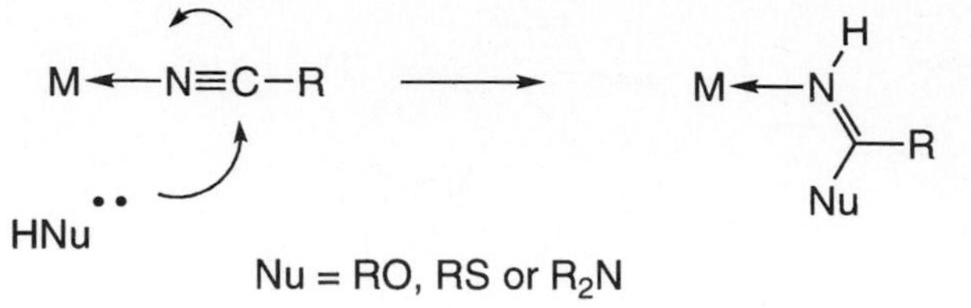

$$M \leftarrow N \equiv C - R \longrightarrow M \leftarrow \overset{H}{N} \underset{Nu}{\overset{}{\diagup}} R$$

HNu

Nu = RO, RS or R_2N

Figure 4-15. General scheme for the attack of nucleophiles on a co-ordinated nitrile.

ysis of 2-cyanopyridine (but not the 3- or 4-isomer) in the presence of copper(II) chloride (Fig. 4-14). The ligand co-ordinates to the metal through the pyridine nitrogen atom, and it is worth noting that 2-cyanopyridine does not react with ethanol in the absence of the metal ion.

A very wide range of nucleophiles other than water and alcohols has been shown to react with co-ordinated nitriles (Fig. 4-15). The reaction with ammonia or amines is a particularly convenient method for the preparation of amidine complexes. Specific examples of these reactions are presented in Fig. 4-16, illustrating the formation of amide, iminoether or amidine complexes.

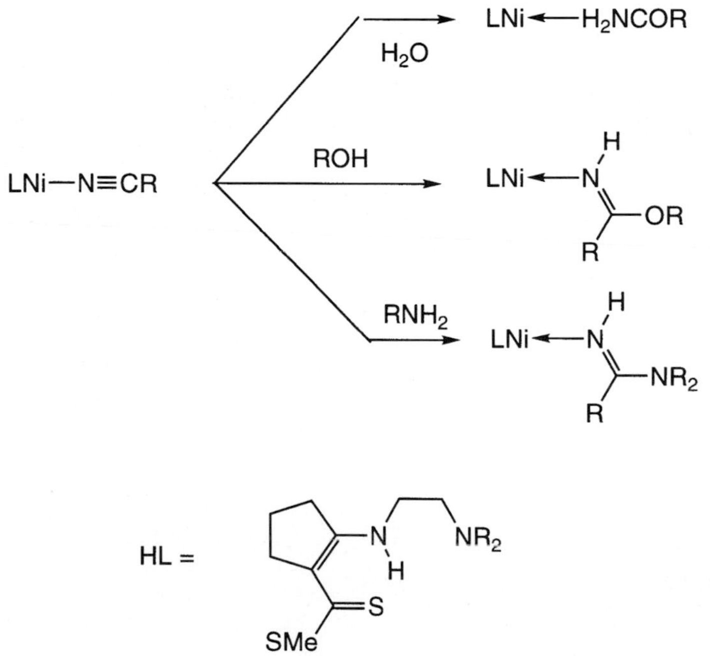

$$LNi \leftarrow H_2NCOR$$

H_2O

ROH

$$LNi - N \equiv CR$$

$$LNi \leftarrow \overset{H}{N} \underset{R}{\overset{}{\diagup}} OR$$

RNH_2

$$LNi \leftarrow \overset{H}{N} \underset{R}{\overset{}{\diagup}} NR_2$$

HL =

Figure 4-16. Reactions of a co-ordinated nitrile with a range of nucleophiles (L = anionic *N,N,S* donor ligand).

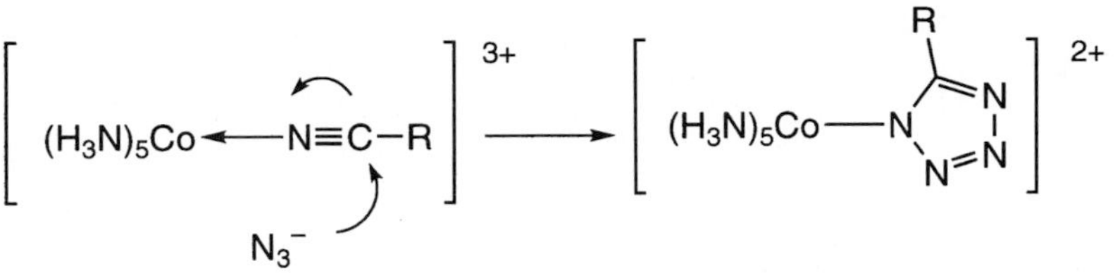

Figure 4-17. The nucleophilic addition of azide to a nitrile to generate a co-ordinated tetrazole. The complex that is initially produced tautomerises to the isomer shown.

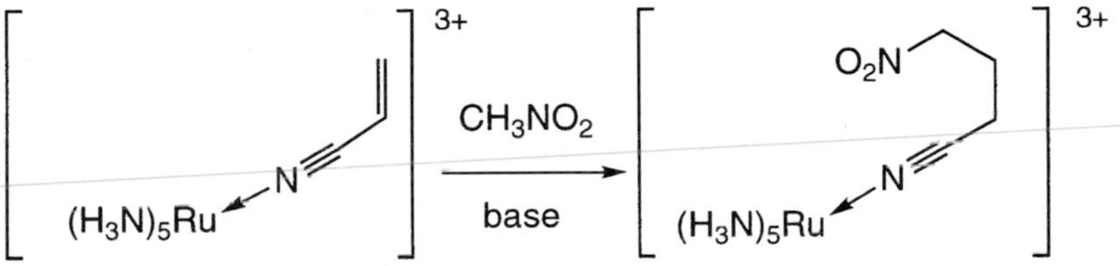

Figure 4-18. The Michael addition of the $O_2NCH_2^-$ anion to co-ordinated acrylonitrile.

Even nucleophiles such as the azide ion react with co-ordinated nitriles, and this provides a method for the preparation of tetrazolate complexes (Fig. 4-17).

The polarisation produced by co-ordination to the metal may be transmitted through a conjugated system. Michael addition reactions of nucleophiles to N-bonded acrylonitrile are known, and provide a convenient method for the preparation of derivatives. A wide range of nucleophiles may be used in these conjugate additions. For example, the anion of nitromethane (generated in situ) reacts with the ruthenium(III) complex $[Ru(NH_3)_5N\equiv CCH=CH_2)]^{3+}$, **4.6**, to yield a complex of 4-nitrobutyronitrile (Fig. 4-18).

Reactions of this type are widely used for the preparation of encapsulating ligands (see Chapter 7).

The reduction of nitriles by the nucleophilic attack of hydride transfer reagents has also been widely investigated and is a process with considerable synthetic potential. The reduction yields amine complexes, and rates are typically about ten thousand times faster than for reduction of the free ligand (Fig. 4-19). Once again, the principal effect appears to be associated with the build-up of positive charge on the ligand.

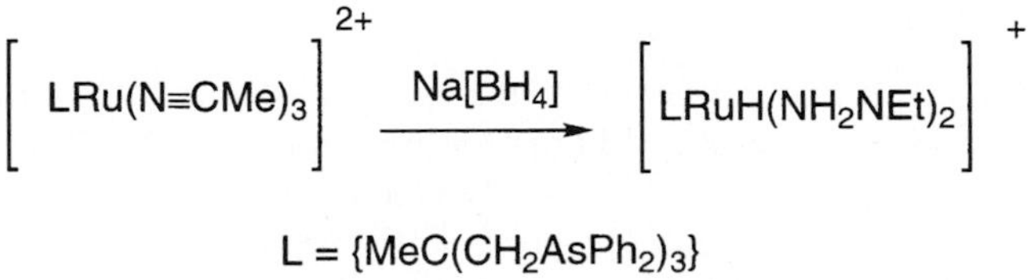

$$L = \{MeC(CH_2AsPh_2)_3\}$$

Figure 4-19. The reduction of a co-ordinated nitrile to a co-ordinated amine.

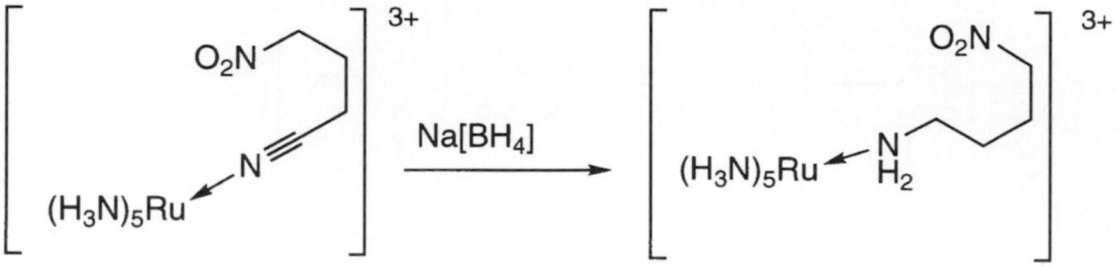

Figure 4-20. The reduction of a co-ordinated nitrile to an amine

    The reduction is relatively selective; for example, the nitrile group of the Michael addition product shown in Fig. 4-18 may be selectively reduced by sodium borohydride to a complex of 4-nitrobutylamine (Fig. 4-20).

4.3 Reactions of Nucleophiles with sp^2 Hybridised Carbon Centres

In the same way that we were primarily concerned with reactions of nitriles in the previous section, we will be concerned with the attack of nucleophiles on imines in this section. Imines, $R_2C=NR'$, are the nitrogen analogues of carbonyl groups, and we saw in Chapter 2 that imines may be stabilised by co-ordination to a metal ion capable of back-donation to the ligand π^*-levels. We shall investigate the synthetic utility associated with the formation of co-ordinated imines in a later chapter. However, it is also possible to *promote* the hydrolysis of the imine by co-ordination to a positively charged metal ion.

4.3.1 Hydrolysis of Imines

The equilibrium between carbonyl compound, amine and imine is metal-ion dependent and the position of the equilibrium may be perturbed by the co-ordination of any of the components to a metal ion (Fig. 4-21).

    In many cases, free imines are hydrolytically unstable. In general, it is difficult to form imines from carbonyl compounds and amines in aqueous solution. This is not always the case, and it is sometimes possible to form conjugated imines which are stabilised by delocalisation in aqueous conditions (Fig. 4-22).

    The balance between stabilisation and activation of the imine towards hydrolysis depends on the relative polarisation of the ligand and the back-donation from the metal, as discussed in Chapter 2. It is very difficult to successfully predict the overall stabilisation or destabilisation of a given imine towards hydrolysis in the presence of a given metal ion. Some imines are stabilised by co-ordination to copper(II), whereas others are destabilised (Fig. 4-23).

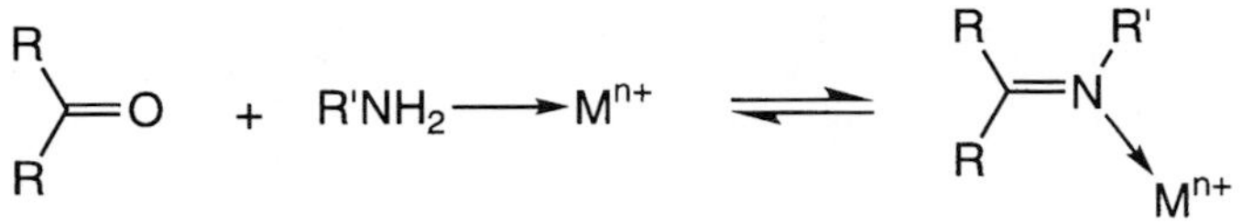

Figure 4-21. Co-ordination of amine or imine to a metal ion may control the position of equilibria involving imines.

The presence of electrons in *d* orbitals, which may be involved in back donation, is not a prerequisite for the stabilisation of an imine by co-ordination; some imines are stabilised by co-ordination to lead(II). The many factors involved (charge on metal, charge on ligand, back-donation, configuration of ligand, stabilisation of products, *etc.*) are interdependent and finely balanced. The formation of a chelated imine complex is an important factor, but once again examples are known in which chelated ligands are either activated or deactivated towards hydrolysis.

In general, the greater the thermodynamic stability of the imine complex, the smaller the tendency towards hydrolysis. The hydrolysis of the imine formed from aniline and benzaldehyde is enhanced 100,000 times in the presence of copper(II). The importance of the electron configuration of the metal ion is seen in the reactions of this same ligand; the imine is *stabilised* with respect to hydrolysis on co-ordination to a d^6 iron(II) centre. This may be partially ascribed to the effective back-donation from the low-spin d^6 centre (Fig. 4-24). In this case, the free imine is reasonably stable to hydrolysis in the absence of metal ions.

Paradoxically, this imine is structurally very closely related to the amidate ester which is *produced* by the ethanolysis of 2-cyanopyridine in the presence of copper(II) (Fig. 4-14)! There is indeed a very fine balance between destabilisation and stabilisation of the co-ordinated imine.

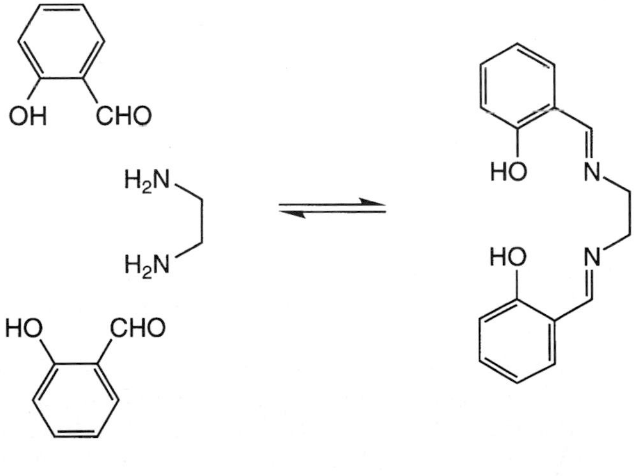

Figure 4-22. The imine formed from the reaction of 1,2-diaminoethane with salicylaldehyde is stable to hydrolysis.

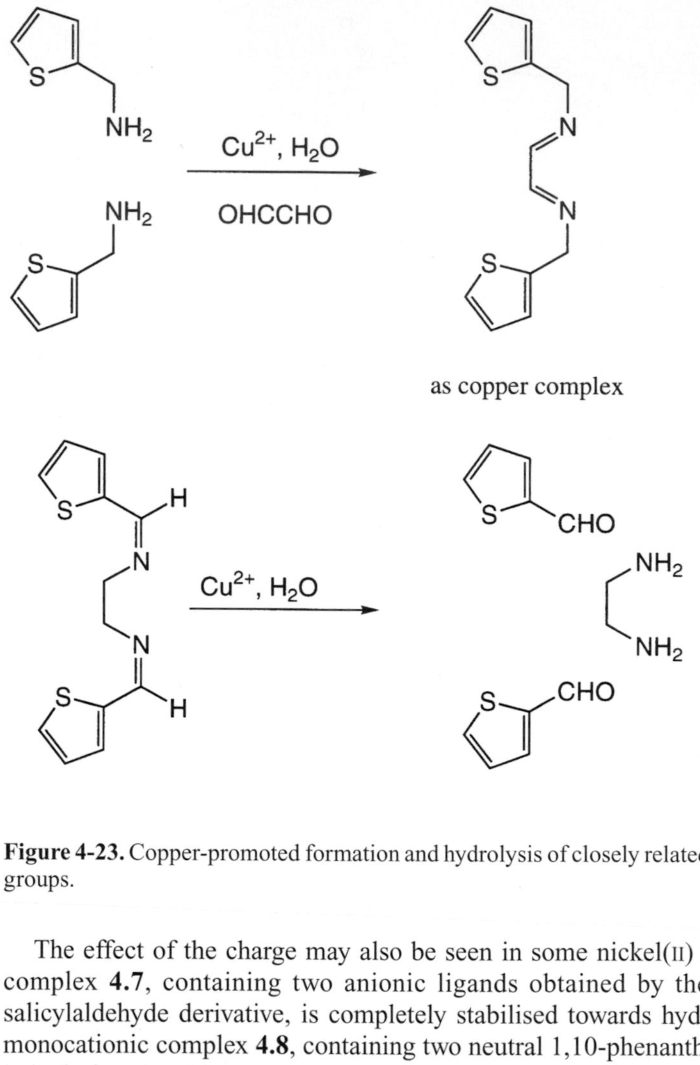

as copper complex

Figure 4-23. Copper-promoted formation and hydrolysis of closely related imines containing thienyl groups.

The effect of the charge may also be seen in some nickel(II) complexes. The neutral complex **4.7**, containing two anionic ligands obtained by the deprotonation of the salicylaldehyde derivative, is completely stabilised towards hydrolysis. In contrast, the monocationic complex **4.8**, containing two neutral 1,10-phenanthroline ligands, is rapidly hydrolysed to the corresponding salicylaldehyde complex (Fig. 4-25). Presumably, the overall positive charge of the complex promotes the attack of the nucleophile upon the electrophilic carbon centre.

The interconversion of imines and the constituent carbonyl compounds and amines is a stepwise process in which the first stage is the formation of an intermediate aminoalcohol or aminol (Fig. 4-26).

Many examples are known in which the co-ordination of an imine to a metal centre *activates* it towards nucleophilic attack by water to yield the aminol (aminoalcohol) or related derivative. In the absence of the metal ion, most aminols either dehydrate to yield imines or collapse to the parent amines and carbonyl compound.

We will consider the reactivity of macrocyclic compounds in more detail in chapter 6, but let us consider here the reaction of a nickel(II) cyclic imine complex with water.

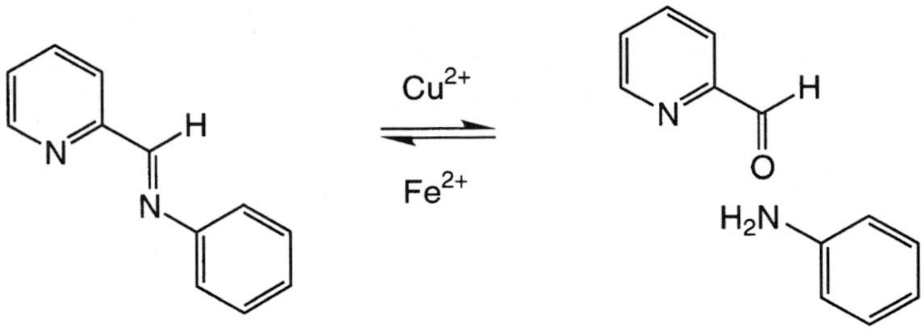

Figure 4-24. The hydrolysis of the imine formed from the reaction of aniline with benzaldehyde. The position of equilibrium is to the left in the presence of iron(II) and to the right in the presence of copper(II).

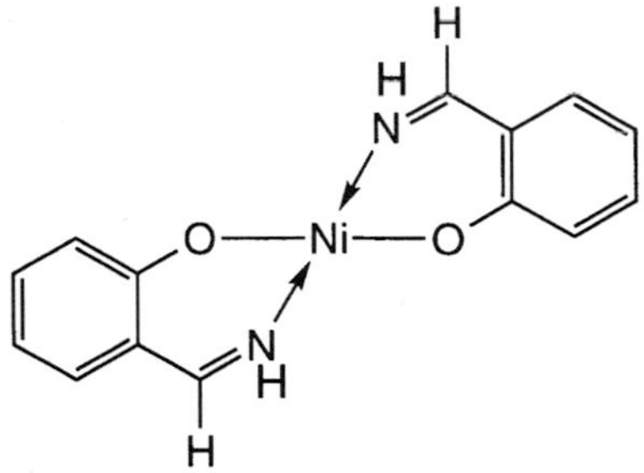

4.7

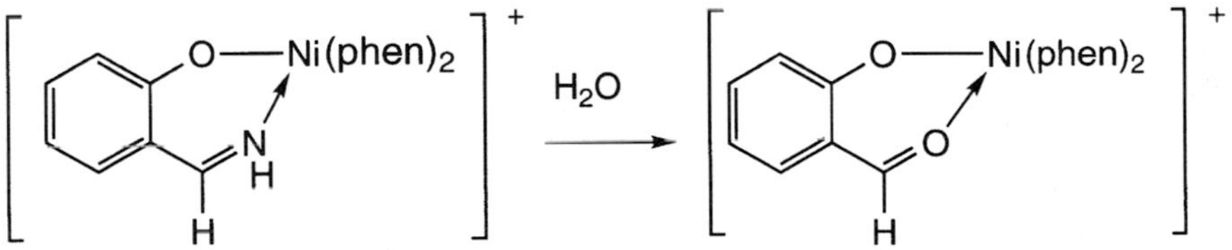

4.8

Figure 4-25. Hydrolysis of the cationic nickel(II) complex **4.8**.

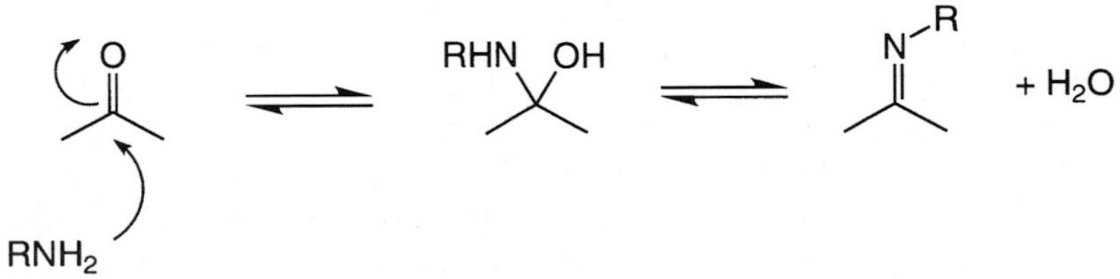

Figure 4-26. The aminoalcohol (aminol) intermediate in the formation of an imine.

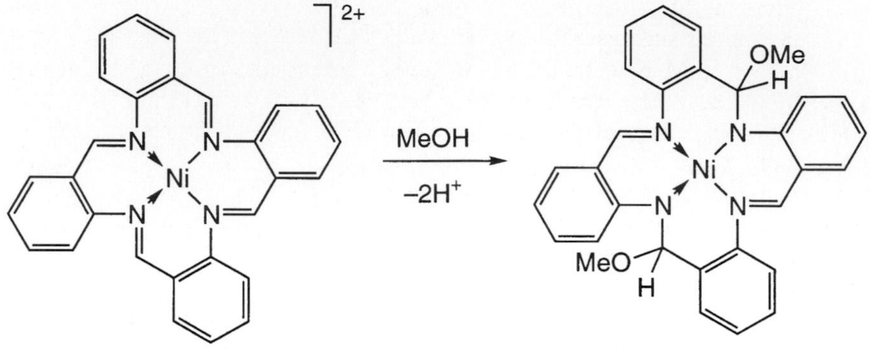

4.9

Figure 4-27. A hole size effect is involved in the alcoholysis of the imine to give the aminol complex.

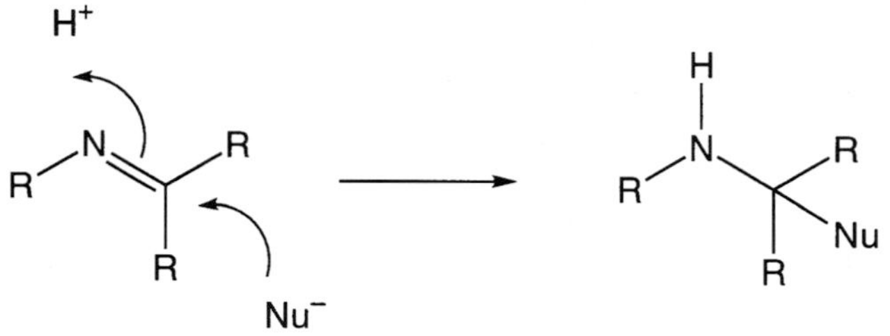

Figure 4-28. The addition of a general nucleophile HNu across an imine C=N double bond.

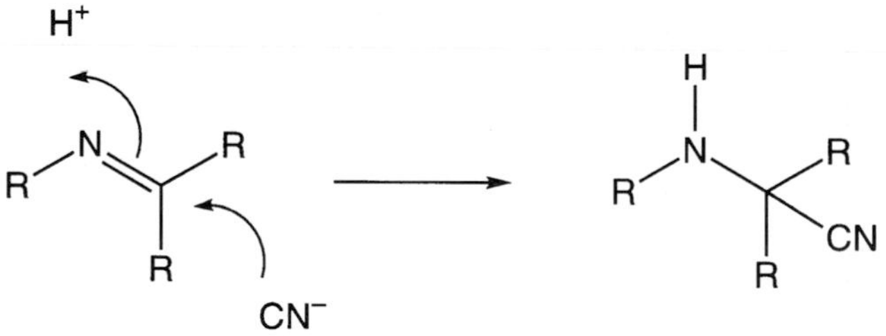

Figure 4-29. The formation of a cyanoamine by the addition of HCN to an imine.

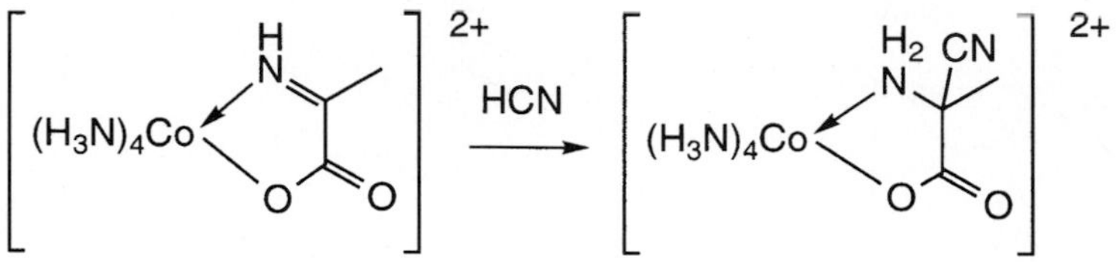

Figure 4-30. The addition of HCN to an imine co-ordinated to a kinetically inert cobalt(III) centre.

The four nitrogen donor atoms define points on a circle of radius r. This radius is fixed in relatively rigid ligands such as **4.9**, and so any metal ion centrally placed within this cavity will have a fixed M-N distance *regardless* of the ionic radius of the metal ion. In the case of **4.9** the cavity is slightly too large for optimal Ni-N distances. This could either result in the labilisation of the complex, or in its activation towards some reaction which might alter the cavity size. The conjugated ligand is rigid and cannot adopt a conformation which gives optimal Ni-N contacts. One of the ways in which the nitrogen donor atoms can be moved closer to the metal ion is by destroying the planarity of the ligand and allowing the donor atoms to 'fold over' the metal ion. This is very readily achieved by the addition of water or an alcohol to the imine to yield an aminol derivative, with the concomitant generation of a tetrahedral sp^3 hybridised centre. A common feature of reactions of this type is the formation of an amido complex of a deprotonated amine, as seen in Fig. 4-27.

Notice that there is also some element of charge control over this reaction – the product is the neutral *bis* adduct containing the deprotonated form of the new ligand rat-

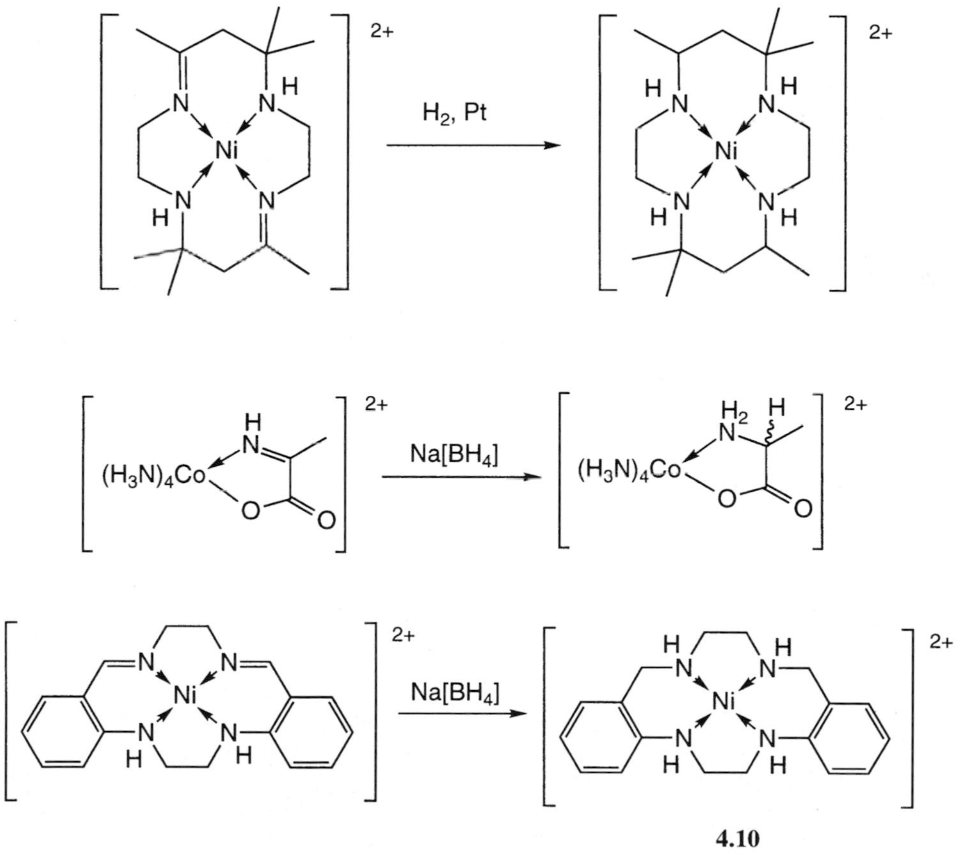

Figure 4-31. The reduction of co-ordinated imines by dihydrogen or sodium borohydride.

her than the dianionic tetrakis(methanol) adduct which would arise by the addition of methanol to each of the imine bonds, although the size of the newly-generated cavity will also dictate the number of nucleophiles that add. Other examples of reactions of this type will be encountered in the co-ordination chemistry of macrocyclic ligands, and in the next chapter when we consider the *formation* rather than the *hydrolysis* of imines.

Co-ordinated imines have also been shown to react with a wide range of other nucleophiles, resulting in a formal addition of HNu across the C=N bond (Fig. 4-28).

A typical example is seen in the addition of hydrogen cyanide to an imine to yield a cyanoamine (Fig. 4-29). Many of these reactions have been used to best advantage in the synthesis of macrocyclic ligands and complexes, and as such are considered in Chapter 6. A simple example of such a reaction is seen in the addition of HCN to the cobalt(III) complex indicated in Fig. 4-30. The starting complex is also readily prepared by a metal-directed reaction.

4.3.2 Reduction of Imines

The reduction of a co-ordinated imine is a reaction of very great synthetic significance in the preparation of macrocyclic ligands, and is also involved in more general Leuckart-type reductions leading to secondary amines. The reduction may be achieved by the use of molecular dihydrogen or a hydride transfer agent such as sodium borohydride (Fig. 4-31). The metal does not appear to play any major electronic role in these reactions, other than stabilising the intermediate imine and co-ordinating the product.

The metal ion does, however, introduce a new subtlety into these reductions. The reduction of the two imine groups in the nickel(II) complex **4.10** is readily achieved with $Na[BH_4]$. The free tetraamine ligand would be expected to exhibit a facile pyramidal inversion at each nitrogen atom, whereas in the nickel(II) complex this inversion is not possible without significant weakening (or breaking) of the Ni-N bonds. In macrocyclic complexes it is very often found that the complex obtained by the reduction of a co-ordinated imine does not possess the same stereochemistry as that obtained by the direct reaction of the free amine with metal ion.

4.3.3 Reactions of Cyanates

The carbon atom of cyanates is activated towards nucleophilic attack upon co-ordination of the nitrogen to a metal ion (Fig. 4-32).

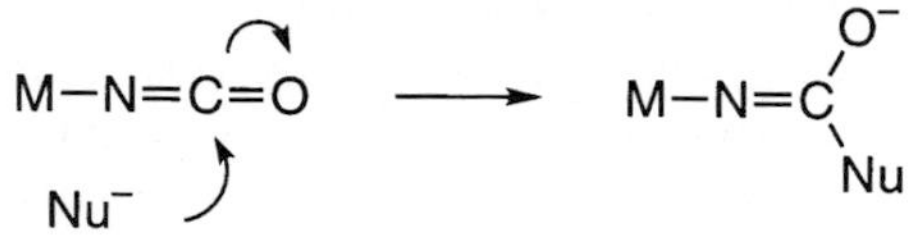

Figure 4-32. Nucleophilic attack upon a co-ordinated cyanate.

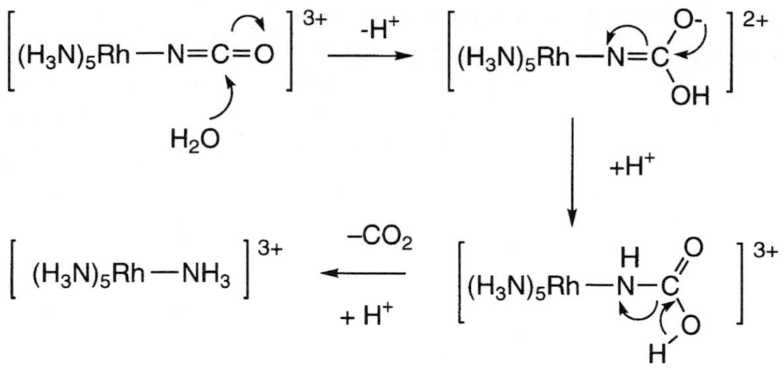

Figure 4-33. The hydrolysis of a rhodium(III) cyanate.

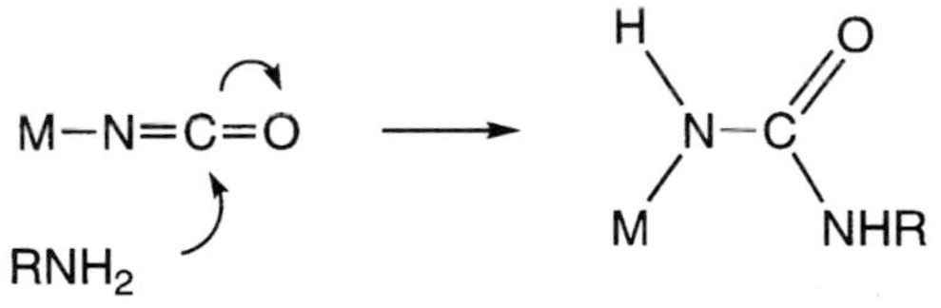

Figure 4-34. The reaction of an amine with a co-ordinated cyanate.

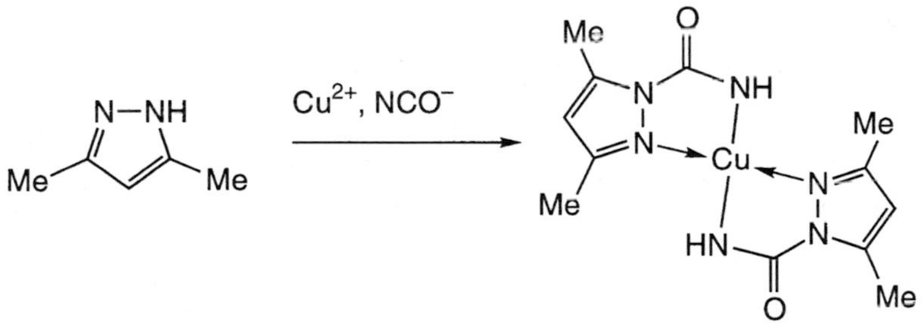

Figure 4-35. The copper(II) mediated reaction of cyanate with 3,5-dimethylpyrazole.

We have already seen an example of this in the nucleophilic attack of water on $[Rh(NH_3)_5(NCO)]^{2+}$, which is produced as an intermediate in the hydrolysis of $[Rh(NH_3)_5(H_2NCONH_2)]^{3+}$. The use of the kinetically inert d^6 rhodium(III) centre ensures that the Rh–N bond remains intact even during cleavage of the C–N bond (Fig. 4-33).

In the case of amine nucleophiles, the products from the reaction with co-ordinated cyanates are carbamates or ureas (Fig. 4-34), and this provides a particularly convenient method for the preparation of carbamate complexes. An example of this behaviour is seen in the reaction of 3,5-dimethylpyrazole with cyanate in the presence of copper(II) salts (Fig. 4-35). Similar reactions are observed with co-ordinated thiocyanates and other heterocumulenes.

4.4 Reactions of Nucleophiles with *sp³* Hybridised Carbon Centres

This type of reaction is related to the well-known S_N2 reaction, widely encountered in organic chemistry. The essential feature of this type of a reaction is the presence of a leaving group attached to the *sp³* carbon centre. The reactions may be controlled in a number of ways. The metal could either co-ordinate to the leaving group and stabilise the anionic species as it is formed, or it could co-ordinate to some other group and increase the δ^+ charge on the *sp³* carbon atom and hence activate it towards the initial attack by the nucleophile (Fig. 4-36).

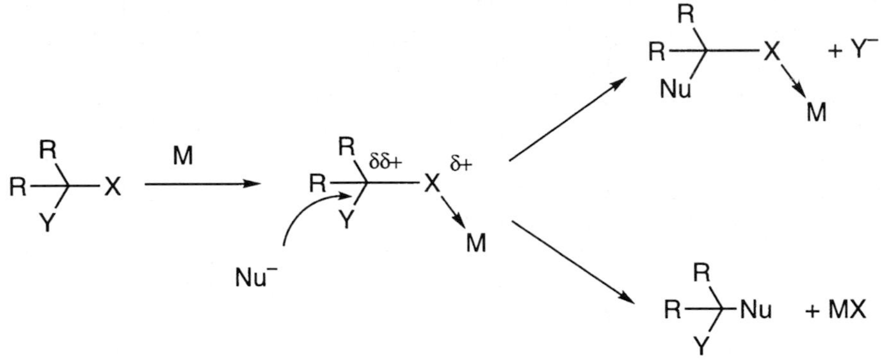

Figure 4-36. Polarisation at an *sp³* hybridised carbon centre by co-ordination of a substituent to a metal. The co-ordinated substituent may also be the leaving group.

4.4.1 Hydrolysis of Thioethers and Related Reactions

The chemistry of organic sulfur compounds does not always parallel that of their corresponding oxygen analogues, and metal-directed chemistry has proved to be of considerable utility in promoting 'copy-cat' reactions. For example, ethers are readily cleaved upon treatment with strong mineral acids, whereas thioethers are unreactive under these conditions (Fig. 4-37).

This behaviour is ascribed to the differences in hardness and softness between oxygen and sulfur. Oxygen is a 'hard' donor and interacts strongly with the 'hard' acceptor H^+. The role of the proton in these reactions is to polarise the C-O bond, and enhance the leaving group ability of the leaving group (Fig. 4-38). In contrast to oxygen, sulfur is 'soft'

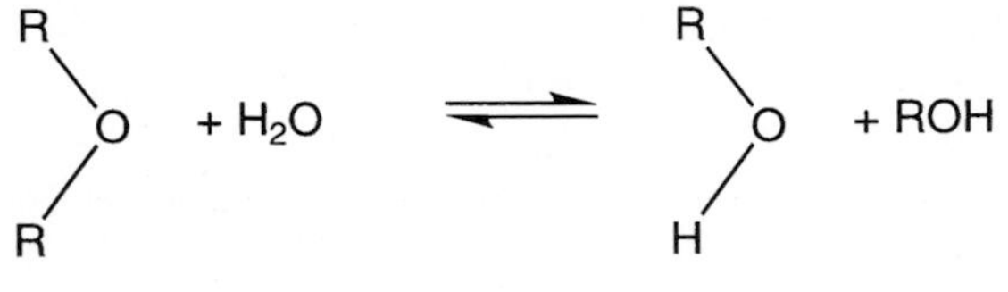

Figure 4-37. The hydrolysis of ethers.

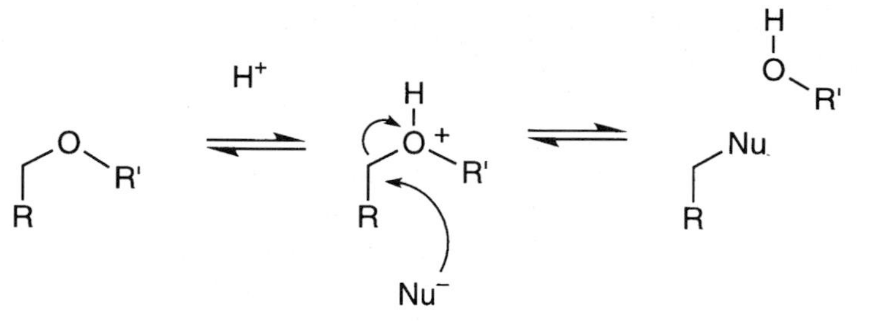

Figure 4-38. The proton acting as a Lewis acid in the hydrolysis of ethers.

and does not interact strongly with protons. Accordingly, protons are not particularly effective Lewis acids for the hydrolysis of thioethers, and these compounds are usually relatively stable to both acid and alkaline hydrolysis

However, the reaction requires only a *general* acid catalyst rather than the specific acid catalyst H^+, and the corresponding reactions of the soft thioether may be better mediated by softer Lewis acids such as Cu^+, Ag^+, Hg^{2+}, Pd^{2+}, Pt^{2+} or Au^{3+}. In many cases the aquated metal ion is the most convenient Lewis acid, but in the case of some metals, particularly the second and third row transition metal ions, the aqua ions are not isolable and other complexes (particularly those with chloride ligands) are equally effective. The role of these softer metal ions as Lewis acids is two-fold. Firstly, the sulfur is co-ordinated to the metal, which increases the polarisation of the C–S bond and enhances the electrophilic character of the carbon, and, secondly, the thiol (or thiolate) leaving group is stabilised by co-ordination (Fig. 4-39).

In the same manner, the hydrolysis of thioesters and thioamides is dramatically (and synthetically usefully) accelerated in the presence of mercury(II) compounds. Notice that a variety of reactions is occurring sequentially in the examples presented in Fig. 4-40.

The use of soft metal ions to direct the course of reactions of sulfur compounds has been utilised in the preparation of nitriles from thioamides. The first step involves the alkylation of the thioamide to give the iminothioester, which is then converted to the nitrile on treatment with mercury(II) salts (Fig. 4-41).

Note that these reactions are reversible, and indeed one of the best methods for the preparation of thioether complexes is by alkylation of a co-ordinated thiolate. In general, the dealkylation reactions require forcing conditions, but in some cases they proceed in an

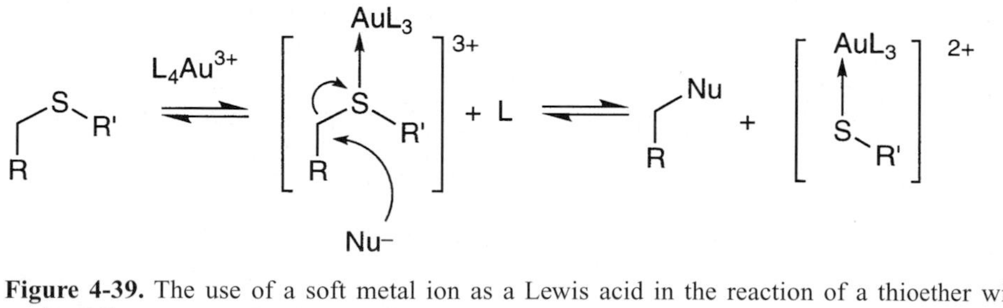

Figure 4-39. The use of a soft metal ion as a Lewis acid in the reaction of a thioether with a nucleophile

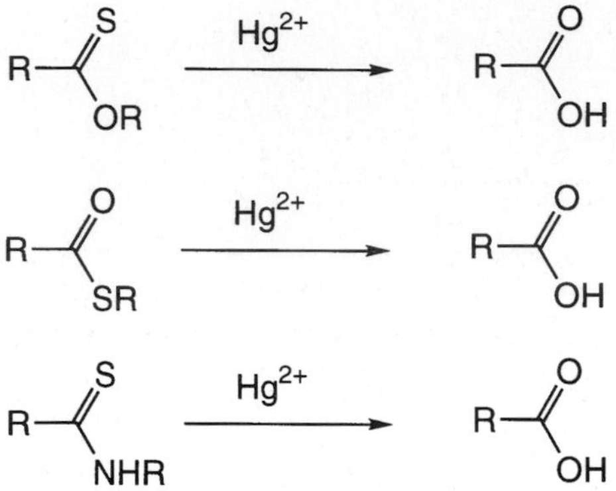

Figure 4-40. Mercury-mediated reactions of thioacid derivatives.

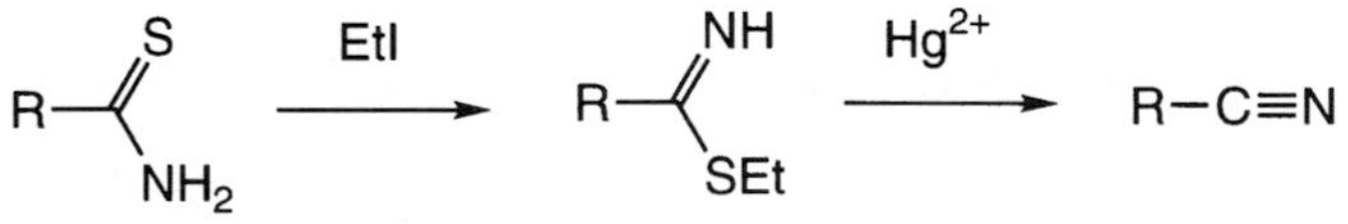

Figure 4-41. A mercury-mediated conversion of a thioamide to nitrile.

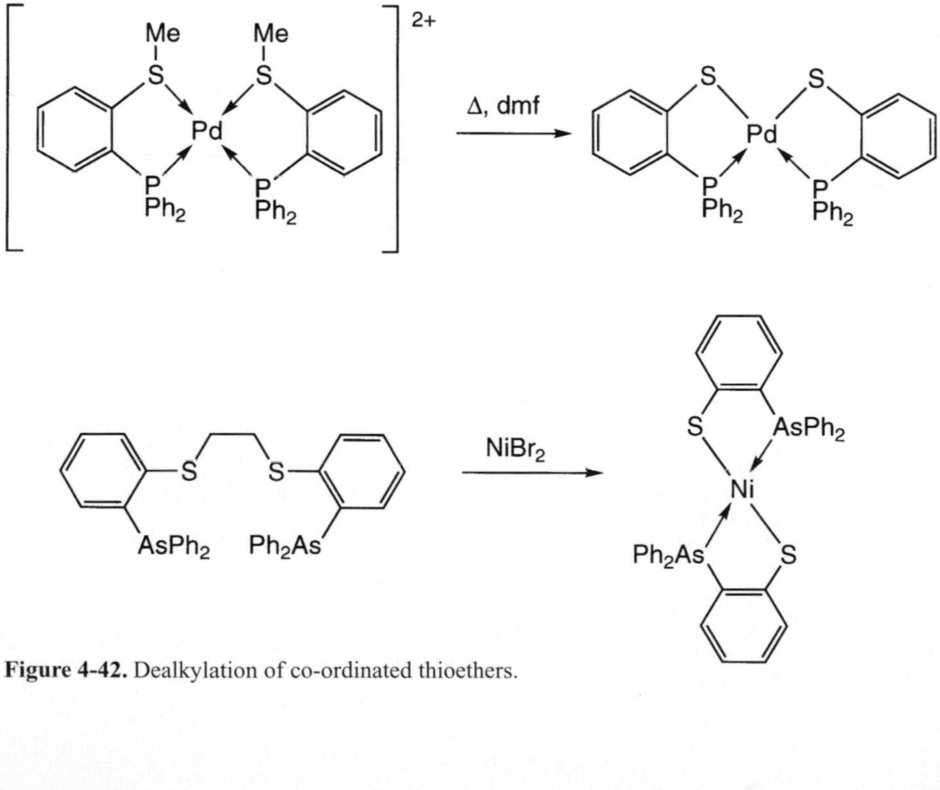

Figure 4-42. Dealkylation of co-ordinated thioethers.

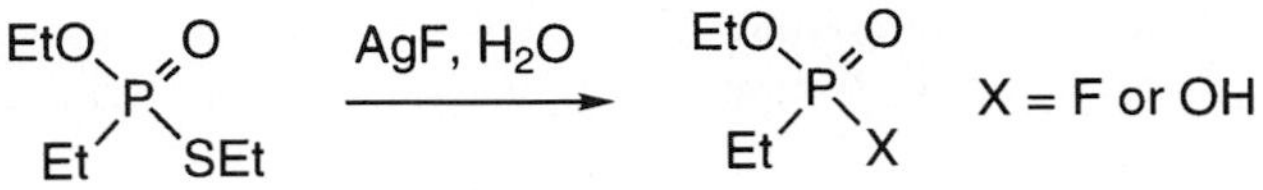

EtO, O   AgF, H₂O   EtO, O
  P=   P=   X = F or OH
Et SEt   Et X

Figure 4-43. A silver-mediated reaction of a thiophosphate.

extremely facile manner. Dimethylformamide has been shown to be a particularly good solvent for these reactions, which may be viewed formally as the transfer of an alkyl group from sulfur to nitrogen. Once again, charge control is of some importance, and the products are the neutral complexes (Fig. 4-42).

These processes are general and any reaction that involves the loss of thiol or thiolate is accelerated by the addition of soft metal cations, especially mercury(II). Thioesters and thiophosphate esters are readily hydrolysed by treatment with aqueous mercury(II) or silver(I) salts. Thus, the reaction of silver fluoride with diethyl ethylthiophosphonate

$$EtO\diagdown P(=O) \diagdown Me \diagup S(CH_2)_2N^iPr_2$$

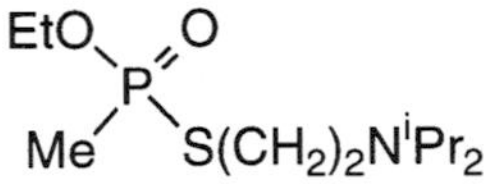

4.11

yields substitution products derived from the displacement of the ethylthiolate group by fluoride or hydroxide. The relevance of these observations is seen from the close structural analogy between diethyl ethylthiophosphonate (Fig. 4-43) and the potent nerve agent VX (**4.11**).

The hydrolysis reactions of thioacetals or thioketals are also accelerated by soft cations such as silver(I) or mercury(II) (Fig. 4-44).

This reaction is widely utilised in organic synthesis, when carbonyl groups may be protected as the thioacetals or thioketals. Unlike acetals or ketals, the thio compounds do not undergo acid catalysed hydration, and may be used in acidic reaction conditions. The metal-directed hydrolysis is rationalised in terms of the soft–soft interaction of the sulfur with the metal cation, in contrast to the hard–soft interaction with a proton. Hydrolysis is readily achieved on treatment with aqueous mercury(II) or silver(I) salts. Once again, the

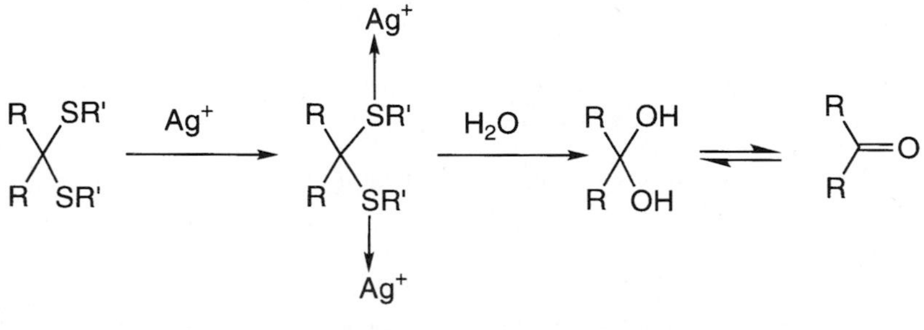

Figure 4-44. Silver-mediated hydrolysis of dithioketals.

role of the metal is mainly to increase the electrophilic character of the sp^3 carbon by co-ordination to the leaving group, but it is also important in stabilising the anionic leaving group.

4.4.2 Arbuzov Reactions

We saw in Chapter 2 that metal ions may exert control over Arbuzov reactions. The Arbuzov reaction involves attack by a nucleophile upon the carbon atom of the alkyl group in an alkoxyphosphine (Fig. 4-45).

In these reactions the polarisation effect of the metal can activate an alkyl group towards nucleophilic attack. The incoming nucleophile is most commonly a halide or pseudohalide, and iodide or thiocyanate are particularly active in these processes. Numerous examples of such reactions are known, which are accelerated by co-ordination of the metal to oxygen. An example is shown in Fig. 4-46 in which the metal stabilises a phosphonate leaving group.

The complexes $[Pt\{P(OMe)_3\}_4]^{2+}$ and $[ClPt\{P(OMe)_3\}_3]^+$ also undergo analogous Arbuzov reactions with halide to yield complexes of $(MeO)_2P(O)H$. However, these particular complexes have been shown to undergo single, double and even triple Arbuzov reactions!

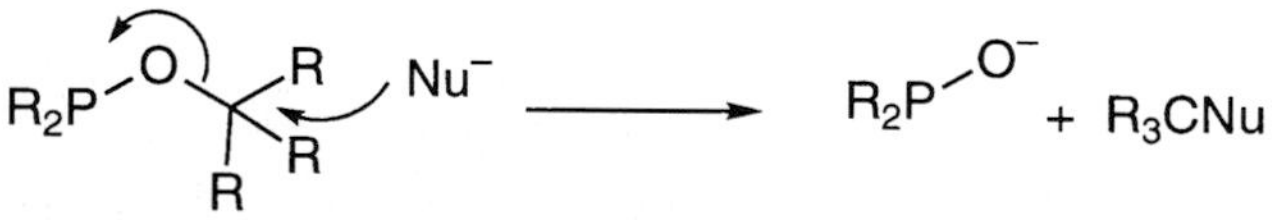

Figure 4-45. The Arbuzov reaction. The final products may be derived from further reactions of those indicated.

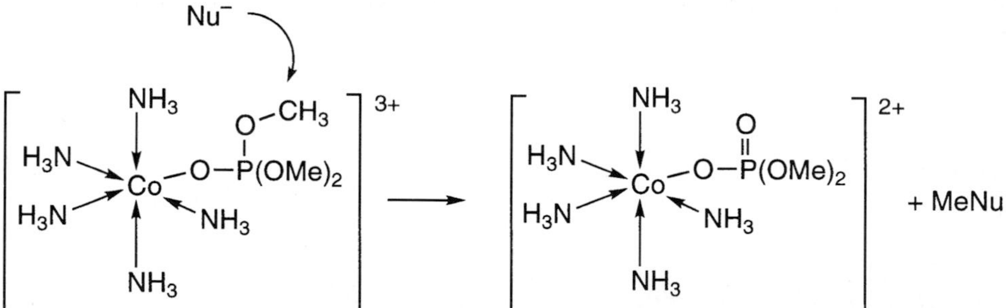

Figure 4-46. A metal-assisted Arbuzov reaction involving a kinetically inert cobalt(III) centre.

$$CoCl_2 + MeSCN \rightleftharpoons [(Co(MeSCN)_2Cl_2] \rightleftharpoons [Me_3S]_2[Co(NCS)_4] + MeCl$$

Figure 4-47. A cobalt-mediated reaction of methyl thiocyanate.

A very similar reaction at a sulfur centre is seen in a thermal reaction of the complex obtained from the reaction of cobalt(II) chloride with methyl thiocyanate (Fig. 4-47). The initial adduct is probably $[Co(MeSCN)_2Cl_2]$, but the mechanism of the thermal rearrangement to give a salt of $[Co(NCS)_4]^{2-}$ is not known.

4.5 Hydrolysis of Phosphate Esters and Related Reactions

The hydrolysis of phosphate esters is accelerated in the same way as the hydrolysis of carboxylates. The principal metal ions involved in these reactions in biological systems are those of Groups I and II, although cobalt(III), copper(II), zinc(II) and nickel(II) complexes have proved to be extremely active in synthetic and mechanistic models. The major role of the metal is to co-ordinate to oxygen and polarise the P-O bond and increase the positive charge on the phosphorus, although a very important secondary function is to increase the leaving group ability of an anionic fragment. With polyphosphates a very important steric role is played, in which chelation to the metal both increases the stability of the complex, thereby favouring ligand polarisation, and also orients the ligand in a conformation favourable for attack by the incoming nucleophile (Fig. 4-48).

As in so many cases, the most unambiguous studies have been made using non-labile cobalt(III) complexes, and a disconcertingly large range of different mechanisms have been demonstrated. The hydrolysis of monodentate *O*-bonded phosphate is observed in the acetyl phosphate complex $[Co(NH_3)_5(O_3POCOMe)]^+$ in which polarisation of the P-O bond activates the phosphate towards attack by external hydroxide or water. The product is $[Co(NH_3)_5(OH)]^{2+}$ and we see the secondary role of the metal in stabilising the hydroxide leaving group. The inert cobalt(III) centre ensures that the Co-O bond remains intact during the P-O bond-cleavage. A variation on this theme is observed if a good leaving group is attached to the phosphorus centre, in which case the product is a cobalt(III) phosphate complex. This may be combined with intramolecular attack by a co-ordinated nucleophile, as seen in Fig. 4-49. Here, a metal-stabilised hydroxide is involved in an intramolecular reaction with a monodentate *O*-bonded ligand to give a co-ordinated didentate phosphate in which a new P-O bond has been formed.

The fine balance between available mechanisms is seen in the hydrolysis of the related complex $[Co(NH_3)_5(O_3POAr)]^+$, which also occurs by an S_N1cb mechanism, with attack on the co-ordinated phosphate by co-ordinated amide. A similar mechanism involving

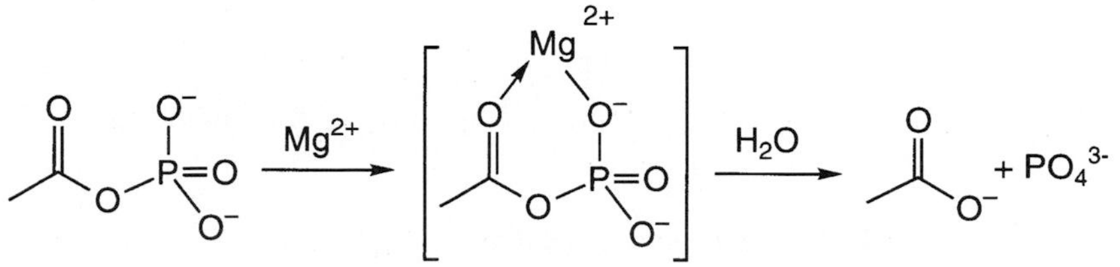

Figure 4-48. Magnesium mediated hydrolysis of a mixed anhydride.

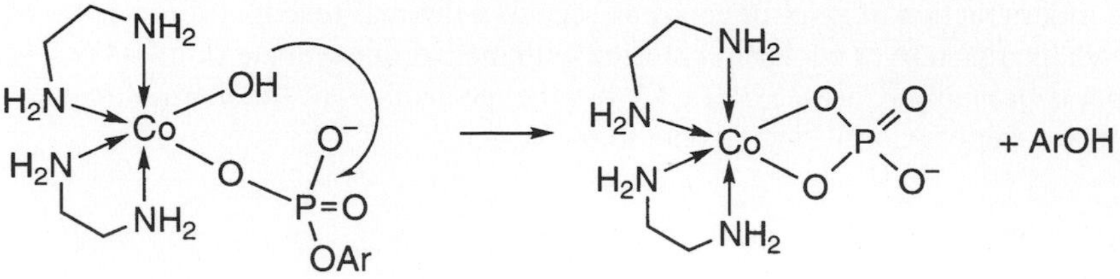

Figure 4-49. Hydrolysis of a monodentate phosphate resulting from intramolecular attack by a co-ordinate hydroxide nucleophile.

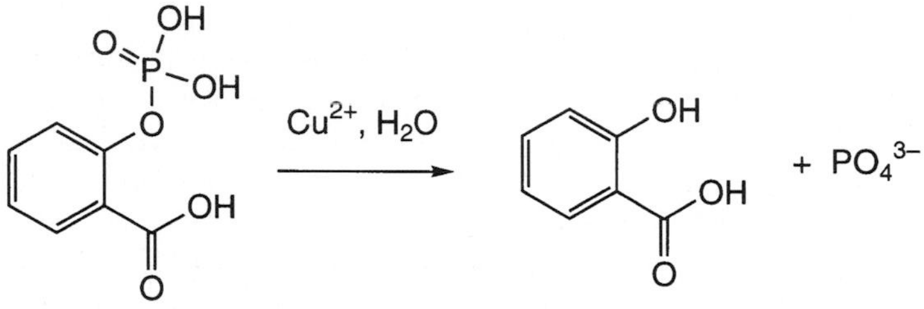

Figure 4-50. The copper-promoted hydrolysis of a phosphate ester.

intramolecular attack by co-ordinated hydroxide is thought to be operative in the hydrolysis of cobalt(III) polyphosphate complexes.

Very rapid hydrolyses of phosphate esters and polyphosphates are observed in the presence of labile metal ions such as copper(II), calcium(II) or magnesium(II) (Fig. 4-50).

Reactions involving the formation and hydrolysis of phosphate and polyphosphate esters are of vital importance in biological systems in which it is found that magnesium ions are almost invariably implicated. The formation and decomposition of adenosine triphosphate are the fundamental reactions involved in energy storage in living systems. In this context, it is perhaps relevant to note that the hydrolysis of ATP is enhanced, albeit in a very modest manner, by some cobalt(III) complexes.

It is found that the hydrolysis of fluorophosphate esters is also accelerated by transition metal ions and complexes. This would be an observation of little general interest, except for the fact that fluorophosphate esters form one of the more commonly encountered types of nerve gases (Fig. 4-51). The hydrolysis of fluorophosphate esters is increased dramatically in the presence of copper(II) and other transition metal complexes, and this sug-

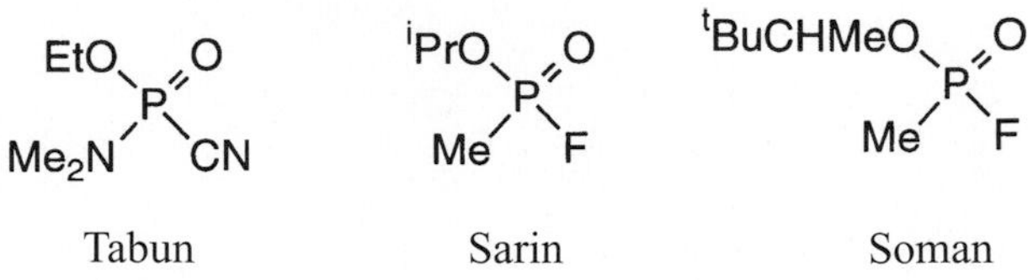

Tabun   Sarin   Soman

Figure 4-51. Some phosphate derivatives used as nerve gases.

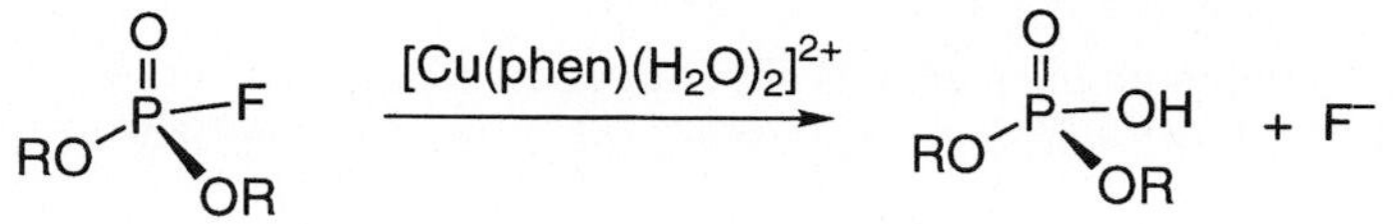

Figure 4-52. The copper(II)-mediated hydrolysis of fluorophosphates.

gests one way by which surfaces contaminated with nerve agents might be rendered harmless by a treatment as simple as spraying with solutions of transition metal complexes (Fig. 4-52).

4.6 Conclusions

In this chapter we have seen a range of reactions in which the key feature is attack by an external nucleophile on a molecule co-ordinated to a metal ion. The ligand may be co-ordinated through a variety of donor atoms and the nucleophile may attack at a range of sites. Many of these reactions very closely mimic those observed in biological systems.

The principal role of the metal ion is to increase the polarisation of the ligand prior to attack by the nucleophile, but a very important secondary role involves the stabilisation of anionic leaving groups.

Suggestions for further reading

1. D.F. Martin, *Prep. Inorg. React.* **1964**, *1*, 59.
   – A short review of cyclic imines.
2. D.L. Leussing in *Metal Ions in Biological Systems*, *Vol. 5,* (ed. H. Sigel), Marcel Dekker, New York, **1976**, p. 1.
   – A relevant chapter in a series devoted to bioinorganic chemistry.
3. A.E. Martell, *Pure Appl. Chem.* **1968**, *17*, 129.
   – An article about the reactions of co-ordinated nitrile ligands
4. D.H. Busch, *Science* **1971**, *171*, 241.
   – More about co-ordinated nitriles
5. B.N. Storhoff, H.C. Lewis Jr., *Coord. Chem. Rev.* **1977**, *23*, 1.
   – A short review dealing with reactions of co-ordinated nitriles
6. S.J. Bryan, P.G. Huggett, K. Wade, J.A. Daniels, J.R. Jennings, *Coord. Chem. Rev.* **1982**, *44*, 149.
   – Another review dealing primarily with the reactions of nitriles.
7. L.F. Lindoy, *Coord. Chem. Rev.* **1969**, *4*, 41.
   – A review dealing with the co-ordination chemistry of diimines.

8. D.P.N. Satchell, *Chem. Soc. Rev.* **1978**, 345.
   – A specialist review dealing with the reactions of sulfur compounds.
9. B.S. Cooperman in *Metal Ions in Biological Systems, Vol. 5*, (ed. H. Sigel), Marcel Dekker, New York, **1976**, p. 79.
   – A review dealing with the metal-directed reactions of phosphates.
10. R.W. Hay, P.J. Morris in *Metal Ions in Biological Systems, Vol. 5*, (ed. H. Sigel), Marcel Dekker, New York, **1976**, p. 173.
    – Another review dealing with phosphates. This is more from the viewpoint of the inorganic mechanistic chemist.
11. H. Sigel, *Coord. Chem. Rev.* **1990**, *100,* 453.
    – A good review of metal-directed phosphate hydrolysis
12. R.W. Hay in *Comprehensive Co-ordination Chemistry, Vol. 6*, (eds G. Wilkinson, R.D. Gillard, J.A. McCleverty), Pergamon, Oxford, **1987.**
    – A good overview of the reactions of co-ordinated phosphate ligands.
13. R.W. Hay in *Reactions of Co-ordinated Ligands, Vol. 2*, (ed. P.S. Braterman), Plenum, New York, **1989.**
    – Rather similar to the reference above.

The continuing series of books edited by Sigel, *Metal Ions in Biology*, Marcel Dekker, provides a good source of data for the interaction of metal ions with biological systems, and many good examples of metal-directed reactions of biological molecules.

5 Stabilisation of Anions and the Reactions of Co-ordinated Ligands with Electrophiles

5.1 General Introduction

In the preceding two chapters we have been primarily concerned with the attack of external nucleophiles upon co-ordinated ligands. In this chapter we will consider the complementary processes which involve the reactions of co-ordinated nucleophilic ligands with electrophiles (usually, but not necessarily, external). We will also be concerned with a number of related processes in which the role of the metal ion in stabilising an anionic or other nucleophilic form of a ligand is thought to be paramount. We shall begin by thinking about some more of the ways in which co-ordination to a metal ion might control the typical reactions of a carbonyl compound.

5.2 Stabilisation of Enolates

We discussed in Chapter 3 the way in which a carbonyl compound may be rendered nucleophilic at either carbon or oxygen by the formation of an enol or an enolate (Figs 3-4 and 3-5), and commented briefly that co-ordination to an electropositive metal centre might have an effect upon the generation and reactivity of such species. Indeed, it seems very unlikely that reactions which involve enols or enolates will not be controlled by co-ordination to a metal centre. The generation of an enolate is likely to be favoured by co-ordination to a positively charged metal centre. In reality, the 'organic' chemistry of enolates is generally that of their alkali or alkaline earth complexes! This follows from the common practice of using alkali metal compounds as the bases for the deprotonation of

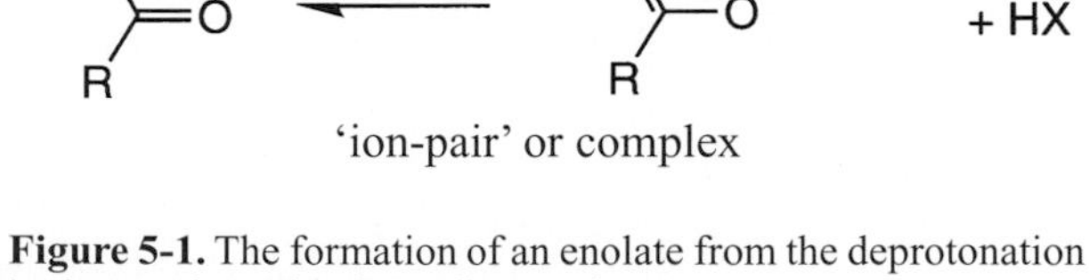

'ion-pair' or complex

Figure 5-1. The formation of an enolate from the deprotonation of a ketone. The enolate is stabilised by interaction with the sodium cation.

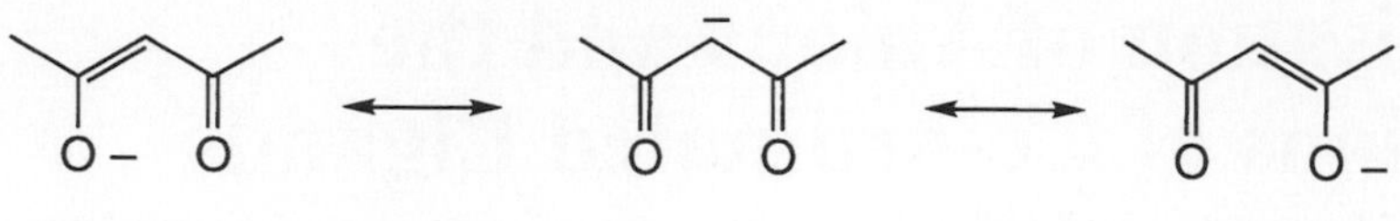

Figure 5-2. The additional resonance forms that may be written for the anion derived from the deprotonation of a 1,3-dicarbonyl compound account for the lowering of the pK_a.

the enol. A better representation of the process shown in Fig. 3-4 is presented in Fig. 5-1, in which the involvement of the sodium ion is made explicit.

It is often rather difficult to assess the effect of the co-ordination of the enolate to the alkali metal ion upon the acidity of free ketones or aldehydes. This is partly due to their high pK_a (usually in the range 20-25) and partly to the fact that the strong bases usually employed for the deprotonation are derivatives of alkali metals. The latter feature means that it is not at all easy to determine the pK_a of the carbonyl compound in the absence of co-ordination effects. To a certain extent, these problems are obviated by considering the behaviour of 1,3-dicarbonyl compounds. The introduction of the second carbonyl group lowers the pK_a of the free ligand to the more accessible region of 8-12. The lowering of the pK_a is associated with the ability to write a larger number of resonance forms, in which the negative charge is placed on the more electronegative oxygen atoms (Fig. 5-2).

The 1,3-diketonate anions which are formed are excellent didentate chelating ligands for transition metals. In general, the formation of a diketonate complex is so favourable that simply treating a metal salt with the 1,3-diketone in the presence of a mild base results in the formation of a complex of the deprotonated ligand. In some cases, it is not necessary to add an external base – another ligand co-ordinated to the metal centre may be capable of acting as the base (Fig. 5-3).

Aza-analogues of oxygen compounds frequently show related reactivity patterns, and this is certainly seen in a comparison of the chemistry of imines and carbonyl compounds. For example, 1,3-diimines are readily deprotonated to yield the 1,3-diketonate analogues. The most frequent consequence of this is that reactions which are expected to yield 1,3-diimine complexes often lead to those of the deprotonated species. This is seen in the formation of the gold(III) complex of a deprotonated macrocycle in the reaction of 1,2-diaminoethane with 2,4-pentanedione in the presence of Na[AuCl$_4$] (Fig. 5-4). The exact sequence of events in this reaction is not known, but note that the product is a square-planar gold(III) complex of the doubly-deprotonated macrocycle, rather than a gold(I) species!

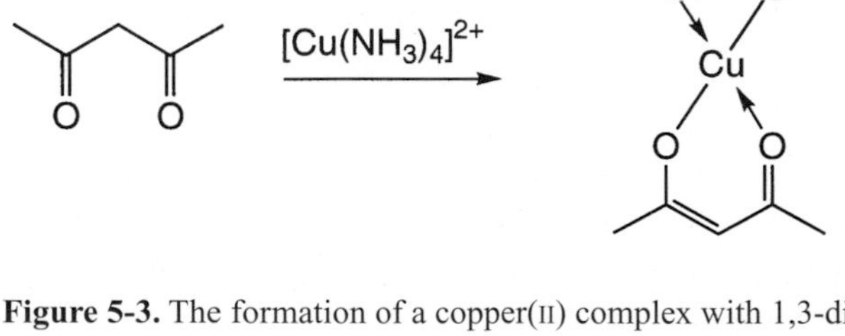

Figure 5-3. The formation of a copper(II) complex with 1,3-diketonate ligands.

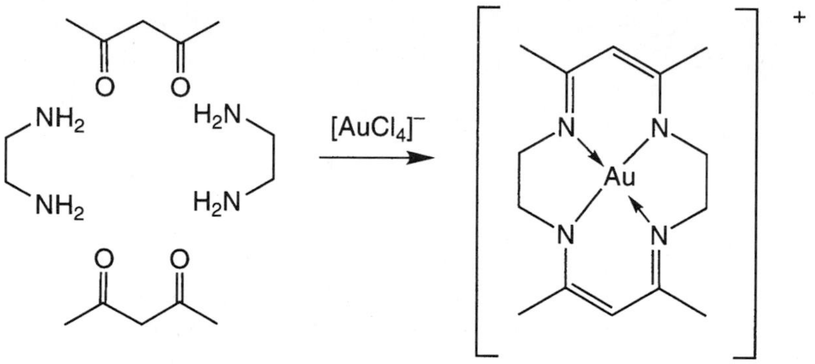

Figure 5-4. The template formation of a gold(III) complex of a dianionic macrocycle

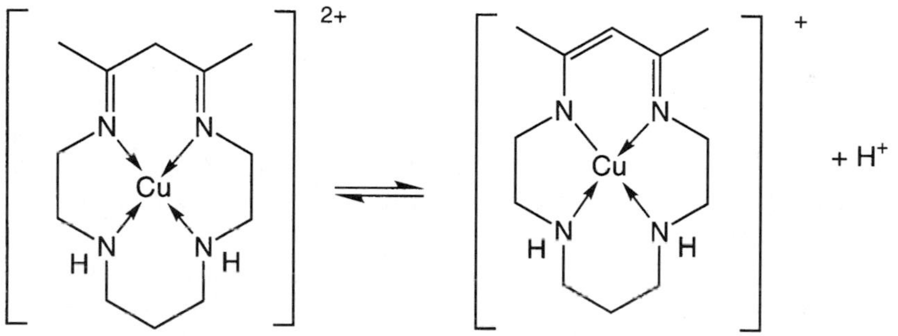

$$pK_a = 9.3$$

Figure 5-5. The deprotonation of a 1,3-diimine ligand co-ordinated to a copper(II) centre

In relatively few cases, separate complexes containing both the protonated and the deprotonated forms of the ligands may be isolated, and it is then possible to directly determine the pK_a values for the co-ordinated ligands (Fig. 5-5).

5.3 Reactivity of Co-ordinated Enolates with Electrophiles

In the same way that the formation of the enolate may be promoted by co-ordination to a metal ion, so may the reactions of that enolate be promoted or modified. Spectacular examples of the reactions of co-ordinated enolates are observed in the electrophilic halogenation of β-ketoesters. These reactions are accelerated dramatically in the presence of copper(II) and other metal ions (Fig. 5-6).

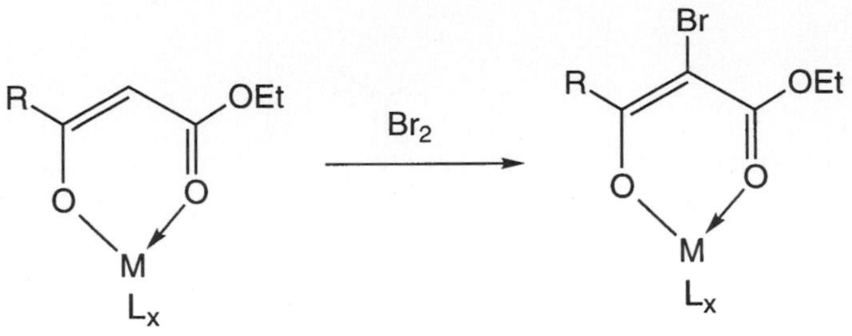

Figure 5-6. The bromination of a co-ordinated β-ketoester

The reaction of the ketoester discussed above is just one example of a more general type of reaction exhibited by co-ordinated 1,3-diketonates. A very extensive series of electrophilic substitution reactions has been described for transition metal 1,3-diketonate complexes. The nucleophilic species is the enolate form of the diketone, which is stabilised by co-ordination to the metal, as we saw earlier. These complexes are often very stable and provide an excellent basis for a systematic study of reactions of co-ordinated ligands with electrophiles. The effects of systematically varying the ligand and the metal may be probed. A wide range of electrophiles and metals have been shown to undergo these reactions and, indeed, the chelated 1,3-diketonate ring has been described as an inorganic analogue of benzene. Electrophilic reactions which have been reported include halogenation, nitration, Friedel–Crafts acylation and Vilsmeier formylation. Some of these reactions have been utilised for the preparation of novel co-ordination compounds which cannot readily be prepared from the free ligands (indeed, some of the free ligands are unknown). Excellent yields of bromo compounds may be obtained by direct halogenation, and this is a useful synthetic method for their preparation. For example, the bromination of the complex $[Rh(F_3CCOCHCOCF_3)_3]$ gives a mixture of the halogenated complexes $[Rh(F_3CCOCHCOCF_3)_{3-n}(F_3CCOCBrCOCF_3)_n]$. Intermediate enolate forms are undoubtedly involved in the related reaction of $[Rh(O_2CCH_2CO_2)_3]^{3-}$ with N-bromosuccinimide to yield the α-brominated species $[Rh(O_2CCHBrCO_2)_3]^{3-}$.

It is possible to form the corresponding nitro compounds by direct nitration of the co-ordinated diketonate (Fig. 5-7), but they are also readily prepared by electrophilic nitration of the corresponding iodo compounds (which are, in turn, obtained by direct iodination) (Fig. 5-8).

The analogy between the behaviour of these diketonate complexes and benzene derivatives is indeed remarkable, and reaction sequences very reminiscent of those observed in organic aromatic chemistry are common. In Fig. 5-9, the reduction of a nitro to an amino derivative is illustrated, a conversion reminiscent of the reduction of nitrobenzene to aniline. It is usually necessary to use kinetically inert d^3 or d^6 metal complexes in sequences of this type.

Once the 3-amino-2,4-diketonate has been prepared, it then shows reactivity typical of an aminobenzene derivative. For example, reaction with nitrous acid and sodium fluoride allows the formation of the 3-fluoro derivative in a Sandmeyer reaction (Fig. 5-10). This may usefully be compared to the reaction sequence from conventional aromatic chemistry

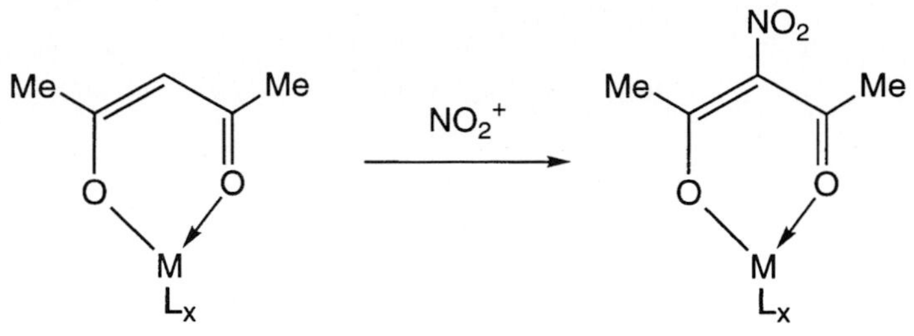

Figure 5-7. The direct nitration of a co-ordinated 2,4-diketonate ligand. The 2,4-diketonate complexes are not usually stable to nitric acid and the nitration is effected by milder reagents, such as a
mixture of copper(II) nitrate and acetic anhydride (which generates N_2O_4). When tris(diketonate)
complexes, such as $[Co(acac)_3]$ are nitrated, mixtures of mono-, bis- and tris- substituted products
are obtained.

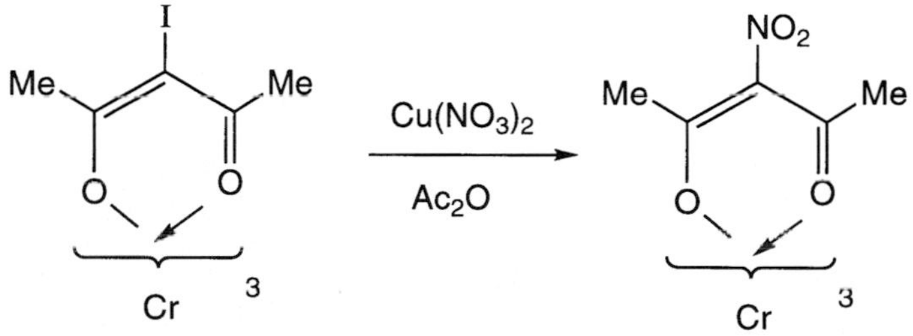

Figure 5-8. Nitrated derivatives of 2,4-diketonates may be obtained by an *ipso* substitution of the
corresponding iodo compound. This allows the preparation of a range of complexes which cannot
be obtained by direct nitration.

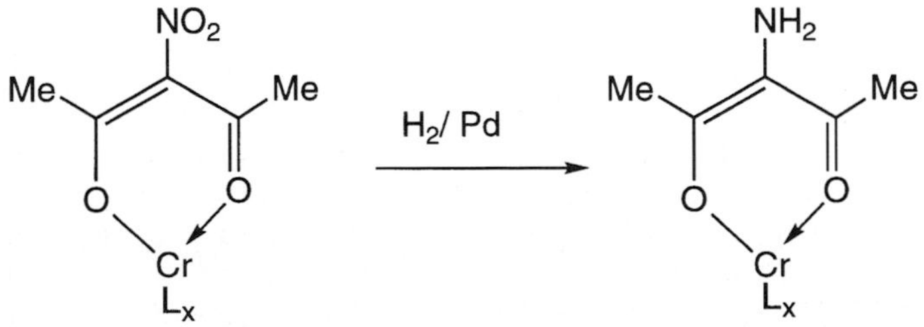

Figure 5-9. The reduction of a co-ordinated 3-nitro-2,4-diketonate to a co-ordinated 3-amino-2,4-
diketonate.

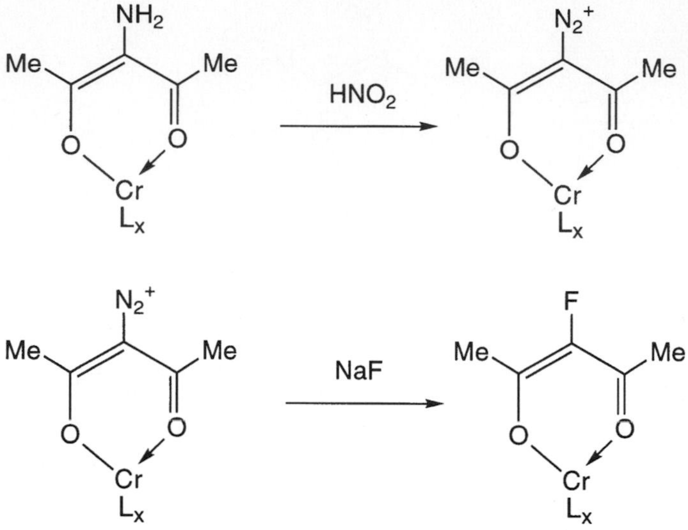

Figure 5-10. A reaction sequence involving a 2,4-diketonate ligand co-ordinated to chromium(III) which allows the introduction of a fluoro substituent.

shown in Fig. 5-11. The scope of these reactions is only limited by the availability of the substituted 1,3-diketonate products. These products have considerable potential as supramolecular building blocks.

However, the reactions of co-ordinated 1,3-dicarbonyl complexes with electrophiles are not always so simple. The reaction of dicyanogen with co-ordinated, and free, 1,3-diketonates has been studied in some detail and results in the formation of a variety of exotic products, such as **5.1**, **5.2** and **5.3**.

A range of other related 2,4-dicarbonyl compounds react with dicyanogen to give products derived from initial electrophilic attack upon the 3-position. For example, the reaction with $C_6F_5NHCOCH_2COR$ is promoted by $[Ni(acac)_2]$, and the major products are the free ligands **5.4** and **5.5**. The copper(II) complexes of **5.4** and **5.5** are obtained directly from the reaction of dicyanogen with copper(II) complexes of $C_6F_5NHCOCH_2COR$.

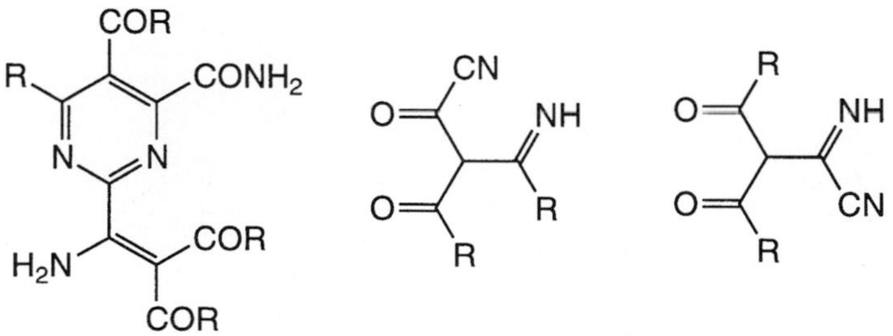

5.1   5.2   5.3

$$ArH \rightarrow ArNO_2 \rightarrow ArNH_2 \rightarrow ArN_2^+ \rightarrow ArF$$

Figure 5-11. A conventional sequence in aromatic chemistry which may be compared to that presented in Figure 5-10.

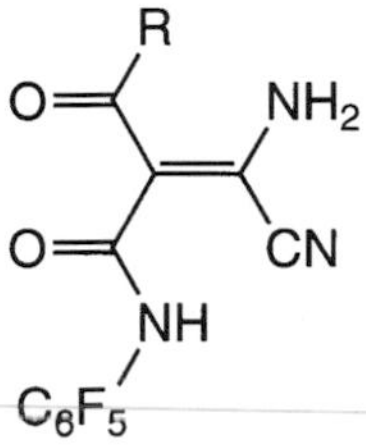

5.4 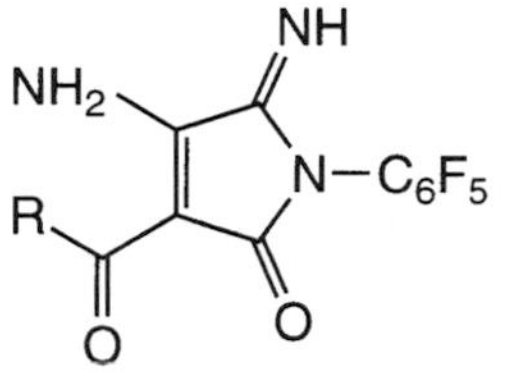   5.5

The analogy between imines and carbonyls was introduced earlier, and just as 1,3-diketonate complexes undergo electrophilic substitution reactions at the 2-position, so do their nitrogen analogues. Reactions of this type are commonly observed in macrocyclic ligands, and many examples arc known. Electrophilic reactions ranging from nitration and Friedel–Crafts acylation to Michael addition have been described. Reactions of 1,3-diimines and of 3-iminoketones are well known. The reactions are useful for the synthesis of derivatised macrocyclic complexes, as in the preparation of the nickel(II) complex of a nitro-substituted ligand depicted in Fig. 5-12.

These 'masked' enolate complexes also react smoothly with carbonyl compounds, in reactions surprisingly reminiscent of the reaction of dimedone with formaldehyde (Fig. 5-13). The dinuclear products are of some interest as models for the dinuclear sites of some metalloproteins.

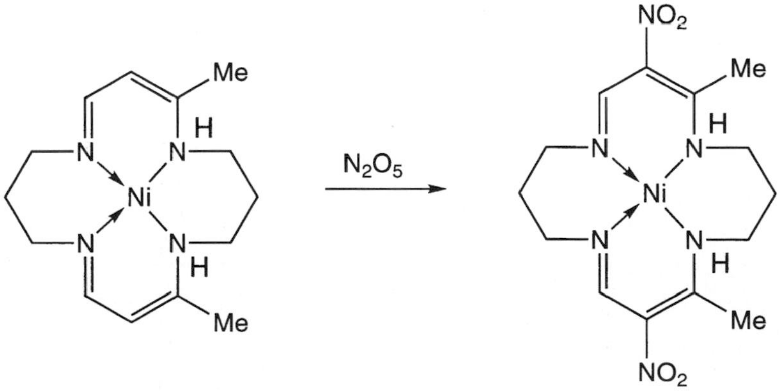

Figure 5-12. The nitration of a nickel(II) macrocyclic complex.

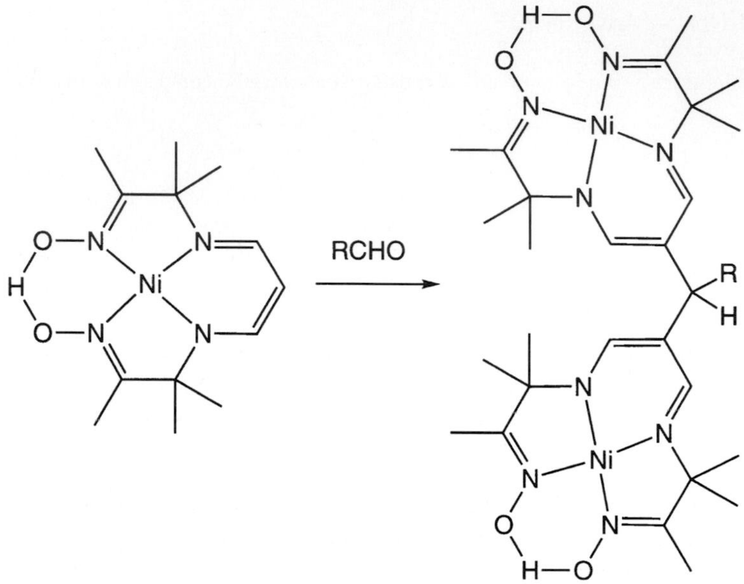

Figure 5-13. The linking together of two *pseudo-macrocyclic* nickel complexes by electrophilic attack of an aldehyde on a diazadiketonate. Both the product and the starting material are neutral and contain nickel(II) centres. Each ring is doubly deprotonated - first at the 1,3-diazadiketonate and second at the dioxime.

These types of reactions have been widely utilised by Jäger for the synthesis of a remarkable range of macrocyclic complexes, which are discussed in a little more detail in Chapter 6. Even mixed *ON* donor ligands related to 1,3-diketones behave as nucleophiles, and bromination of 3-iminoketonates has been achieved using *N*-bromosuccinimide (Fig. 5-14).

Some related reactions have been reported for some co-ordinated 1,2-diimines. A typical electrophilic substitution reaction of this type is observed in the bromination of the

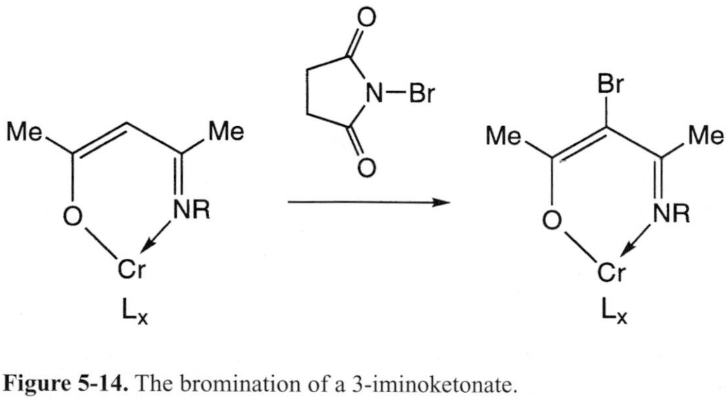

Figure 5-14. The bromination of a 3-iminoketonate.

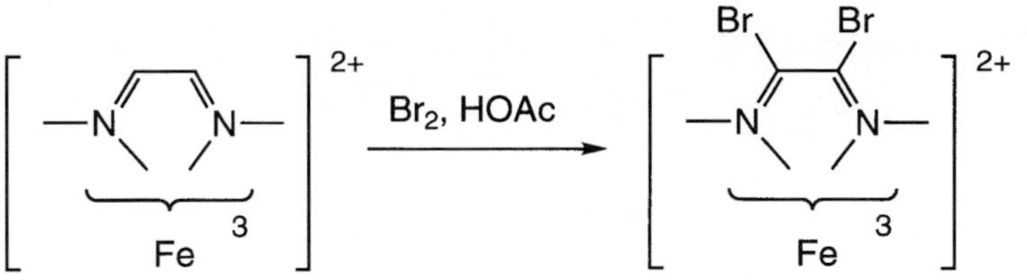

Figure 5-15. The bromination of a 1,2-diimine. Although the ligand is neutral, co-ordination to the metal centre allows a smooth reaction with the electrophile.

tris(2,5-diazahexa-2,4-diene)iron(II) cation, which proceeds smoothly to give a complex of the dibromo derivative (Fig. 5-15). The free dibromo ligand is only accessible with extreme difficulty. We will have more to say about this system later, but suffice it to say that the combination of the kinetically inert d^6 iron(II) centre and the strong field ligands are of considerable importance.

    Returning to the main theme in this section, another case where chelation to a metal centre controls reactions involving enolates is seen in complexes of amino acid derivatives. Amino acids are commonly found in metal complexes as the chelated anions in which the carboxylate oxygen and the amino group are co-ordinated to the metal. The co-ordinated amino acid anion could be in the keto (**5.6**) or enolate (**5.7**) form.

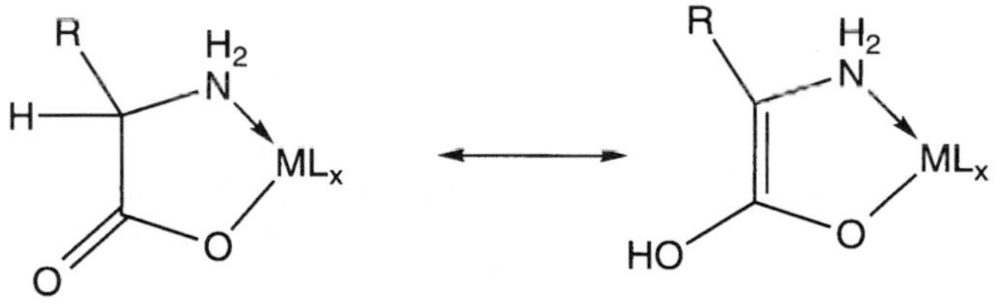

5.6   **5.7**

    These co-ordinated amino acids are, thus, activated to electrophilic attack at the β-position. One of the simplest manifestations is seen in the exchange of the methylene hydrogen atom for deuterium or tritium (Fig. 5-16).

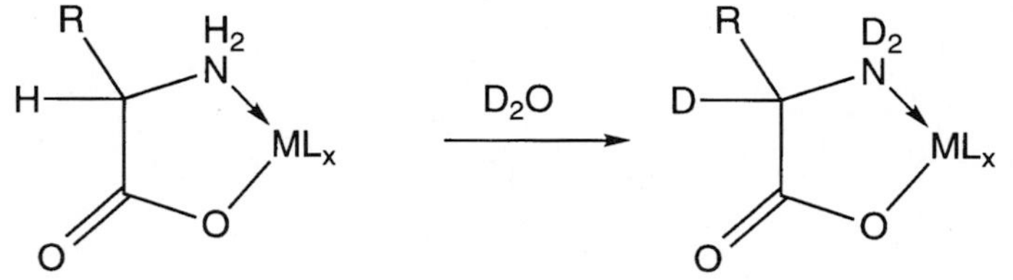

Figure 5-16. The deuteration of a co-ordinated amino acid. Note that usually exchange of the amino hydrogen atoms also occurs.

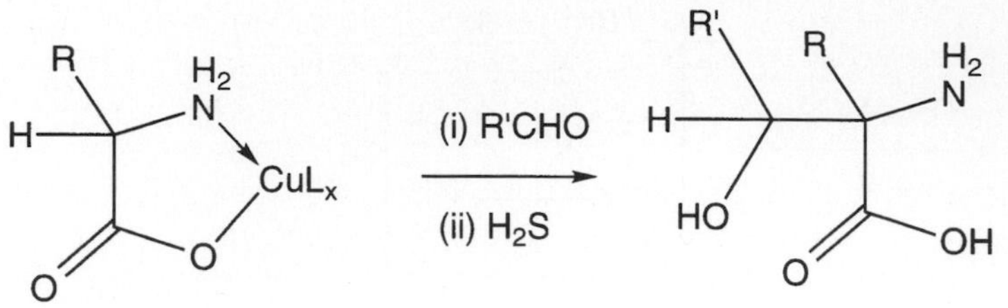

Figure 5-17. The reaction of an aldehyde with a co-ordinated amino acid provides a useful method for functionalisation of the ligand. The precipitation of copper sulfide in the final step is a convenient way of obtaining the free ligand.

This is interesting, but far more useful chemical transformations may be achieved. For example, reaction with an aldehyde R'CHO results in a condensation to give a pendant R'CHOH group (Fig. 5-17).

The final step in the reaction involves demetallation of the organic product by H_2S with the resultant loss of copper(II) sulfide. The reaction is not quite as simple as it appears, and the intermediate copper(II) complex which is demetallated is not of the expected

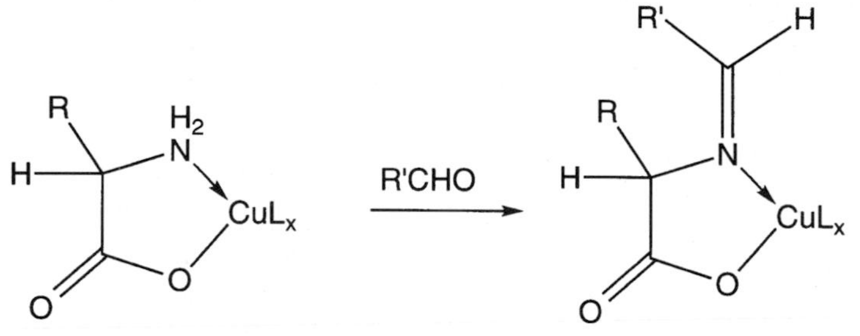

5.8

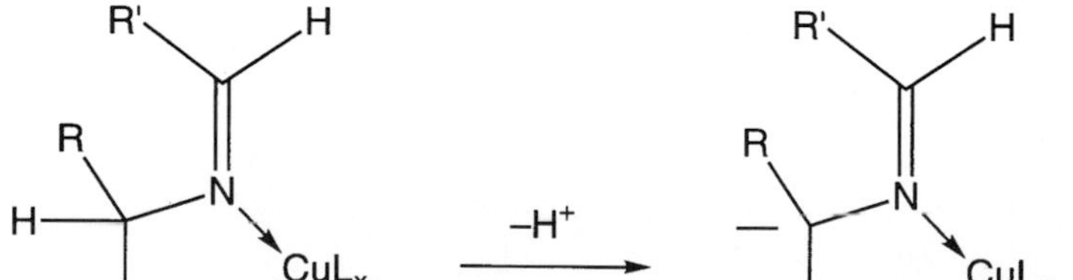

Figure 5-18. The first steps in the reaction depicted in Figure 5-17 involve the formation of an imine, rather than direct attack of the aldehyde upon the carbon.

simple hydroxyalkyl ligand, but of an oxazolidine. The initial reaction involves the formation of an imine **5.8** by reaction of the aldehyde with the co-ordinated amino group. This is a 1,3-diketonate analogue, which is reasonably acidic and is readily deprotonated (Fig. 5-18).

The next steps involve attack of the deprotonated ligand upon a second equivalent of the electrophilic aldehyde. This generates an alkoxide, which undergoes an intramolecular nucleophilic attack upon the imine to give a co-ordinated oxazolidinene, **5.9** (Fig. 5-19).

In the final stage of the reaction, demetallation by hydrogen sulfide is accompanied by ring opening of the oxazolidine and hydrolysis of the imine. This reaction is of some synthetic utility; for example, in the case of R = H and R' = Me, the overall sequence corresponds to a conversion of the amino acid glycine (**5.10**) to threonine (**5.11**). The use of a chiral octahedral complex instead of a copper(II) centre allows the reaction to proceed in a stereospecific manner, and the reaction of *L*-[Co(en)$_2$(H$_2$NCH$_2$CO$_2$)]$^{2+}$ with acetaldehyde yields a threonine complex in low enantiomeric excess. One of the most convenient syntheses of threonine does indeed utilise the reaction of a copper glycine complex with acetaldehyde; after demetallation with hydrogen sulfide, threonine is obtained in 95 % yield.

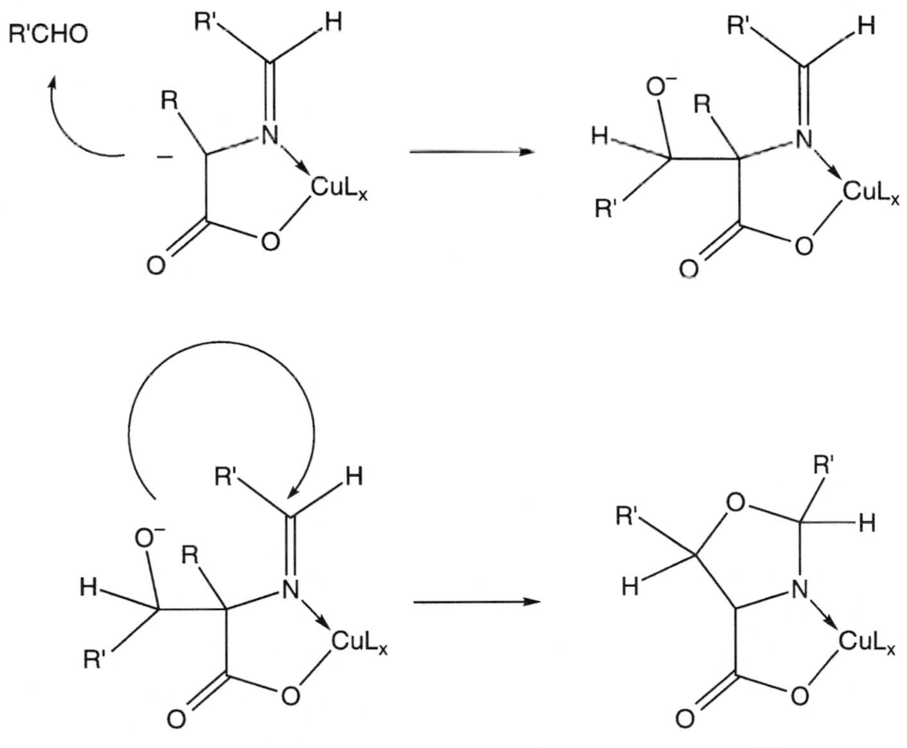

Figure 5-19. The formation of an oxazolidine intermediate.

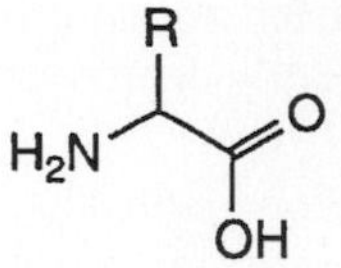

5.10   R = H

5.11   R = CH(OH)Me

5.12   R = (CH₂)₃NH₂

5.13   R = (CH₂)₃NHCONH₂

In some cases, the function of the metal ion is more to deactivate alternative sites of reaction than to activate a particular atom towards attack by an electrophile. A good example of this is seen in the transamination reaction of ornithine (**5.12**) with urea. Co-ordination of the ornithine to copper(II) results in the formation of a five-membered chelate ring, leaving the amino group of the 3-aminopropyl substituent as the most nucleophilic site in the complex. Reaction of this complex with urea results in a transamination process and the formation of the copper(II) complex of the substituted urea, which is the amino acid citrulline (**5.13**) (Fig. 5-20). The complex may be demetallated to yield the free amino acid in respectable yields.

The stabilisation of an enolate (intermediate or product) is also important in the decarboxylation reaction of β-ketoacids. The decarboxylation of such compounds is facile, and is the key to the synthetic utility of ethyl acetoacetate and diethyl malonate. The mechanism of decarboxylation involves the formation of an enol (Fig. 5-21), and so is expected to be subject to metal ion control.

The decarboxylation of simple β-ketoacids, such as acetoacetic acid, is not metal promoted (Fig. 5-22) - this is in part due to formation of the chelate complex, which is in the enolate form. Mechanistic studies have indicated that the enol or enolate is *inactive* in the decarboxylation reaction. The mechanism indicated in Fig. 5-21 is not applicable to the metal complex.

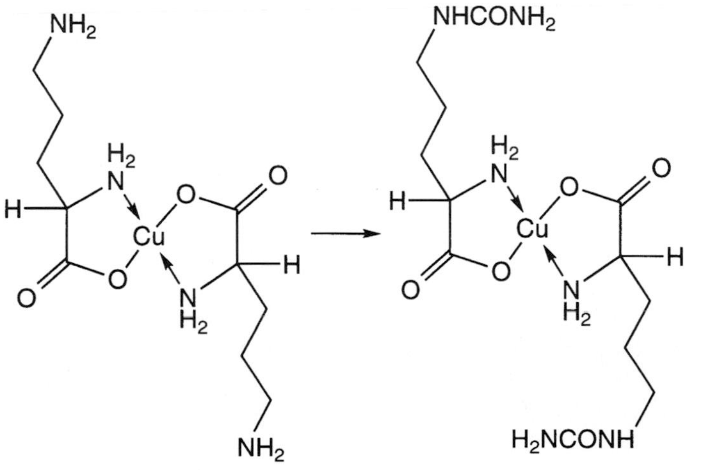

Figure 5-20. The conversion of co-ordinated ornithine to co-ordinated citrulline on reaction with urea.

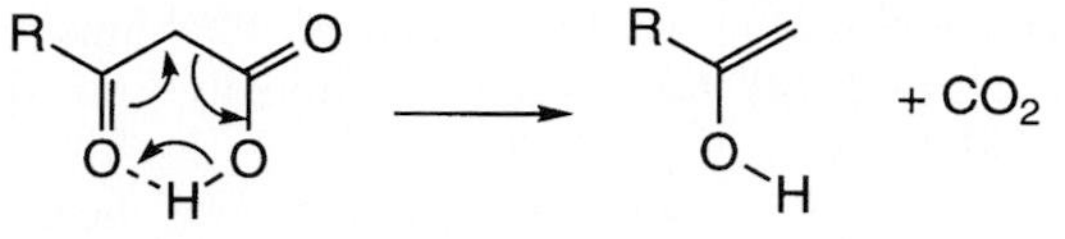

Figure 5-21. The concerted process for the decarboxylation of a β-ketoacid. The carboxyl hydroxy group is hydrogen bonded to the carbonyl group. The product is an enol which usually tautomerises to the desired ketonic product.

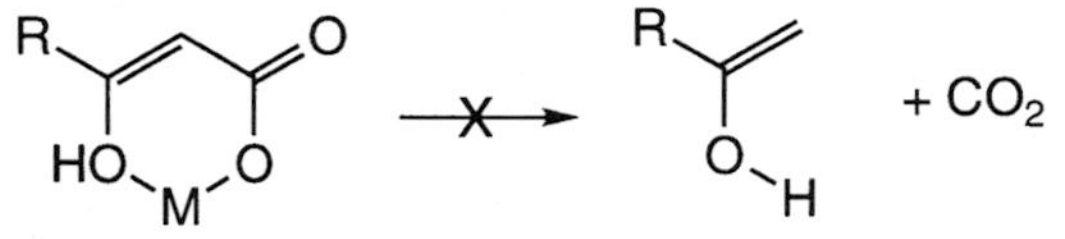

Figure 5-22. The decarboxylation of acetoacetate is not metal ion promoted. The ligand is present in an enolic form.

However, if a further donor group is introduced, a chelate may be formed that does *not* involve the carboxylate group to be lost. In these cases, the decarboxylation is drama- tically enhanced in the presence of metal ions. This is exactly the situation which pertains with oxalacctic acid, which undergoes a facile metal-promoted decarboxylation (Fig. 5-23). The rate of decarboxylation of oxalacetic acid is accelerated some ten thousand times in the presence of copper(II) salts. The metal ion is thought to play a variety of roles, inclu- ding the stabilisation of the enolate that is produced after loss of carbon dioxide.

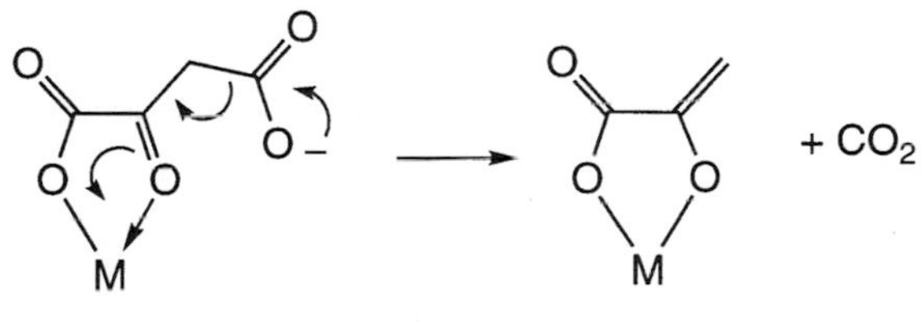

Figure 5-23. The rate of decarboxylation of oxalacetic acid is accelerated many thousands of times in the presence of metal ions.

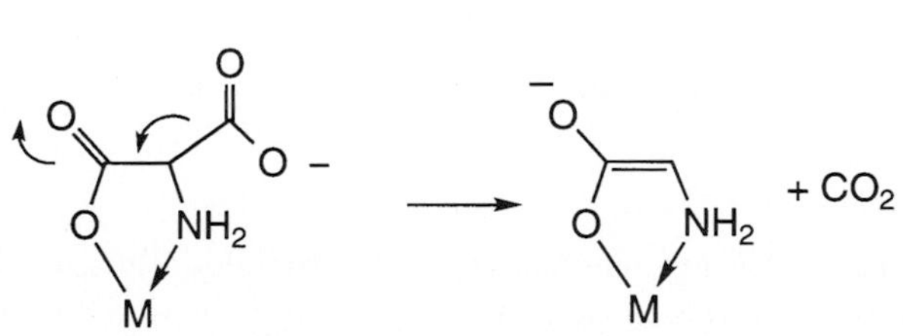

Figure 5-24 . The decarboxylation of a co-ordinated aminomalonate ligand also initially produces a co-ordinated enolate.

A very closely related decarboxylation reaction which is accelerated by metal ions is observed with aminomalonate complexes (Fig. 5-24). Once again, the initially-formed product is in the enolate form.

These types of decarboxylation reaction may be used for the stereoselective metal-directed synthesis of amino acids when combined with the use of chiral complexes. Thus, chiral cobalt(III) complexes containing the didentate ligand $(HO_2C)_2CMeNH_2$ may be decarboxylated in high yield to give alanine $(HO_2CCHMeNH_2)$ complexes in good enantiomeric excess.

5.4 Reactions of Co-ordinated Amines

In principle, co-ordination of an amine to a metal ion renders an amine non-nucleophilic as the lone-pair of the nitrogen atom is involved in the metal–ligand bond. However, co-ordinated amines may still show nucleophilic properties. This can be achieved in one of two ways. The simplest mechanism invokes an equilibrium between the free and the co-ordinated amine (Fig. 5-25). If this equilibrium is set-up, the free amine may react with appropriate electrophiles. The product of this reaction may then react with the metal ion to form a complex of the new ligand. Clearly, in these cases where the product may co-ordinate to the metal, it is necessary to make detailed kinetic studies of the rates of the reaction of free and co-ordinated ligand to distinguish between processes involving ligand dissociation and those involving reaction of the *co-ordinated* ligand.

$$[M \leftarrow NH_2R]^{n+} \rightleftharpoons M^{n+} + RNH_2$$

Figure 5-25. A ligand dissociation reaction may allow small amounts of free amine to be present in solution. This free amine may then react with electrophiles.

Such equilibria as Fig. 5-25 allow the generation of small but controlled amounts of a free amine in solution. This effective reduction in the nucleophilicity of the co-ordinated amine may be used to good advantage. For example, the alkylation of co-ordinated amines rarely proceeds beyond the monoalkylated stage. In contrast, the reactions of the free amines usually proceed further to give a mixture of polyalkylated products (Fig. 5-26).

$$RNH_2 + R'X \rightarrow RNHR' + RNR'_2 + [RNR'_3]X$$

Figure 5-26. The alkylation of a primary amine gives mixtures of mono- and disubstituted products, together with the quaternary ammonium salt. This occurs because the electron releasing effects of the alkyl groups are such that the introduction of more alkyl groups makes the nitrogen more nucleophilic.

However, even 'non-labile' amine complexes may react with electrophiles. These can react by an alternative mechanism which involves the deprotonation of the co-ordinated amine to generate a negatively-charged amido ligand (Fig. 5-27). These two limiting pathways are clearly very closely related, merely varying in the precise point along the reaction co-ordinate at which N–H bond cleavage becomes dominant.

$$[M \leftarrow NH_2R]^{n+} \rightleftharpoons [M \leftarrow NHR]^{(n-1)+} + H^+$$

Figure 5-27. The deprotonation of a co-ordinated amine provides another way of rendering the amine nucleophilic. The sp^2 hybridised nitrogen atom bears a lone pair which may be used in reactions with electrophiles.

In many cases it is not known unambiguously which of these two mechanisms is operative. The pathway involving ligand deprotonation is most favoured by high oxidation state metal ions, and is relatively well-established for complexes of metal ions such as platinum(IV), where appropriate intermediates may be isolated, and for cobalt(III), where there is very convincing kinetic evidence for the involvement of such deprotonated intermediates. The careful design of experiments and the selection of the correct complexes is crucial in this area of study.

Perhaps the simplest reaction to envisage is the alkylation of a co-ordinated amine. These reactions are well-known and usually occur under strongly basic conditions. It is most likely that these reactions involve deprotonated amido intermediates, and are considered in that context. As we have seen in Chapter 2, the acidity of an amine proton should increase upon co-ordination to a metal centre, and with the charge on that metal. As a consequence, we might expect to see new types of reaction products derived from the amido ligand, particularly with high oxidation state metal complexes. The former effect is indeed the case, and dramatic reduction of the pK_a of ammonia and amines is observed upon co-ordination to a metal ion (Table 5-1).

Table 5-1. The pK_a values of some metal-amine complexes.

Complex	pK_a	Complex	pK_a
$[Ru(NH_3)_6]^{3+}$	12.4	cis-$[Pt(NH_3)_4Cl_2]^{2+}$	9.8
$[Ru(NH_3)_6]^{2+}$	7.9	$[Pt(NH_3)_6]^{4+}$	7.2
$[Pt(NH_3)_5(NH_2)]^{3+}$	10.1	$[Pt(en)_3]^{4+}$	5.5
$[Co(NH_3)_6]^{3+}$	>14		

The enhancement of the acidity of the amine is not limited to sites which are directly co-ordinated to the metal, but may also be transmitted for considerable distances through the molecule. For example, the complex bis(di-2-pyridylamine)palladium(II) undergoes a facile deprotonation (Fig. 5-28). This reaction also serves to exemplify the importance of charge control over the level of protonation. The deprotonation yields the *neutral* doubly deprotonated complex.

A direct consequence of the enhanced acidity of co-ordinated amines is seen in the reactions with chlorine in aqueous solution of some platinum(IV) complexes. In these reactions the nucleophilic attack of an intermediate amido complex upon chlorine leads to the formation of dichloroamido complexes (Fig. 5-29).

Perhaps of more synthetic utility is the alkylation of co-ordinated amines. As we illustrated in Fig. 5-26, the attempted alkylation of free amines usually results in the formation of numerous products. This is ascribed to the greater nucleophilicity of the alkylated amines with respect to the starting material. As an example, the reaction of 1,2-diaminoethane with iodomethane yields three major products (Fig. 5-30).

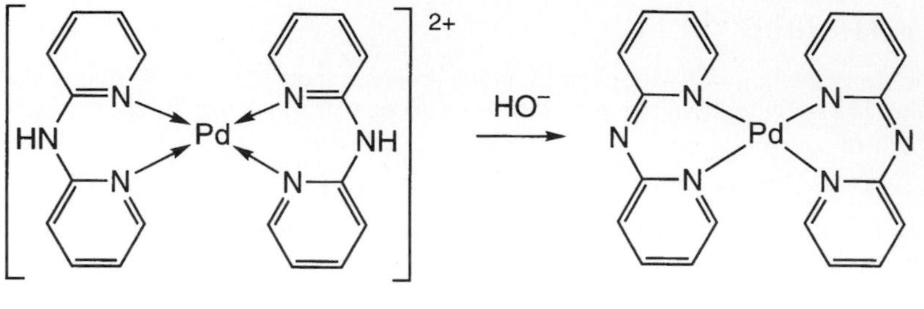

Figure 5-28. The deprotonation of a co-ordinated di-2-pyridylamine ligand, in which the site of deprotonation is not directly bonded to the metal.

$$[L_5Pt-NH_3]^{n+} \rightleftharpoons [L_5Pt-NH_2]^{(n-1)+} \rightleftharpoons [L_5Pt-NCl_2]^{(n-1)+}$$

Figure 5-29. The reaction of a deprotonated amine ligand with chlorine gives dichloramido complexes.

In contrast, co-ordination of 1,2-diaminoethane to a metal may sufficiently deactivate it that monoalkylation or dialkylation may be achieved in a selective manner. For example, $[Rh(en)_3]^{3+}$ salts react with iodomethane to give complexes containing the N,N'-dimethyl-1,2-diaminoethane ligand (Fig. 5-31). The two alkyl groups are introduced to different nitrogen atoms. In the case of the free ligand, the introduction of the first methyl group rendered the nitrogen to which it was bonded *more* nucleophilic, and hence more reactive with the alkylating agent. In consequence, the free ligand introduces the second (and third) methyl groups at the same carbon atom. In contrast, the effect of introducing the first methyl group in the metal complex is to *increase* the stability of the Rh–N bond to that nitrogen to which it is bonded. The result is that it is the other nitrogen atom in the ligand which is more nucleophilic towards the second alkylation step. This represents a very subtle way of controlling the reactivity of amine ligands.

Charge effects may also play an extremely important role in controlling the reactions of co-ordinated amines with electrophilic reagents. This is very clearly seen in the alkylation reactions of nucleophilic sites remote from the metal. On electrostatic grounds we would expect the reaction of positively charged complexes with electrophiles to be less favoured than the reaction of neutral or anionic complexes, and this is indeed the case. Consider the attempted alkylation of the non-co-ordinated isoquinoline rings in the copper(II) complexes **5.14** and **5.15**. Compound **5.14** is derived from salicylaldehyde and

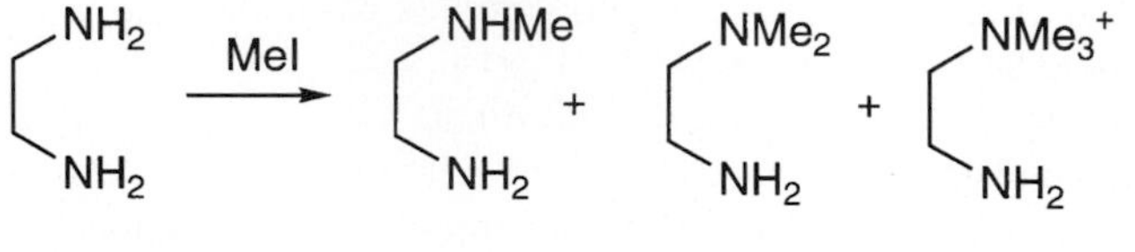

Figure 5-30. The methylation of 1,2-diaminoethane with iodomethane gives mixtures of mono-, di- and trimethylated products.

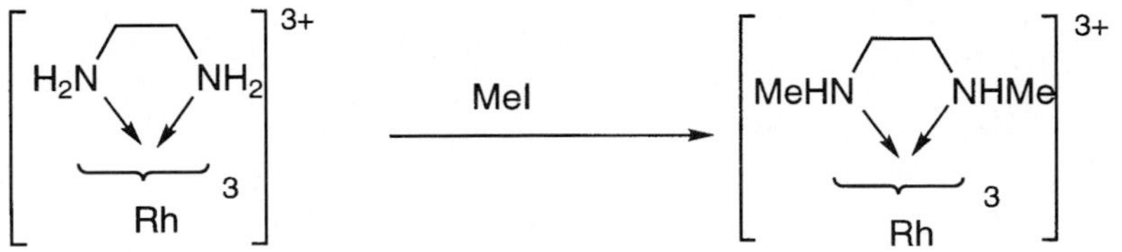

Figure 5-31. Methylation of the [Rh(en)$_3$]$^{3+}$ cation yields complexes of *N,N'*-dimethyl-1,2-diaminoethane.

deprotonation of the phenolic hydroxy groups means that the complex is neutral. In contrast, **5.15** contains a neutral ligand and is dicationic. It is found that only the neutral complex **5.14** may be alkylated, whilst the 2-pyridinecarbaldehyde derivative **5.15** is inert. Once again, the metal ion is exerting a subtle influence over the reactivity of a site which is spatially remote from the site of co-ordination.

However, this is not to say that it is impossible to alkylate cationic complexes. The reaction of the ruthenium(II) complex [Ru(**5.16**)$_2$]$^{2+}$, in which only the three chelating nitrogen atoms of the 2,2':6',2"-terpyridine moiety are co-ordinated to the metal, with iodomethane in acetonitrile gives the alkylated product [Ru(**5.17**)$_2$]$^{4+}$ in near quantitative yield.

In addition to the charge control over the reaction discussed above, there is also a marked element of conformational control over alkylation reactions. This is seen clearly in the methylation of the nickel(II) complex of the tetraaza macrocyclic ligand, cyclam (Fig. 5-32). Reaction of the nickel complex with methylating agents allows the formation of a *N,N',N",N"'*-tetramethylcyclam complex. In this product, each of the four nitrogen atoms is four-co-ordinate and tetrahedral, and specific configurations are associated with each. Of the four methyl groups in the product, two are oriented above the square plane about the nickel, and two below it.

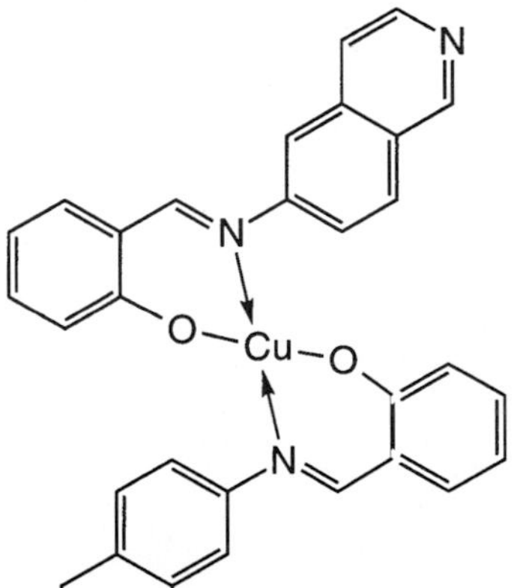

5.14 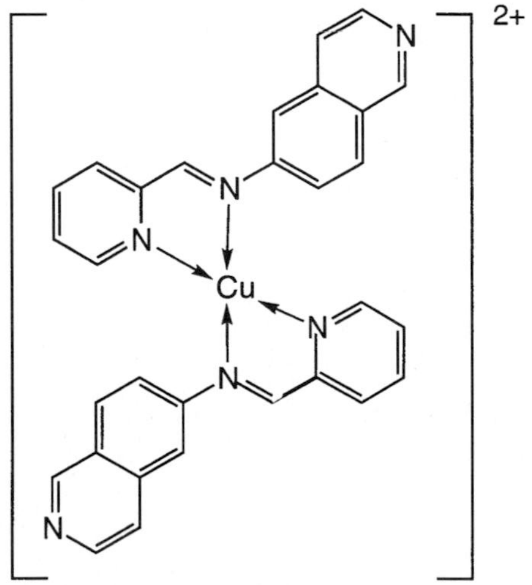   5.15

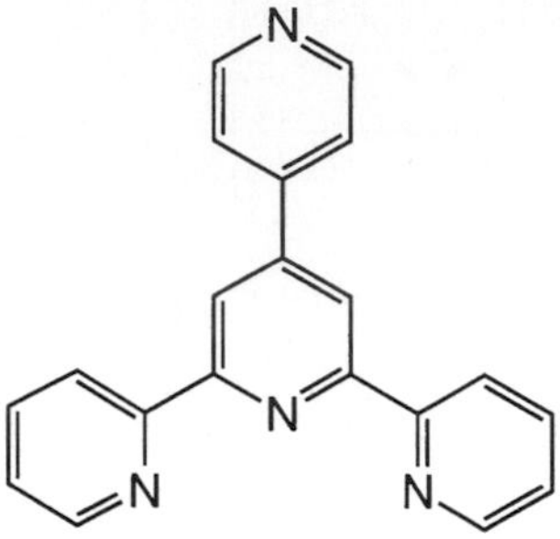

5.16 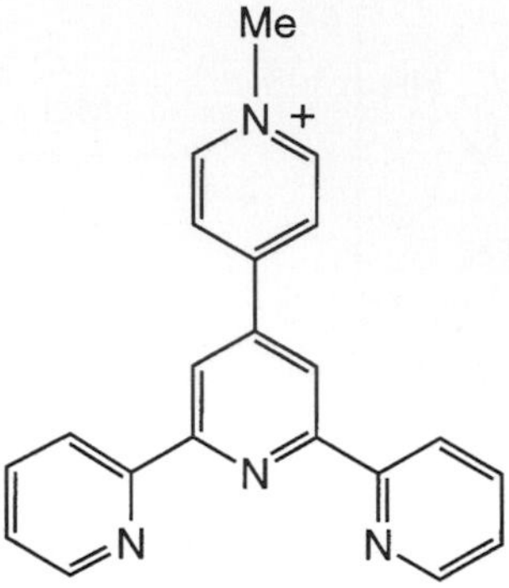   **5.17**

In the nickel(II) complex of the N,N',N'',N'''-tetramethylcyclam, the sp^3 conformation about each nitrogen atom prevents inversion and allows the isolation of specific conformers. What is remarkable is that the direct reaction of nickel(II) salts with the free N,N',N'',N'''-tetramethylcyclam ligand gives a different conformer (Fig. 5-33) in which the four methyl groups are all on the same face with respect to the square plane about the nickel centre. Notice that this also results in changes of the relative orientation of the cyclohexane-like chelate rings.

Even very mild electrophiles may become involved in these reactions. Although triethylamine does not normally react with dichloromethane, in the presence of certain platinum(II) salts quaternisation occurs and the salt $[Et_3NCH_2Cl]_2[PtCl_4]$ may be isolated. The precise mechanism of these reactions is not known, but it seems likely that electrophilic attack upon a co-ordinated triethylamine is the key step. Notice that a tertiary amine, which cannot undergo competing deprotonation reactions, is involved in the reaction.

Ligand deprotonation may result in some interesting ambiguities in oxidation state, which ultimately allow us to consider new patterns of reactivity. We begin by considering the iron(III) complex of a macrocyclic amine **5.18**. Deprotonation of the amino group in the complex gives an iron(III)-amido complex **5.19** (Fig. 5-34).

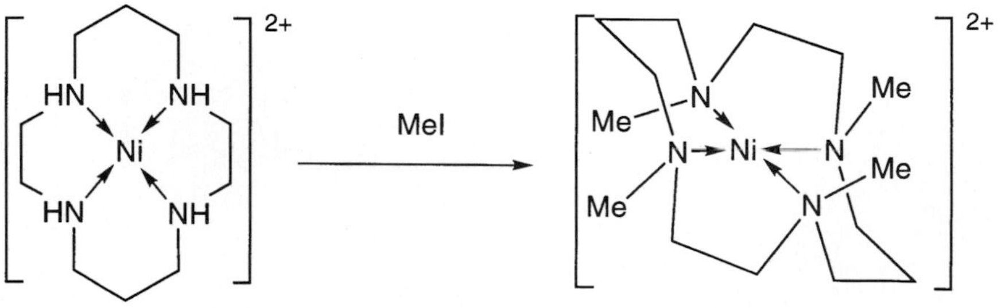

Figure 5-32. The methylation of a nickel(II) cyclam complex to give one specific conformer of the N,N',N'',N'''-tetramethylcyclam complex.

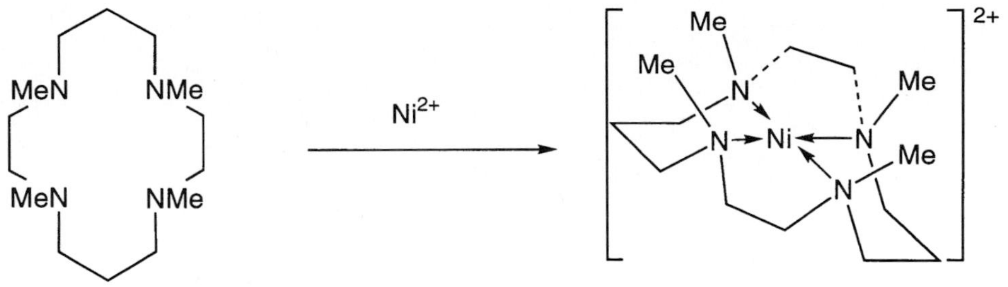

Figure 5-33. The co-ordination of nickel(II) to *N*,*N'*,*N''*,*N'''*-tetramethylcyclam gives a different conformer to that obtained from the methylation of nickel(II) cyclam complexes.

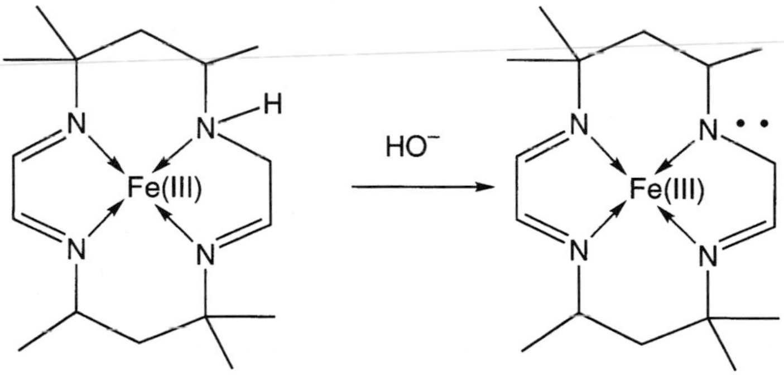

5.18   **5.19**

Figure 5-34. The deprotonation of an iron(III) complex of a macrocyclic amine to give an iron(III)-amido complex. The presence of the lone pair of electrons (negative charge) on the deprotonated nitrogen atom is emphasised (••).

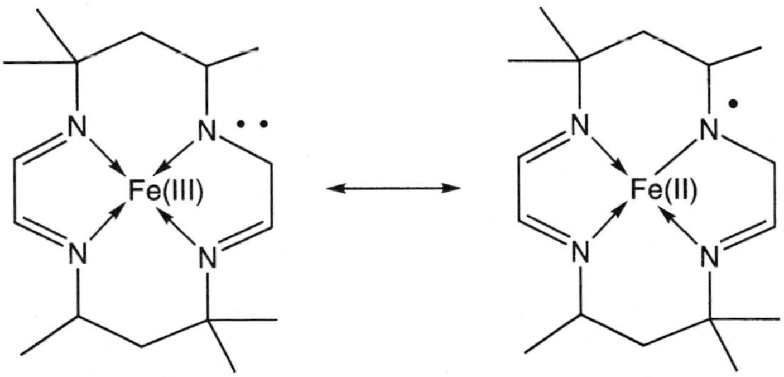

5.20

Figure 5-35. Tautomeric forms of the iron(III)–amido complex generated in Figure 5-34. The iron(II) complex **5.20** is simply a tautomeric form of **5.19**, in which an electron has been transferred from nitrogen to the iron(III) centre.

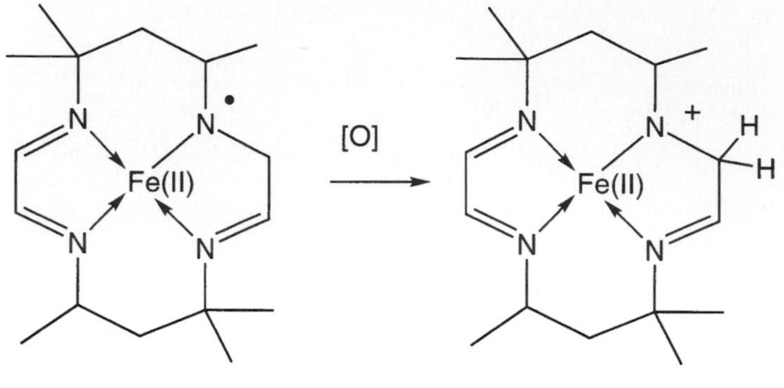

5.21

Figure 5-36. Oxidation of **5.20** yields an iron(II)-nitrenium complex **5.21** by the loss of the electron from the radical nitrogen centre. The nitrenium centre has only six electrons on the nitrogen and is electron deficient.

However, the iron(III)–amido complex could also be written as an iron(II) amido radical complex **5.20**. These two representations are, of course, simply limiting valence bond descriptions of the complex (Fig. 5-35). The difference simply involves the transfer of one electron from the lone pair on the nitrogen in **5.19** to the iron(III) centre.

The oxidation of **5.20** (by air, nitric acid, iron(III) etc.) results in electron loss from the nitrogen-centred radical and the formation of an iron(II) nitrenium species **5.21** (Fig. 5-36), which then undergoes deprotonation to yield an iron(II) complex of a tetraene **5.22** (Fig. 5-37).

Similar reactions have been observed with a range of macrocyclic ligands and metal ions. The precise products of the reactions are not always as simple as one might expect, although the overall result is the partial or complete oxidation of the ligand. Further examples of this type of reaction are discussed in Chapters 6 and 10.

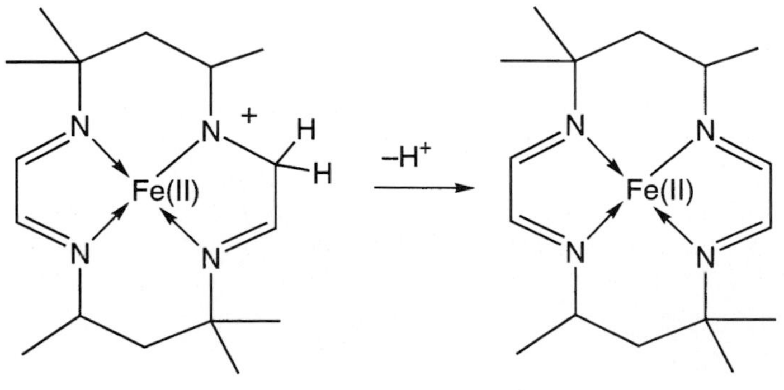

5.22

Figure 5-37. Proton loss from **5.21** generates an iron(II) complex of a tetraimine macrocyclic ligand.

5.4.1 Amine Deprotonation and the S_N1cb Mechanism

The substitution reactions of octahedral metal complexes (Fig. 5-38) have been the subject of intensive investigation over the past forty years, with complexes of the non-labile ions such as chromium(III) and cobalt(III) playing a vital role in these studies.

$$[ML_6]^{n+} + X \rightleftharpoons [ML_5X]^{n+} + L$$

Figure 5-38. The prototypical ligand substitution reaction in octahedral complexes. In principle, the reaction could proceed by associative or dissociative mechanisms.

The bulk of the evidence which has been accumulated (mainly for cobalt(III) complexes) is in favour of a mechanism in which bond-breaking of the M-L bond, as opposed to M-X bond formation, is dominant in the transition state. The investigation of inorganic reaction mechanisms is complicated by the fact that purely associative (S_N2) or purely dissociative (S_N1) pathways are not commonly adopted. The kinetic behaviour and rate equations derived for such reactions are complex, but are in general accord with a dissociative type of mechanism. We are not concerned with the details of the discussion (and indeed the arguments) that have led to these conclusions, but we will linger for a short while to consider one of the red herrings encountered along the way.

Whilst the majority of the available evidence may be interpreted in terms of a dissociative type of mechanism for substitution, there is a substantial number of reactions for which the kinetic data indicate a purely associative S_N2 mechanism. In particular, it was found that reactions of cobalt(III) ammines in basic conditions obeyed pure second-order kinetics (Fig. 5-39).

$$[Co(NH_3)_5Cl]^{2+} + OH^- \rightleftharpoons [Co(NH_3)_5(OH)]^{2+} + Cl^-$$

$$\text{rate} = k[(Co(NH_3)_5Cl^{2+})][OH^-]$$

Figure 5-39. The reaction of some cobalt(III) complexes with hydroxide obey second-order kinetics.

Although this was at first thought to indicate that an associative mechanism was indeed operative in these reactions, over the years a body of further data accumulated to suggest that this was not the case. It is now clear that the deprotonation of a co-ordinated amine is the key step in this mechanism, which is based upon dissociative ligand loss from an amido intermediate. This process is known as the S_N1cb mechanism, and was mentioned briefly in Chapter 2, and illustrated in Fig. 2-13. The first step involves the deprotonation of the co-ordinated amine (Fig. 5-40).

$$[Co(NH_3)_5Cl]^{2+} + OH^- \rightleftharpoons [Co(NH_3)_4(NH_2)Cl]^+ + H_2O$$

Figure 5-40. The first step in the reaction of cobalt(III) ammine complexes with hydroxide often involves deprotonation of an ammine ligand to give an amido species.

$$[Rh(NH_3)_6]^{3+} + D_2O/OD^- \rightleftharpoons [Rh(ND_3)_6]^{3+}$$

Figure 5-41. Ammine ligands co-ordinated to a kinetically inert metal centre may undergo rapid base-catalysed deuterium exchange reactions.

Although these amido complexes can only be isolated in very rare circumstances, the base-catalysed deuteration of ammine ligands is readily observed (Fig. 5-41) and provides good circumstantial evidence for this process. The rate of base catalysed deuteration is considerably faster than that observed in neutral D_2O and is also of comparable rate, or faster than, the overall substitution process.

The next steps in the reaction involve the formation of a five-co-ordinate complex by loss of a halide ion from the amido intermediate (Fig. 5-42). This species is stabilised by π-bonding between the metal centre and the amido ligand. It is convenient to consider the electron count at the metal centre for each of the species involved. In the starting complex we have a total of 18 electrons (9 from the cobalt, 10 from the five ammine ligands, 1 from the chloride and -2 for the overall dipositive charge).[1] The first formed amido complex also possesses 18 electrons (9 from the cobalt, 8 from the four ammine ligands, 1 from the chloride, 1 from the amido ligand, and -1 for the charge). In the five-co-ordinate species we apparently have a 16-electron centre (9 from the cobalt, 8 from the four ammines, 1 from the amido ligand and -2 for the charge). However, the lone pair of electrons on the amido ligand may also become involved in forming a π-bond with the metal, thus transforming the ligand into a 3-electron donor. The result is to permit an 18-electron count at the metal centre in the intermediate generated in Fig. 5-42.

$$[Co(NH_3)_4(NH_2)Cl]^+ \rightleftharpoons [Co(NH_3)_4(NH_2)]^{2+} + Cl^-$$

Figure 5-42. The six-co-ordinate amido complex undergoes chloride loss to generate a five-co-ordinate intermediate.

This five-co-ordinate complex then undergoes a rapid reaction with the incoming ligand (water or some other species) to generate a new six-co-ordinate amido complex, which becomes protonated to generate the observed product of the reaction (Fig. 5-43).

$$[Co(NH_3)_4(NH_2)]^{2+} + H_2O \rightleftharpoons [Co(NH_3)_4(NH_2)(H_2O)]^{2+}$$

$$[Co(NH_3)_4(NH_2)(H_2O)]^{2+} \rightleftharpoons [Co(NH_3)_4(NH_3)(HO)]^{2+}$$

Figure 5-43. The final steps in the S_N1cb mechanism

---

[1] There are a number of different ways used for the counting of electrons in transition metal complexes. In this book I have adopted a *neutral atom* formalism, in which the metals are treated as neutral, neutral ligands as two electron donors and charged ligands such as chloride as one electron donors. For a further discussion of this topic see *Transition Metal Chemistry*, M.Gerloch and E.C. Constable, VCH, Weinheim, **1994**.

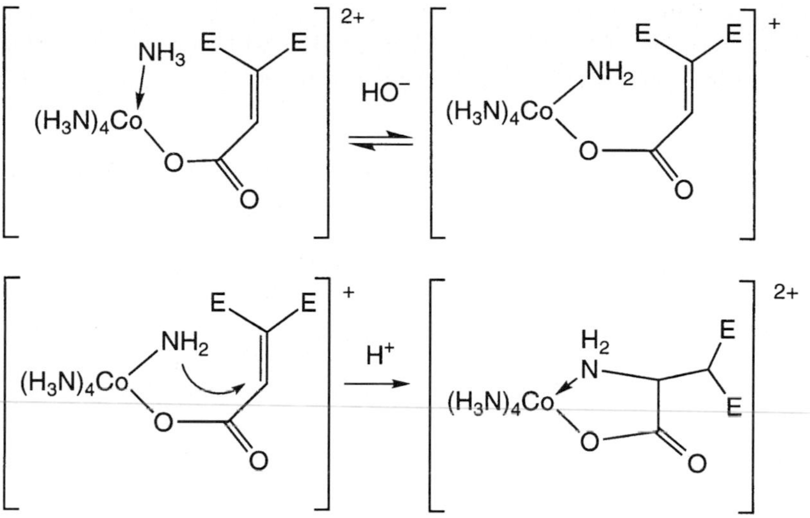

Figure 5-44. A co-ordinated amido ligand is formed by the deprotonation of an ammine; in this sequence of reactions it is then involved in an intramolecular conjugate addition to give an *NO*-co-ordinated amino acid derivative (E = CO$_2$Et).

This is an important mechanism, and we have seen the consequences of attack by an intramolecular nucleophile (ligand) in earlier chapters. A particularly interesting example is seen in the intramolecular Michael addition of a co-ordinated amide at a cobalt(III) centre to yield an amino acid derivative (Fig. 5-44).

In the case of complexes containing macrocyclic amines, it is often possible to isolate the intermediate deprotonated amido complexes. In the next section, we will consider some very important aspects of co-ordination chemistry, in which reactions of co-ordinated amine and/or amide play a crucial role.

5.5 Formation and Reactivity of Imines

Ligands containing the R$_2$C=NR' imine grouping have played a major role in the development of contemporary co-ordination chemistry and are a common feature in the highly structured molecular architectures which are regularly encountered. To a co-ordination chemist concerned with 'normal' oxidation state compounds (i.e., those in the +2 or +3 oxidation states), the π-acceptor imine ligand provides an *N*-donor analogue to the ubiquitous carbonyl ligand found in organometallic compounds. Accordingly, much of the remainder of this chapter is concerned with the metal-directed preparation, stabilisation and reactivity of imines. We have discussed the hydrolysis of co-ordinated imines in Chapter 4, and further aspects of the preparation and reactivity of cyclic imines will be found in

Chapter 6. As discussed in Section 4.3.1, imines are formally related to carbonyl compounds by the addition of amine and elimination of water, and we will begin by considering the ways in which metal ions may control this process.

5.5.1 Formation by Condensation of an Amine with a Carbonyl Compound

At the beginning of this chapter we considered the ways in which co-ordination to a metal ion might control the reactions of a carbonyl compound. We considered the possible fates of the tetrahedral intermediate formed by the attack of a nucleophile upon the carbonyl carbon atom. In the case of a nucleophile such as ammonia or a primary amine another pathway leading to an imine is open.

Even if the imine may not be isolated, the transient species may sometimes be trapped by reaction with a suitable nucleophile. This is the basis of the reductive amination reaction in which an amine is formed from the reaction of ammonia with a carbonyl compound in the presence of a reducing agent such as sodium borohydride or formate. Use of a primary or secondary amine results in the specific formation of secondary or tertiary amines respectively (Fig. 5-45). This synthetic method allows the preparation of high yields of amines, in contrast to the unselective and uncontrollable reaction of alkylating agents with amines. A specific example involving the preparation of α-phenylethylamine from acetophenone is presented in Fig. 5-46.

The position of the equilibrium between imine and carbonyl may be perturbed by interaction with a metal ion. We saw in Chapter 2 how back-donation of electrons from suitable orbitals of a metal ion may stabilise an imine by occupancy of the π^* level. It is possible to form very simple imines which cannot usually be obtained as the free ligands by conducting the condensation of amine and carbonyl compounds in the presence of a metal ion. Reactions which result in the formation of imines are considered in this chapter even in cases where there is no evidence for prior co-ordination of the amine nucleophile to a metal centre. Although low yields of the free ligand may be obtained from the metal-free reaction, the ease of isolation of the metal complex, combined with the higher yields, make the metal-directed procedure the method of choice in many cases. An example is presented in Fig. 5-47. In the absence of a metal ion, only low yields of the diimine are obtained from the reaction of diacetyl with methylamine. When the reaction is conducted in the presence of iron(II) salts, the iron(II) complex of the diimine (**5.23**) is obtained in good yield.

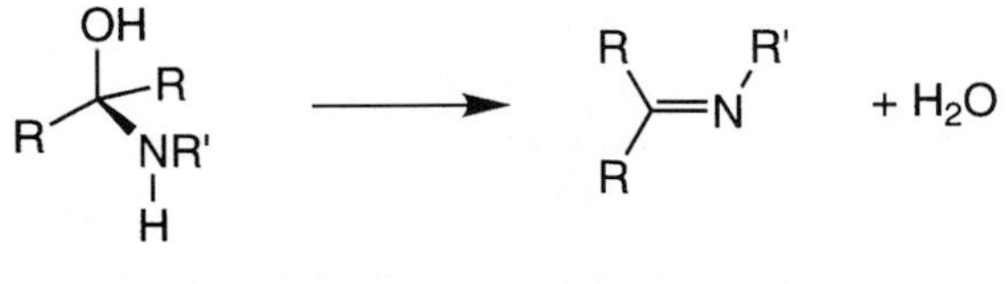

Figure 5-45. The formation of an imine by the dehydration of an aminol. Unless the imine group is conjugated with the R and R' groups, it is usually necessary to use special methods for the removal of water, otherwise the equilibrium lies far to the left.

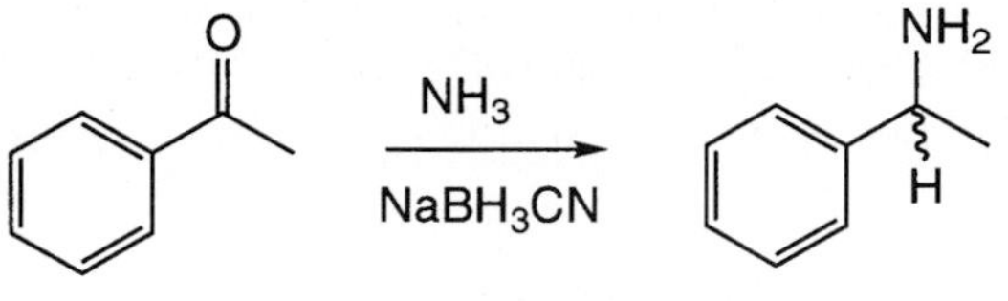

Figure 5-46. The reductive amination of a ketone involves the trapping of a transient imine.

It should be emphasised that many different factors are of importance in reactions of this type, and this should be borne in mind even when we concentrate upon one principal feature. In the reaction shown in Fig. 5-47, the formation of the conjugated product is also important, since the hydrolysis of comparable non-conjugated bisimines is often drama-tically accelerated by the addition of metal ions. The stabilisation of the product by che-lation is probably an additional factor in these reactions. There is also an electronic effect operating. The low-spin d^6 iron(II) centre will give a maximum ligand field stabilisation and back-donation to the strongly π-accepting diimine will be at a maximum at this elec-tron configuration. The α,α'-diimine functionality is a recurrent structural motif in ligands which exhibit a high specificity for iron(II) and other d^6 metal centres, and is pre-sent in ligands such as 2,2'-bipyridine and 1,10-phenanthroline.

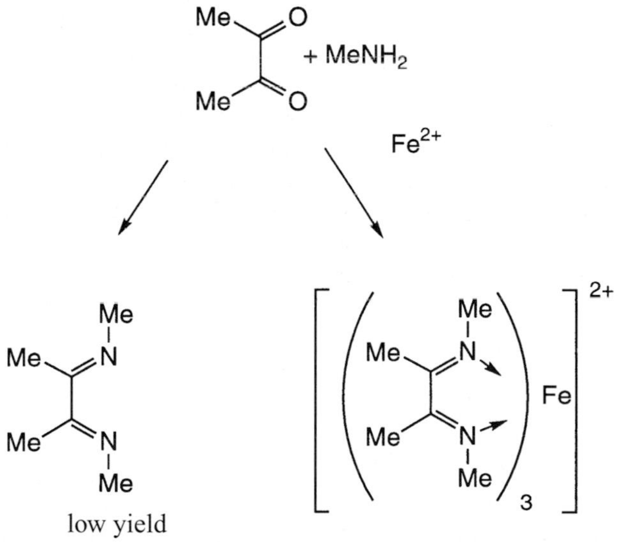

5.23

Figure 5-47. The condensation of diacetyl with methylamine in the presence of iron(II) yields the iron(II) complex of the conjugated diimine. In the absence of metal ion, little of the imine is obtai-ned. A number of factors are of importance in this reaction, including the strong back-bonding bet-ween the π-acceptor diimine and the low-spin d^6 metal centre.

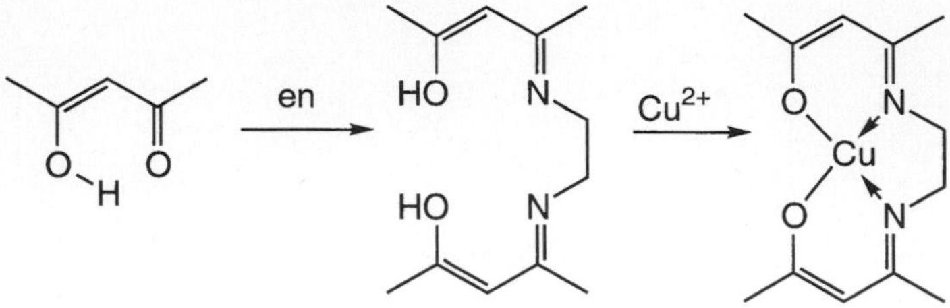

Figure 5-48. Co-ordinated 1,3-diketonates do not always react with amines. If this is the case, an alternative strategy may be adopted.

It is not usually possible to form mixed *NO*-donor ligands by direct reaction of co-ordinated 1,3-diketonates with amines. In part, this is due to the delocalised charge of the formally anionic ligand rendering the diketonate less prone to attack by a nucleophile. This deactivation towards attack by nucleophiles should be contrasted with the facile reactions with electrophiles which have been discussed in Section 5.3. It is possible, however, to form complexes of conjugated *NO*-donor ligands by direct reaction of the metal-free, 1,3-dicarbonyl with amine, followed by co-ordination (Fig. 5-48).

The design of polydentate ligands containing imines has exercised many minds over many years, and imine formation is probably one of the commonest reactions in the synthetic co-ordination chemist's arsenal. Once again, the chelate effect plays an important role in stabilising the co-ordinated products and the majority of imine ligands contain other donor atoms that are also co-ordinated to the metal centre. The above brief discussion of imine formation will have shown that the formation of the imine from amine and carbonyl may be an intra- or intermolecular process. In many cases, the detailed mechanism of the imine formation reaction is not fully understood. In particular, it is not always clear whether the nucleophile is metal-co-ordinated *amine* or *amide*. Some intramolecular imine formation reactions at cobalt(III) are known to proceed through amido intermediates. A particularly useful intermediate (**5.24**) in metal-directed amino acid chemistry is

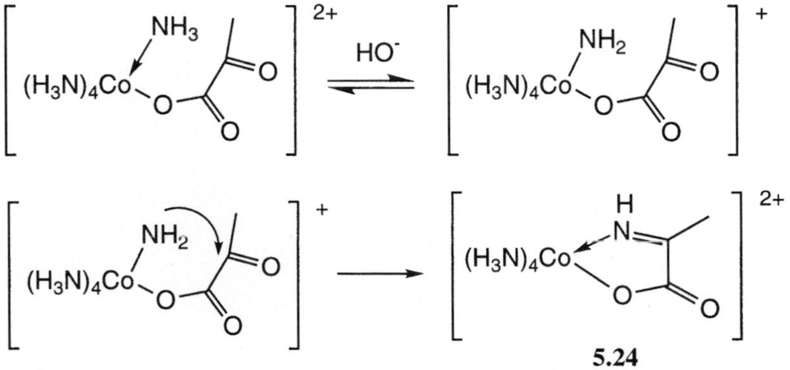

5.24

Figure 5-49. The formation of a co-ordinated imine by the intramolecular reaction of an amido ligand with pyruvate at a kinetically inert cobalt(III) centre.

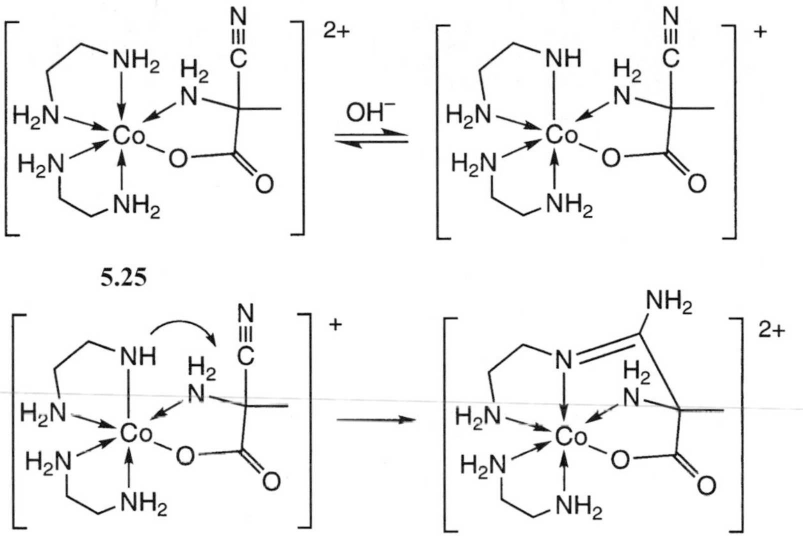

5.25

Figure 5-50. Intramolecular attack by a deprotonated 1,2-diaminoethane ligand upon a nitrile group.

generated by the treatment of a cobalt(III) pyruvate complex with base (Fig. 5-49). In the related complex with two 1,2-diaminoethane ligands, nucleophilic addition of hydrogen cyanide gives the intermediate complex **5.25**. Treatment of this nitrile complex with base results in the generation of an amido intermediate by the deprotonation of the en, which then undergoes an internal reaction to yield amidate (Fig. 5-50).

In recent years, it has been shown that co-ordinated phosphines may also undergo reactions with carbonyl compounds. This is well exemplified in the reactions of $[(MeHPCH_2CH_2PHMe)_2Pd]^{2+}$ (Fig. 5-51). The reaction with formaldehyde yields a complex of an open-chain hydroxymethyl substituted ligand, the same species that is obtained from reaction of the free ligand. This is the phosphorus analogue of the aminol intermediate in imine formation. It is extremely unusual to obtain $RP=CR_2$ systems in the absence of sterically demanding substituents.

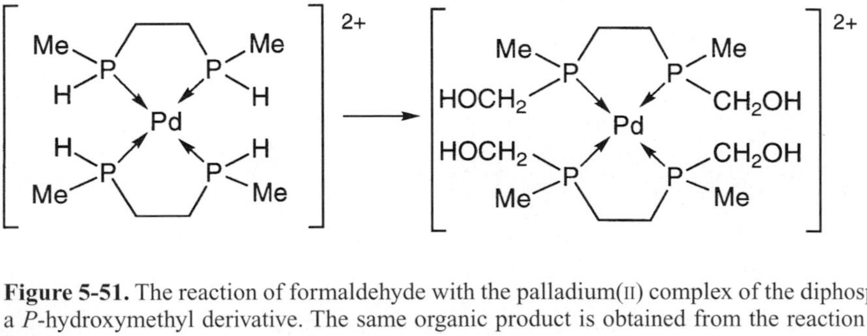

Figure 5-51. The reaction of formaldehyde with the palladium(II) complex of the diphosphine gives a *P*-hydroxymethyl derivative. The same organic product is obtained from the reaction of the free ligand.

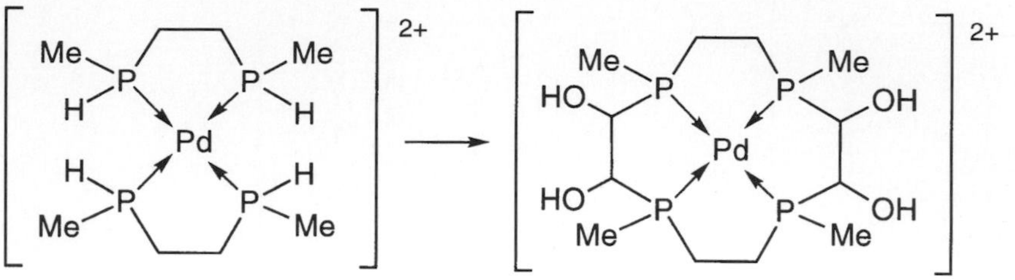

Figure 5-52. The reaction of the palladium diphosphine complex with diacetyl gives a palladium complex of a macrocyclic P_4 donor ligand.

The reaction of the palladium complex with diacetyl is rather more interesting and yields the palladium(II) complex of a novel tetraphospha macrocyclic ligand. Note that the phosphorus analogue of an aminol rather than an imine is once again obtained, representing the general instability of $P=C$ bonds in the absence of sterically hindering substituents (Fig. 5-52).

5.5.2 Transimination

One of the paradoxes of metal–imine chemistry is the observation that in many cases the imine is stabilised with respect to nucleophilic attack by water upon co-ordination, but is still prone to attack by amines. We saw in Chapter 4 how the hydrolysis of imines may be either promoted or inhibited by co-ordination to a metal, and we also saw a number of examples involving nucleophilic attack on an imine by a variety of other nucleophiles. A special case of such a nucleophilic attack involves another amine. The consequence is a transimination reaction, as indicated in Fig. 5-53. Presumably, intermediates of type **5.26** are involved. The procedure is of some synthetic use for the preparation of imine complexes (Fig. 5-54).

Another interesting example of metal-directed chemistry involving the stabilisation and reactivity of imines is seen in the reaction of pyridoxal with amino acids. This reaction is at the basis of the biological transamination of amino acids to α-ketoacids, although the involvement of metal ions in the biological systems is not established. The reaction of pyridoxal (**5.27**) with an amino acid generates an imine (**5.28**), which is stabilised by co-ordination to a metal ion (Fig. 5-55).

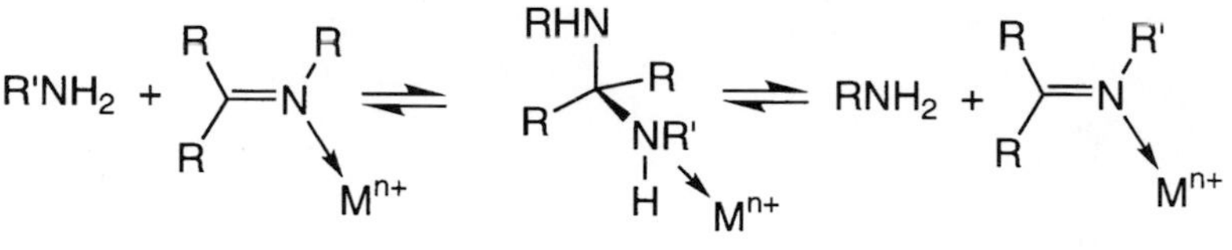

5.26

Figure 5-53. The transimination of a co-ordinated imine by reaction with an amine.

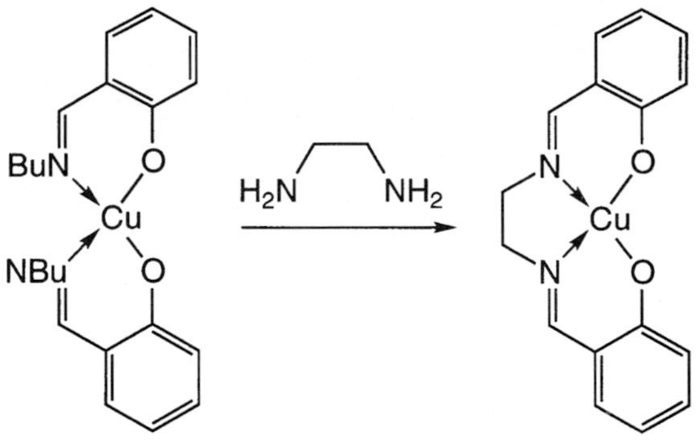

Figure 5-54. Transimination may provide a method for the preparation of imines which are not rea-dily accessible by other methods. This reaction illustrates a way of making *NO* donor ligands with-out the need for nucleophilic attack of amine on a co-ordinated 1,3-diketonate.

The complex is additionally stabilised by co-ordination of the phenoxide, and possib-ly the carboxylate, to the metal ion, illustrating the utility of chelating ligands in the study of metal-directed reactivity. We saw in the previous section the ways in which a metal ion may perturb keto–enol equilibria in carbonyl derivatives, and similar effects are observed with imines. The metal ion allows facile interconversion of the isomeric imines. The first step of the reaction is thus the tautomerisation of **5.28** to **5.29** (Fig. 5-56). Finally, the metal ion may direct the hydrolysis of the new imine (**5.29**) which has been formed, to yield pyridoxamine (**5.30**) and the α-ketoacid (Fig. 5-57).

The kinetically inert cobalt(III) complex of **5.31** has been investigated as a model for metal-directed reactions of this type. The ligand is the imine derived from the condensa-tion of glycine with pyridoxal. The initial stages involving the breaking of the C–H bonds have been studied and the two diastereotopic protons of the CH_2 group have been found to undergo deuterium exchange at different rates.

Transamination reactions of this type have found some synthetic application. The syn-thesis of the nickel(II) complex of a macrocycle indicated in Fig. 5-58 clearly involves

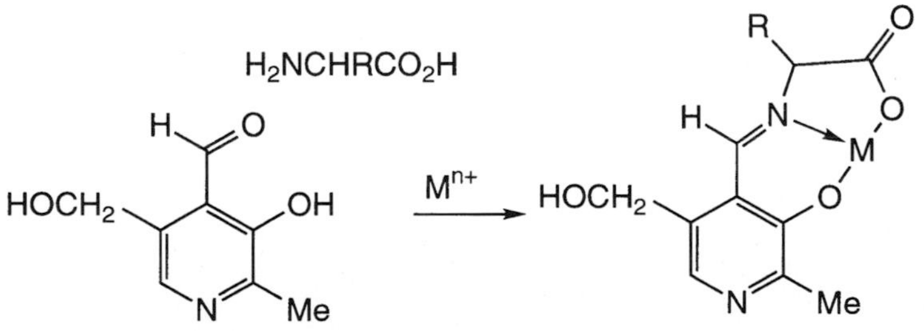

5.27   **5.28**

Figure 5-55. The reaction of an amino acid with pyridoxal in the presence of a metal ion gives a metal complex of an intermediate imine.

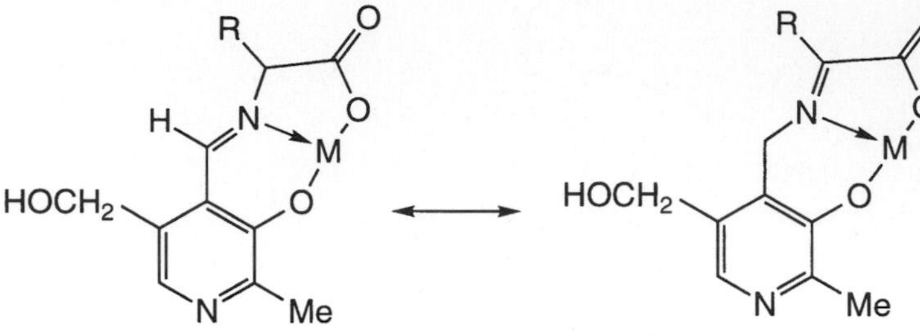

5.28   **5.29**

Figure 5-56. The key step in the metal-directed transimination reaction involves the interconversion of tautomeric imines.

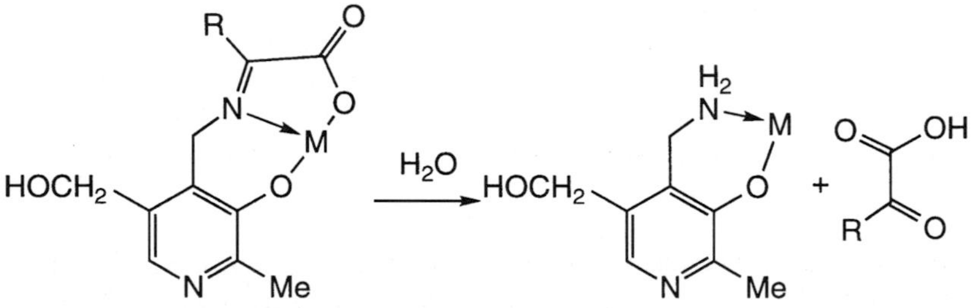

5.29   **5.30**

Figure 5-57. The final step in the transimination involves metal-directed hydrolysis of the new imine to give pyridoxamine and the ketoacid.

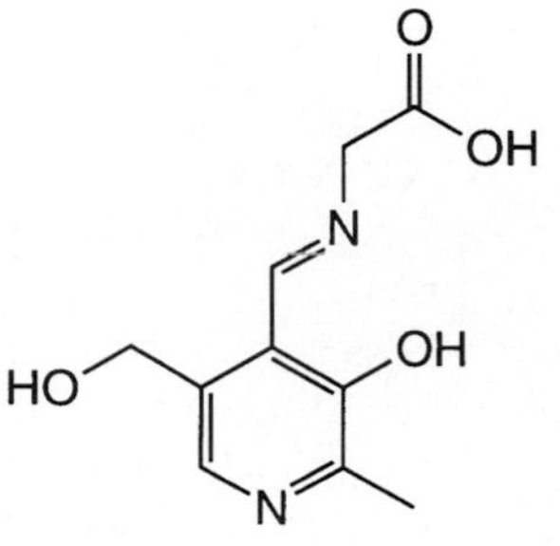

5.31

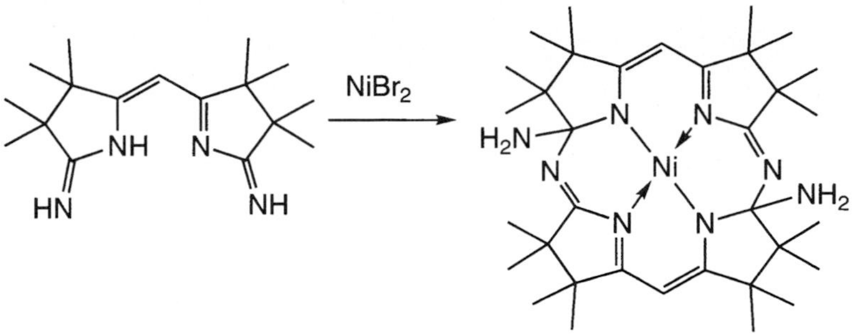

Figure 5-58. The formation of a nickel complex of a tetraza macrocyclic ligand from an acyclic precursor

attack at the carbon of an imine, followed by sequential protic shifts. The facile assembly of the tetraaza macrocyclic ligand about a metal ion has obvious biological connotations and raises interesting questions regarding the origin of porphyrin pigments in early organisms. Eschenmoser and others have made elegant use of metal-directed in reactions in the preparation of such pigments.

5.6 Reactions of Co-ordinated Cyanide

It has been known for many years that the reaction of cyanide ion with alkylating agents shows a dependency on the presence of metal ions. The classic application of this is the formation of *nitriles* by reaction with potassium cyanide and *isonitriles* from the reaction with silver cyanide (Fig. 5-59).

MeI + KCN → MeCN +KI

MeI + AgCN → MeNC + AgI

Figure 5-59. The reaction of alkylating agents with potassium cyanide gives nitriles, whilst the reaction with silver cyanide gives isonitriles.

This type of selectivity was explained in terms of the relative hardness and softness of the metal ions in Chapter 2. Rather more dramatic manifestations of this selectivity are observed if the reactions of co-ordinated cyanide ion (as opposed to ionic 'cyanide') are investigated. The co-ordination of cyanide to a transition metal ion is usually through the softer carbon end. A simple steric effect predicts that the nitrogen atom should be more available as a nucleophilic centre, although we should also consider the electron distribution within the metal-cyanide bonding. The reactions of hexacyanoferrates and other co-ordinated cyanides with alkylating agents have been widely investigated. Alkylation of the cyanide to yield a co-ordinated isonitrile occurs (Fig. 5-60).

$$[(NC)_5Fe-C \equiv N]^{4-} + MeOSO_3Me \rightarrow [(NC)_5Fe-C=NMe]^{3-}$$

$$[(NC)_5Fe-C=NMe]^{3-} \rightarrow \rightarrow [Fe(CNMe)_6]^{2+}$$

Figure 5-60. The reaction of hexacyanoferrate(II) salts with alkylating agents gives iron(II) complexes of isonitriles.

5.7 Reactions of Co-ordinated Water or Hydroxide

We saw in Chapter 2 that co-ordination of a water molecule to a metal ion modifies the pK_a and can make the water considerably more acidic. This stabilisation of the hydroxide anion is rationalised in terms of transfer of charge from the oxygen to the metal in the co-ordinate bond. Some typical pK_a values of co-ordinated water molecules are given in Table 5-2.

Table 5-2. The pK_a values of some aqua complexes.

Complex	pK_a	Complex	pK_a
$[Pd(H_2O)_4]^{2+}$	1.4	$[Fe(H_2O)_6]^{3+}$	2.2
$[Al(H_2O)_6]^{3+}$	5.0	$[Cu(H_2O)_6]^{2+}$	8.0
$[Zn(H_2O)_6]^{2+}$	9.0	$[Co(H_2O)_6]^{2+}$	9.7
$[Ni(H_2O)_6]^{2+}$	9.9	$[Mn(H_2O)_6]^{2+}$	10.6

A co-ordinated hydroxide ligand will still possess some of the nucleophilic properties of free hydroxide ion, and this observation proves to be the basis of a powerful catalytic method, and one which is at the basis of very many basic biological processes. In general, hydrolysis reactions proceed more rapidly if a water nucleophile is replaced by a charged hydroxide nucleophile. This is readily rationalised on the basis of the increased attraction of the charged ion for an electrophilic centre. However, in many cases the chemical properties of the substrate are not compatible with the properties of the strongly basic hydroxide ion. This is exactly the situation that biological systems find themselves in repeatedly. For example, the uncatalysed hydration of carbon dioxide is very slow at pH 7 (Fig. 5-61).

In the laboratory the process would be accelerated by the addition of an alkali metal hydroxide; this both accelerates the reaction by replacing the neutral nucleophile by an anionic one and also perturbs the equilibrium towards the right-hand side by Le Chate-

$$CO_2 + H_2O \rightleftharpoons HCO_3^- + H^+$$

Figure 5-61. The hydration of carbon dioxide is very slow. Biology has found very effective metal-mediated ways to speed up this process.

lier's principle. However, biological systems are not compatible with strongly alkaline aqueous solutions, as anyone who has spilt sodium hydroxide solution on him or herself can testify. Accordingly, we are left with the challenge of attempting to generate signifi-cant concentrations of hydroxide ion at physiological pH (≈ 7). This is achieved by chan-ging the pH of the water molecule by co-ordination to a metal ion. In the case of the en-zyme responsible for the hydration of carbon dioxide, carbonic anhydrase, the water molecule is co-ordinated to a zinc ion and exhibits a pK_a of about 7.

5.7.1 Intramolecular Attack by Co-ordinated Hydroxide

It is very often extremely difficult to demonstrate that a metal-co-ordinated hydroxide ion is involved in a particular reaction. Studies of kinetic behaviour provide one of the most powerful tools for the determination of reaction mechanisms. It is not, however, always easy to distinguish between intra- and intermolecular attack of water or hydroxide. The most unambiguous studies have been made with non-labile cobalt(III) complexes, and we will open this discussion with these compounds.

We saw in Chapter 3 that the hydrolysis of *chelated* amino acid esters and amides was dramatically accelerated by the nucleophilic attack of external hydroxide ion or water and that cobalt(III) complexes provided an ideal framework for the mechanistic study of these reactions. Some of the earlier studies were concerned with the reactions of the cations $[Co(en)_2Cl(H_2NCH_2CO_2R)]^{2+}$, which contained a monodentate amino acid ester. In many respects these proved to be an unfortunate choice in that a number of mechanisms for their hydrolysis may be envisaged. The first involved attack by external hydroxide upon the monodentate *N*-bonded ester (Fig. 5-62). This process is little accelerated by co-ordina-tion in a monodentate manner.

The second mechanism requires a preliminary displacement of chloride by the oxygen of the ester to give a chelated complex which may be attacked by external hydroxide as seen in Chapter 3. In practice, the displacement of chloride from cobalt(III) is very slow and this mechanism proceeds by the S_N1cb mechanism, in which loss of chloride ion is aided by deprotonation of the amine. The first step involves deprotonation of the en ligand followed by chloride loss to give a five co-ordinate intermediate (Fig. 5-63).

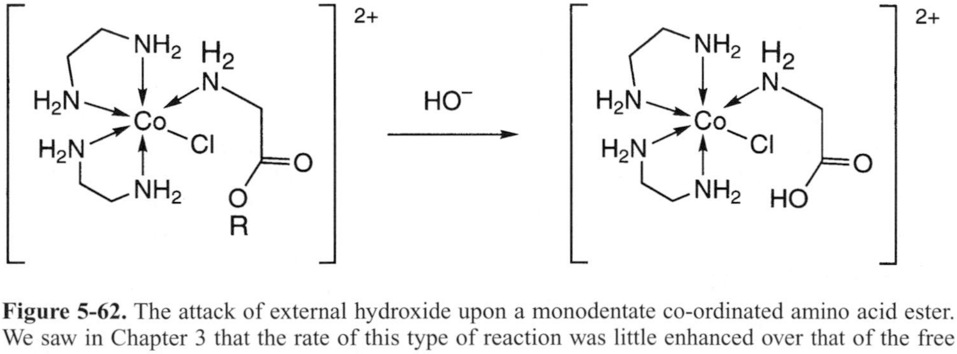

Figure 5-62. The attack of external hydroxide upon a monodentate co-ordinated amino acid ester. We saw in Chapter 3 that the rate of this type of reaction was little enhanced over that of the free ester.

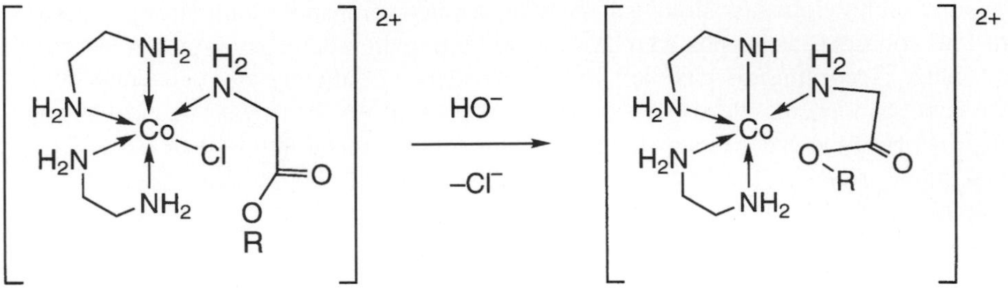

Figure 5-63. The formation of a five co-ordinate intermediate by the loss of chloride ion in an $S_N 1cb$ mechanism involving the deprotonation of the en ligand.

This is followed by a rapid process in which the oxygen atom of the ester becomes co-ordinated to the metal (Fig. 5-64).

The final step involves attack upon the chelated ester by external hydroxide (Fig. 5-65), in a process which we showed in Chapter 3 was dramatically accelerated by co-ordination.

The other mechanism which may be adopted also proceeds by an initial $S_N 1cb$ mechanism to form the five-co-ordinate intermediate, but the fate of this species differs. Instead of attack by an intramolecular ligand (the ester group), attack by an external ligand (water or hydroxide) results in the formation of a hydroxy complex (Fig. 5-66). In this hydroxy compound the ester may undergo attack by the intramolecular hydroxide to yield the co-ordinated carboxylate. This is, of course, the same intermediate that we observed in Fig. 3-8.

The intramolecular attack of hydroxide upon the monodentate ester (Fig. 5-67) would give the same observed product as the other mechanism.

How may we distinguish between pathways that involve external attack by hydroxide and those that involve co-ordinated hydroxide? There is a considerable accumulation of data that suggest the two latter pathways are the most important (i.e., attack of external hydroxide upon monodentate amino acid ester is not greatly accelerated). The attack by external hydroxide may be studied independently and accurate rate constants may be determined for insertion in the composite rate equation with the two competing processes. In some cases it is possible to detect the five-co-ordinate and the other intermediates. Finally, some elegant labelling studies have provided very strong evidence for the exi-

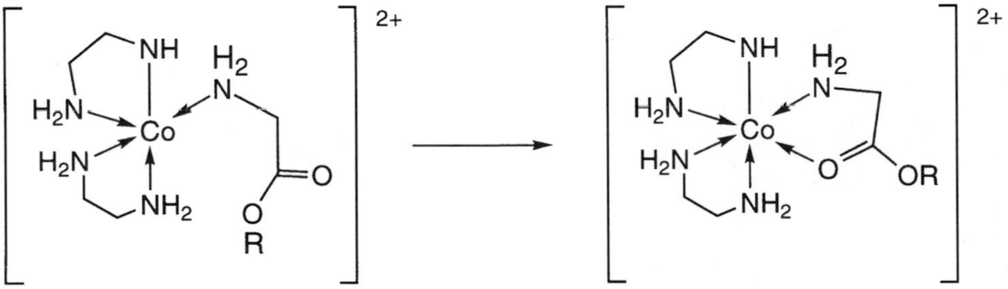

Figure 5-64. The five co-ordinate intermediate is rapidly trapped by chelation to the amino acid ester.

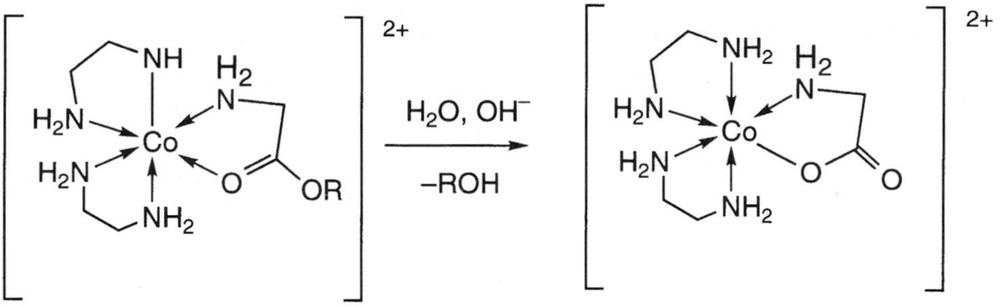

Figure 5-65. Attack upon the chelated ester by external hydroxide is rapid.

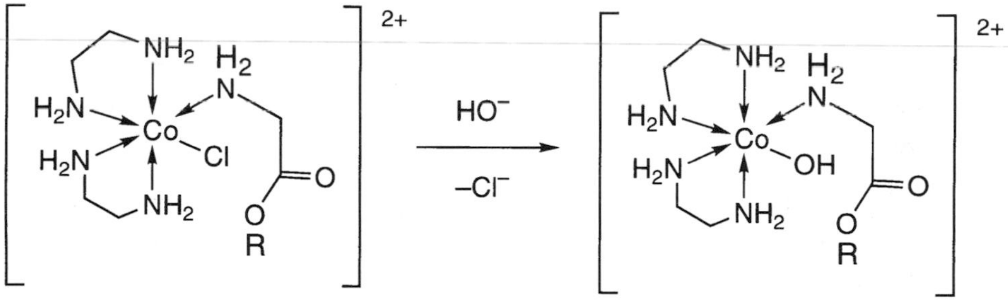

Figure 5-66. The trapping of the five co-ordinate intermediate can give a hydroxy complex with a monodentate amino acid ester ligand.

stence of the two competing mechanisms. The beauty of the labelling experiments relies upon the cobalt(III) centre being non-labile - thus, an ^{18}O-labelled hydroxide bonded to the metal will not exchange to any significant extent with bulk water or hydroxide. Conducting the reaction with labelled complex in unlabelled water will allow a direct determination of the relative importance of the two mechanisms (Fig. 5-68).

Although the above discussion has concentrated upon the hydrolysis of amino acid esters, very similar mechanisms have been demonstrated for the hydrolysis of amino acid amides. A very wide range of intramolecular reactions of this type are now known to occur by intramolecular attack by hydroxide, with most having been demonstrated at non-labile

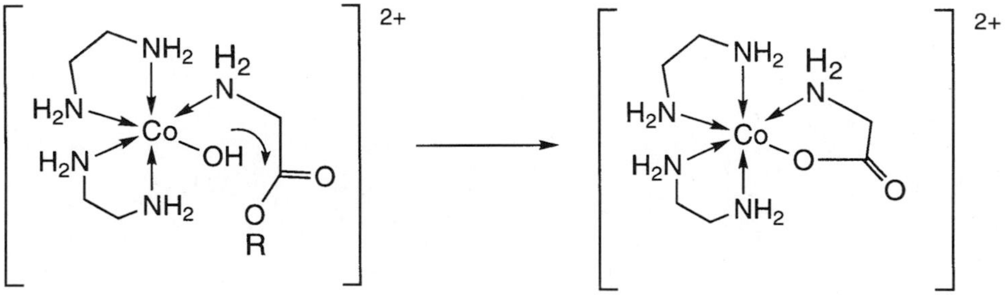

Figure 5-67. The formation of the same chelated amino acid anion is observed whichever mechanism is adopted.

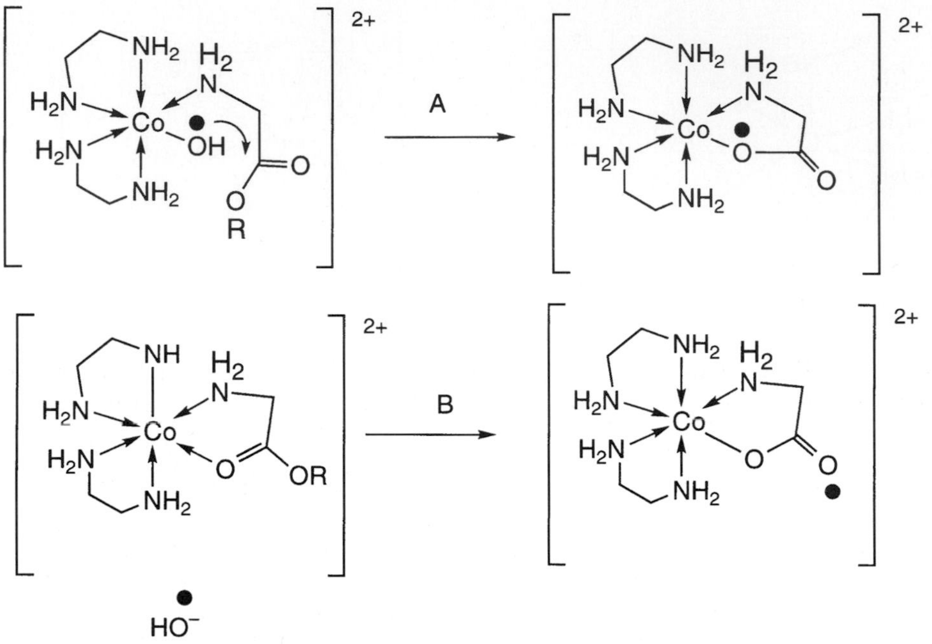

Figure 5-68. The labelling experiment that distinguished between the various pathways for hydrolysis of amino acid esters. The site of the label may be determined by IR spectroscopy or other methods. Pathway A involves co-ordinated hydroxide nucleophile and pathway B, external hydroxide. Both pathways are found to be important for cobalt(III).

cobalt(III) or rhodium(III) centres. Even co-ordinated alkyl halides may undergo intramolecular reaction with co-ordinated hydroxide (Fig. 5-69).

Similar mechanisms may be proposed for the hydrolysis of amino acid esters and amides co-ordinated to labile metal centres such as copper(II) or nickel(II), although mechanistic studies at these centres are much more difficult to perform in view of the rapidity of ligand exchange processes. Further complications arise from the formation of insoluble or colloidal suspensions of metal oxides and hydroxides at higher pH values. In general,

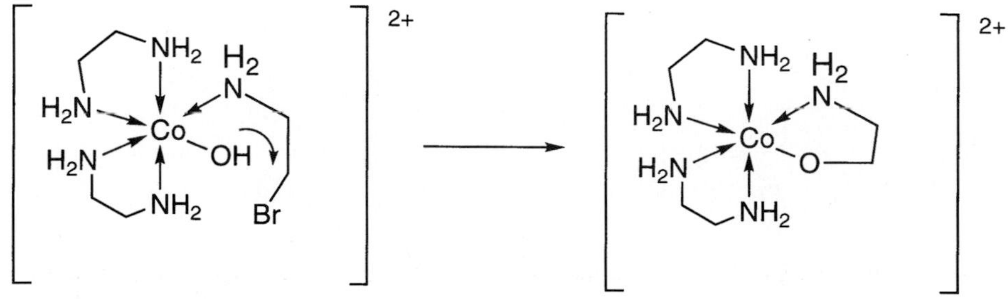

Figure 5-69. The displacement of halide from an alkyl halide in an intramolecular process.

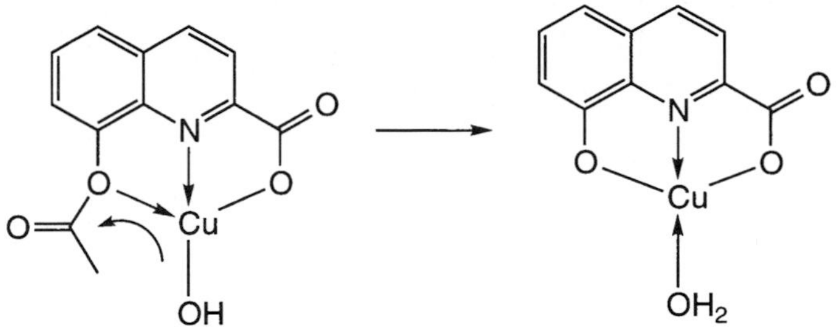

5.32

Figure 5-70. The hydrolysis of the ester **5.32** is accelerated by copper(II) salts. The initial step is the formation of the chelated copper(II) complex, followed by intramolecular attack of co-ordinated hydroxide upon the co-ordinated ester group.

the favoured mechanisms appear to involve attack by external hydroxide on chelated ligand, although examples of both types of limiting mechanism are known.

If we consider the hydrolysis of esters and amides other than those of amino acids there is more evidence for the involvement of co-ordinated hydroxide. Particular attention has been centred on the hydrolysis of esters of chelating heterocyclic ligands. For example, there is very convincing evidence suggesting that the hydrolysis of the ester in the copper(II) and zinc(II) complexes of 8-acetoxyquinoline-2-carboxylate (**5.32**) involves attack by co-ordinated hydroxide (Fig. 5-70). Once again, the metal plays a dual role in activating the carbonyl to nucleophilic attack by co-ordination to the oxygen atom and also in stabilising the anionic leaving carboxylate group by co-ordination.

Another elegant example of the importance of attack by an intramolecular nucleophile is seen in the hydrolysis of some cyclic amides. The ligand **5.33** has been designed to act as a tridentate donor, in which the pyridine nitrogen atom, the carboxylate and the tertiary amino nitrogen are involved in bonding to a metal. Chelation of the metal ion to this ligand in a tridentate manner results in the metal centre being placed in such a position that a water or hydroxide co-ordinate to it is ideally situated for nucleophilic attack upon the *non-co-ordinated* amide group. Very dramatic enhancements of the rate of hydrolysis of **5.33** are observed with some metal ions (Fig. 5-71).

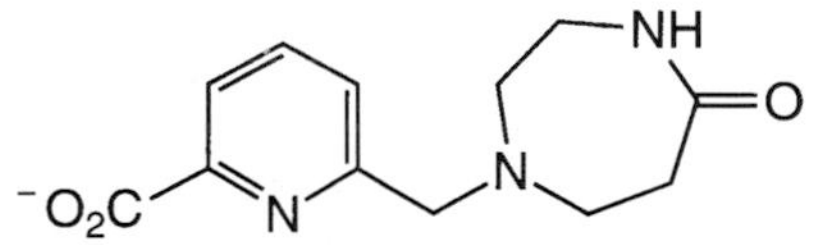

5.33

A related mechanism is almost certainly observed in the almost instantaneous methanolysis of some pivalamide derivatives in the presence of copper(II) chloride. For example, ligand **5.34** forms a five co-ordinate complex with copper(II) chloride. When

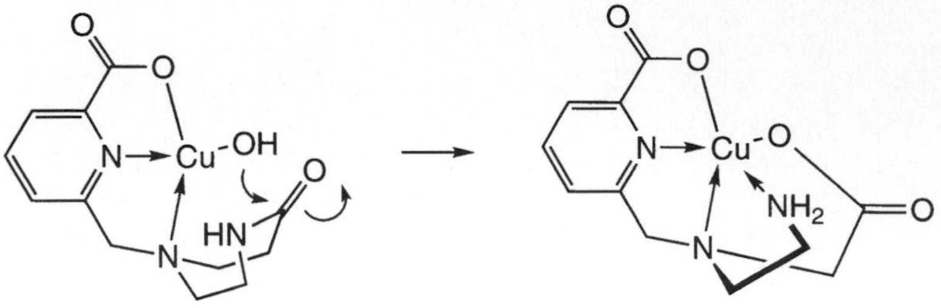

Figure 5-71. The hydrolysis of the tridentate ligand **5.33** is accelerated by co-ordination to a metal ion. The two reaction involves intramolecular attack by co-ordinated hydroxide.

this reacts with methanol, deacylation of the ligand occurs to give methyl pivalate and the copper(II) complex of a secondary amine (Fig. 5-72).

The hydrolysis of phosphate esters by both inter- and intramolecular hydroxide has been discussed in Chapter 4.

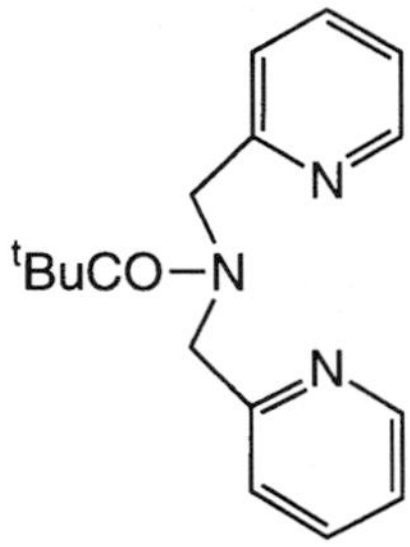

5.34

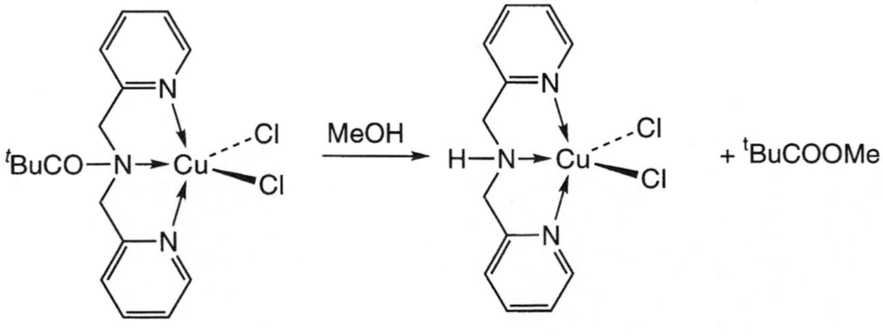

Figure 5-72. The methanolysis of the amide **5.34** is almost instantaneous in the presence of copper(II) chloride. It is likely that the reaction involves attack by a co-ordinated methanol molecule.

5.7.2 Intermolecular Attack by Co-ordinated Hydroxide

In the preceding section we discussed the use of co-ordinated hydroxide as an intramolecular nucleophile. It could also act as a nucleophile to an external electrophile. Over the
past few decades, there has been considerable interest in the nucleophilic properties of
metal-bound hydroxide ligands. One of the principal reasons for this relates to the widespread occurrence of Lewis acidic metals at the active site of hydrolytic enzymes. There
has been a lively discussion over the past thirty years on the relative merits of mechanisms
involving nucleophilic attack by metal-co-ordinated hydroxide upon a substrate or attack
by external hydroxide upon metal-co-ordinated substrate. As we have shown above, both
of these mechanisms are possible with non-labile model systems.

However, we may also design model systems to study the reactions of co-ordinated
hydroxide with external electrophiles. The simplest models utilise non-labile complexes
with a single hydroxide ligand, such as $[M(NH_3)_5(OH)]^{2+}$ (M = Co or Rh). Various electrophiles have been shown to react with such metal-bound hydroxide ligands, and some
of these reactions are indicated in Fig. 5-73.

As in most cases, the involvement of co-ordinated hydroxide in reactions involving
more labile metal ions is much more difficult to demonstrate, although, of course, it is
labile metal ions such as zinc(II) which are often involved in these Lewis acid-directed
reactions in biological systems. However, the use of macrocyclic or polydentate ligands is
a valuable tool in biomimetic chemistry. The chelate effect allows us to form complexes
with metal ions, such as zinc(II), in which the polydentate ligand is relatively non-labile
(in the same way that a protein is a non-labile ligand). An example of this is seen in the
five-co-ordinate zinc complex of a tetraaza macrocyclic ligand, **5.35**, which has been suggested as a good model for the zinc centre of carbonic anhydrase. The co-ordinated water

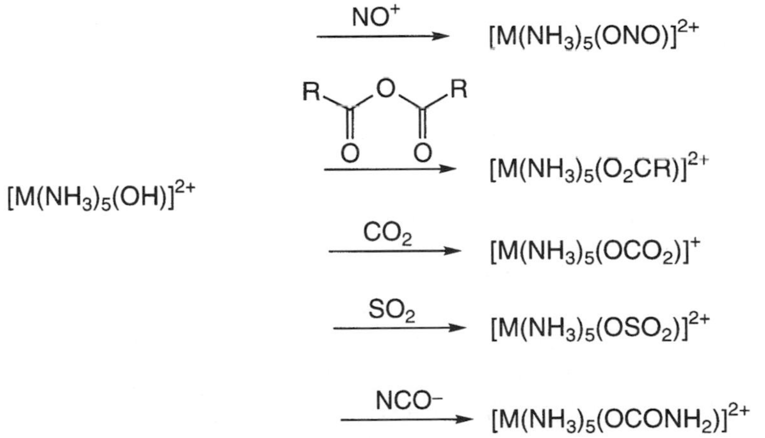

Figure 5-73. Hydroxide complexes of non-labile d^6 metal centres undergo reactions with a selection
of electrophiles. Of particular note is the specific method for the formation of monodentate *O*-bon-
ded complexes with nitrito, carboxylato, carbonato and sulfito ligands.

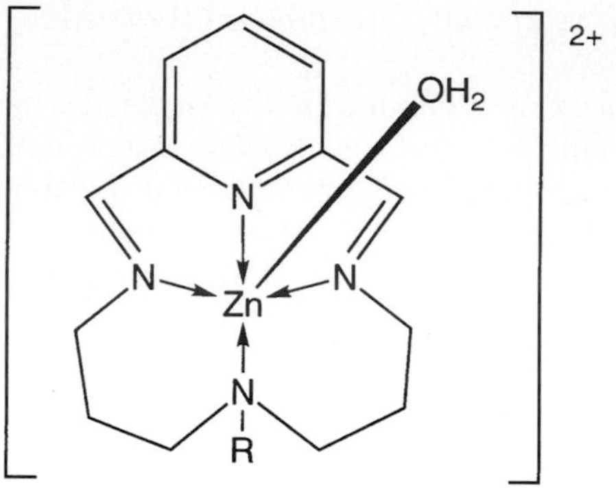

5.35

molecule has a pK_a around 8.5; the hydroxo complex is a reasonable catalyst for the hydration of carbon dioxide, and other reactions of carbonyl compounds.

5.8 Reactions of Co-ordinated Thiolate

One of the simplest and widely used methods of forming C–S bonds involves nucleophilic attack of a thiolate on a suitable C-centred electrophile such as an alkyl halide (Fig. 5-74). Co-ordinated thiolate ligands behave as nucleophiles in exactly the same manner, and the method has been extensively used for the preparation of thioethers and their metal complexes. The method has been particularly commonly utilised in the formation of macrocyclic ligands in templated syntheses (see Chapter 6).

Although this type of reactivity has been most generally used for the synthesis of complexes of macrocyclic ligands, it is a general method for the preparation of cyclic and open-chain thioethers. A typical example is seen in the benzylation of a nickel(II) bis(thiolate) complex (Fig. 5-75). The dinuclear metal complex is smoothly alkylated at sulfur to give a dinickel(II) complex of the thioether. Use of the metal complex prevents competing reactions involving alkylation of the pyridine nitrogen atom from occurring. Indeed, the use of the metal ion to *control* the reactivity of ambidentate ligands such as this is of great importance.

Figure 5-74. The nucleophilic displacement of a leaving group by a thiolate is one of the commonest methods for the formation of C–S bonds.

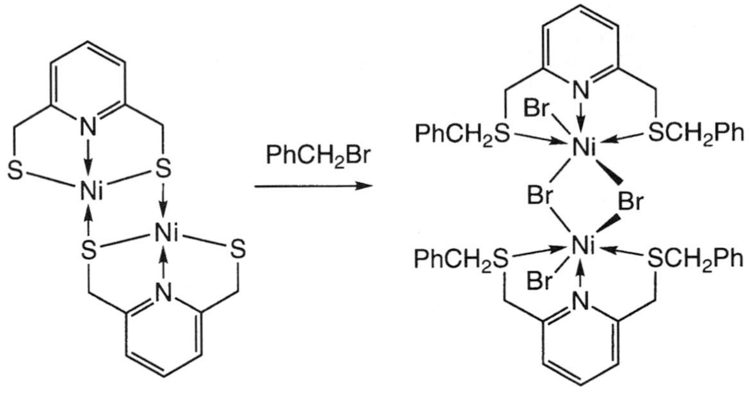

Figure 5-75. The alkylation of a nickel(II) thiolate complex gives a thioether complex. There is no competitive alkylation of the pyridine nitrogen atom.

The selectivity observed in some of these alkylation reactions is great, and it is usually possible to selectively alkylate only the sulfur in a co-ordinated aminothiolate. A typical example is seen in the methylation of bis(2-aminoethylthiolato)nickel(II) with methyl iodide to yield the bis(methylthioethylamine)nickel(II) cation (Fig. 5-76).

In general, these reactions are better characterised with chelated thiolate ligands rather than with monodentate thiolates, although many reactions giving rise to monodentate co-ordinate thioethers are known. A typical example is shown in Fig. 5-77, in which an iron(II) thiolate is converted to a thioether complex.

As mentioned above, reactions of this type have been widely used in the synthesis of macrocyclic ligands. Indeed, some of the earliest examples of templated ligand synthesis involve thiolate alkylations. Many of the most important uses of metal thiolate complexes in these syntheses utilise the *reduced* nucleophilicity of a co-ordinated thiolate ligand. The lower reactivity results in increased selectivity and more controllable reactions. This is exemplified in the formation of an N_2S_2-donor ligand by the condensation of biacetyl with the nickel(II) complex of 2-aminoethanethiol (Fig. 5-78). The electrophilic carbonyl reacts specifically with the co-ordinated amine, to give a complex of a new diimine ligand. The beauty of this reaction is that the free ligand *cannot* be prepared in a metal-free reac-

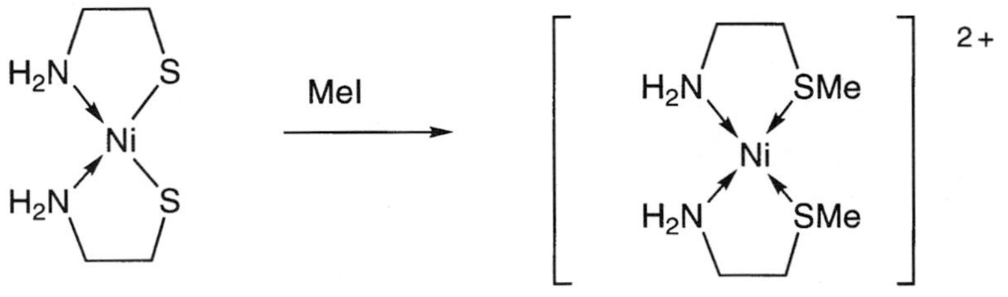

Figure 5-76. The alkylation of an aminothiolate can occur very selectively at the sulfur atom.

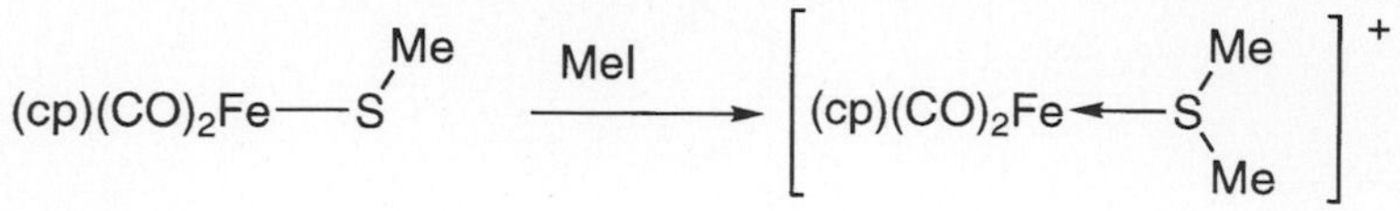

Figure 5-77. The alkylation of a co-ordinated methylthiolato ligand.

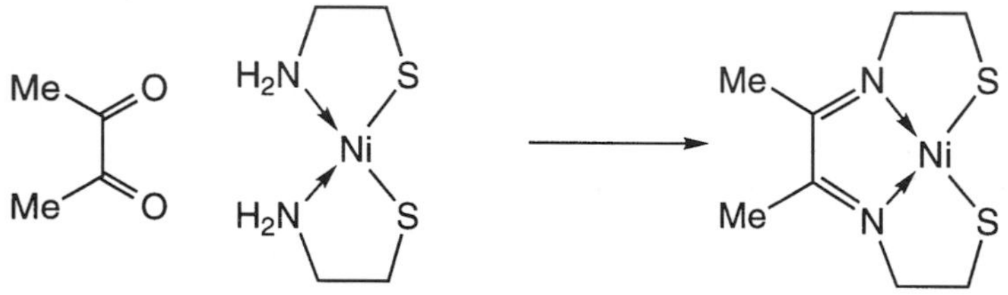

Figure 5-78. The nickel(II) complex of 2-aminoethanethiol reacts smoothly with biacetyl at nitrogen to give a diimine ligand.

tion of biacetyl with 2-aminoethanethiol. The sulfur of thiol or thiolate is too nucleophilic for the free ligand to be isolated and cyclic products, resulting from intramolecular attack of sulfur on the imine, are obtained (Fig. 5-79).

Interestingly, the reaction shown in Fig. 5-79 is reversed in the presence of nickel(II) or other metal ions, which promote a rearrangement to the tetradentate open-chain form

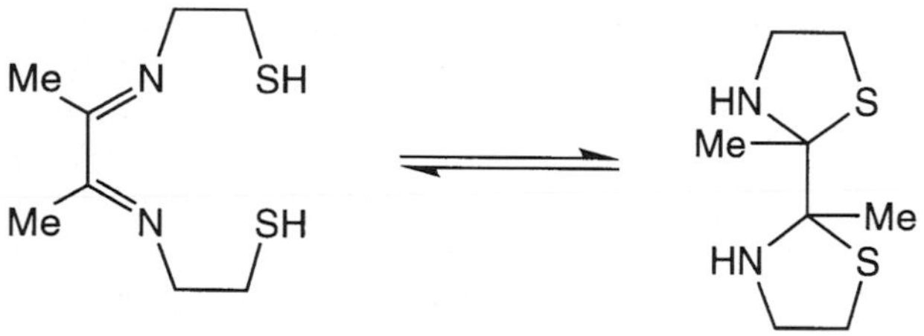

Figure 5-79. Attempts to prepare the ligand of Figure 5-78 in a metal-free reaction of biacetyl with 2-aminoethanethiol are unsuccessful. The thiol sulfur is too nucleophilic, and attacks the imine in an intramolecular process.

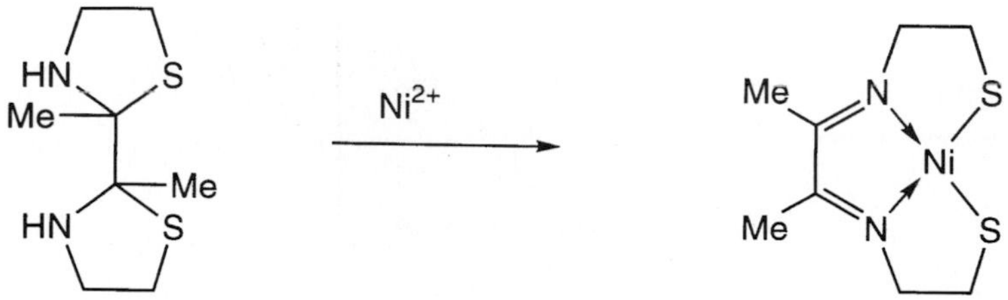

Figure 5-80. A metal-triggered ring opening can give the diimine ligand.

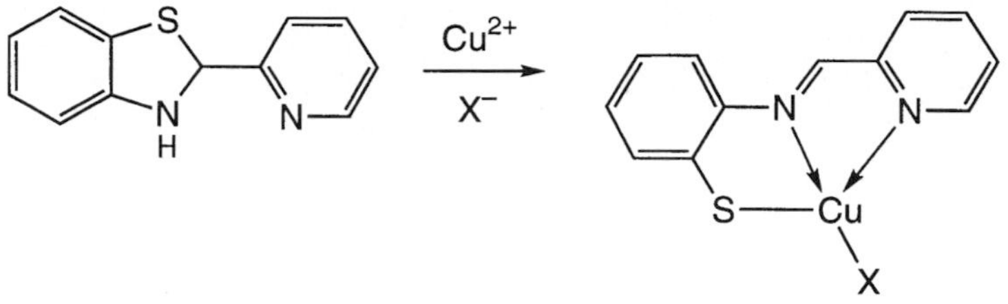

5.36

Figure 5-81. The copper-mediated ring-opening of 2-(2-pyridyl)benzothiazoline gives the complex of an iminothiol ligand.

(Fig. 5-80). In some way, the ligand retains a memory of the metal-binding form. This is the basis of a method of metal-ion triggered switching.

This metal-directed ring-opening of thiazolines, oxazolines and related compounds is a widespread reaction. In many cases, equilibria are set up between the heterocyclic compound and the imine. The position of these equilibria are metal-ion dependent. A typical example is seen in Fig. 5-81. The reaction of 2-(2-pyridyl)benzothiazoline (**5.36**) with copper(II) salts leads to the formation of complexes of the ring-opened tridentate form.

More extensive rearrangements may also occur, and in Fig. 5-82 the ring opening of the bis(thiazoline) **5.37** by a methylthallium(III) complex is demonstrated.

Many exotic electrophiles have been shown to react with co-ordinated thiolate; for example new disulfide bonds may be formed by reaction with S_2Cl_2. The nickel(II) complex of a very unusual tetrasulfide macrocyclic ligand may be prepared by this method (Fig. 5-83). Notice that this reaction utilises the nickel complex of the N_2S_2 ligand prepared by a metal-directed reaction in Fig. 5-78.

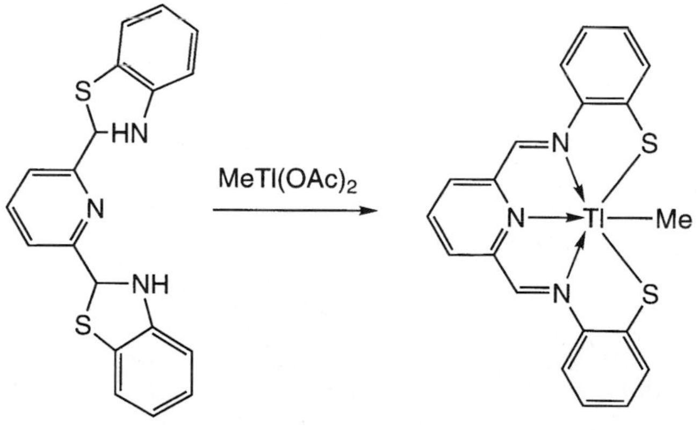

5.37

Figure 5-82. The ring opening of both thiazoline rings in **5.37** leads to the formation of a pentadentate N_3S_2 ligand.

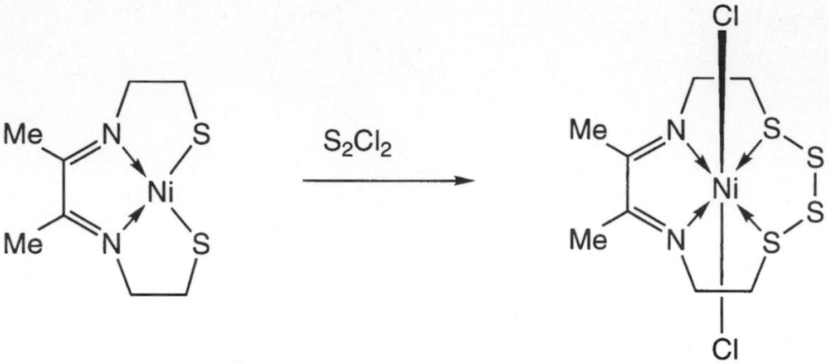

Figure 5-83. The formation of a macrocyclic complex by the formation of S-S disulfide bonds. The co-ordination of the halide leaving groups to the metal is common in reactions of this type.

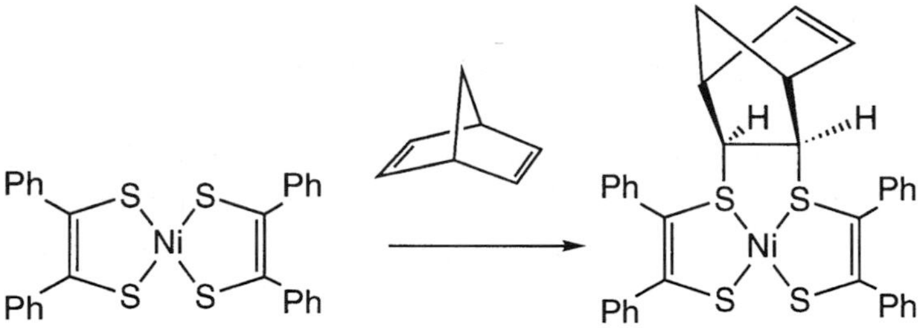

Figure 5-84. The reaction of norbornadiene with a dithiolene complex results in the formation of new C-S bonds.

Even an olefin may be sufficiently electrophilic to react with co-ordinated thiolate, and some nickel dithiolene complexes have been shown to react smoothly with norbornadiene (Fig. 5-84). Naturally, the dithiolene complexes also react with more conventional electrophiles, such as methyl iodide (Fig. 5-85).

There is some evidence that the alkylation of co-ordinated thiolate is a reversible process, and there are a number of examples known in which a co-ordinated thioether is dealkylated to yield a co-ordinated thiolate. In general, these reactions are rather sluggish, but occur in hot dmf solution. Whether the mechanism involves simple thermal dealkylation, or the intermediacy of the dmf as an alkyl group acceptor is not clear.

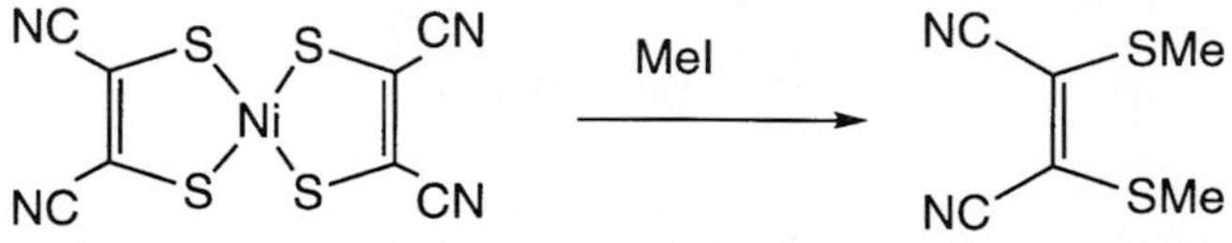

Figure 5-85. The alkylation and demetallation of a nickel dithiolene complex with methyl iodide.

5.9 Other Reactions of Esters and Amides with Co-ordinated Nucleophiles

Transesterification or transamination are metal-directed reactions which are commonly encountered. We have discussed transamination processes in Section 5.5.2 and also in Chapter 3. The key step involves the attack of a co-ordinated alcohol or alkoxy group upon the carbonyl of a co-ordinated ester or amide. Many Lewis acidic metal ions have been shown to be effective catalysts for transesterification reactions; for example, heating diethyl picolinate with copper(II) salts in methanol results in rapid and clean transesterification (Fig. 5-86). In the absence of the metal ion, the rate of reaction is vanishingly slow.

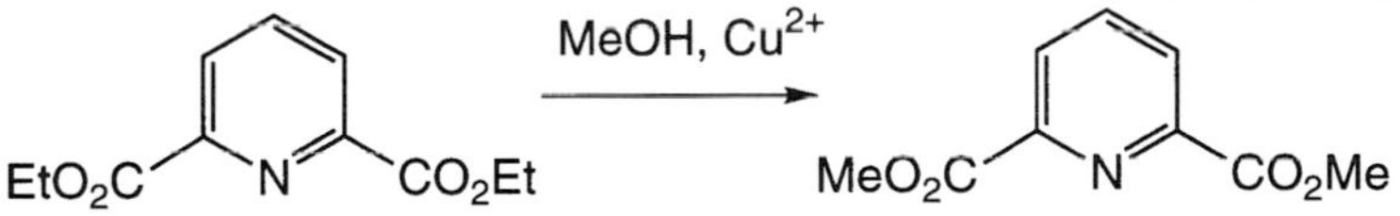

Figure 5-86. The transesterification of the chelating ligand diethyl picolinate is enhanced upon co-ordination to a metal. Reaction with methanolic copper(II) salts gives the methyl ester in a very rapid process.

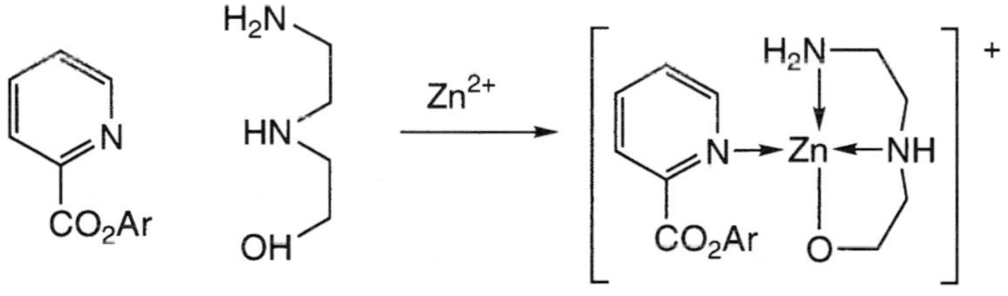

Figure 5-87. The formation of an initial complex in which the incoming nucleophile and the substrate are co-ordinated to a metal complex (Ar = 4-nitrophenyl).

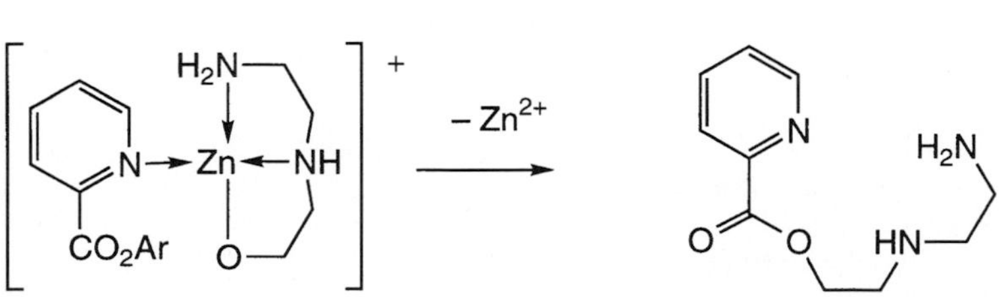

Figure 5-88. The second step of the reaction, involving intermediate **5.38**, yields the transesterification product (Ar = 4-nitrophenyl).

In this reaction, the incoming alkoxide group is almost certainly co-ordinated to the metal centre in the key stages of the reaction. A particularly elegant demonstration of this is seen in the zinc(II) catalysed transesterification of 4-nitrophenyl picolinate with $HO(CH_2)_2NH(CH_2)_2NH_2$ (Figs. 5-87 and 5-88). In the initial step of the reaction, an intermediate complex **5.38** may be isolated (Fig. 5-87).

5.10 Conclusions

This has been a long chapter in which we have seen a very wide range of reactions, all of which are basically controlled by the ability of a positively charged or electropositive metal centre to stabilise anionic ligands. The effect may be expressed by stabilisation of an incipient nucleophile, by reducing the reactivity of a nucleophile, or by stabilising an anionic leaving group. We will see many more examples of reactions of this type in the next chapter.

Suggestions for further reading

1. A. Pasini and L. Casella, *J. Inorg. Nucl. Chem.* **1974**, *36*, 2133.
   – A review covering the reactions of amino acid derivatives
2. R.W. Hay, P.J. Morris in *"Metal Ions in Biological Systems"*, *Vol. 5*, (ed. H. Sigel), Marcel Dekker, New York, **1976**, p. 173.
   – A good review of the metal-mediated reactions of amino acid derivatives from the inorganic chemists viewpoint.
3. D.A. Buckingham in *"Biological Aspects of Inorganic Chemistry"*, (eds. A.W. Addison, W.R. Cullen, D. Dolphin, B.R. James), Wiley, New York, **1976.**
   – A stimulating discussion of the chemistry cobalt(III) complexes of amino acid derivatives
4. R.W. Hay in *Comprehensive Co-ordination Chemistry*, *Vol. 6*, (eds. G. Wilkinson, R.D. Gillard, J.A. McCleverty), Pergamon, Oxford, **1987.**
   – A good review of metal-mediated reactions of amino acid derivatives and other biomolecules.
5. R.W. Hay in *Reactions of Co-ordinated Ligands*, *Vol. 2*, P.S. (ed. Braterman), Plenum, New York, **1989.**
   – Very similar to the above.
6. D.F. Martin, *Prep. Inorg. React.* **1964**, *1*, 59.
   – Mainly about imines.
7. N.F. Curtis, *Coord. Chem. Rev.* **1968**, *3*, 3.
   – Mainly about imine macrocycles.

6 Cyclic Ligands and the Template Effect

6.1 What is a Macrocycle?

Macrocyclic ligands have played an important part in the development of modern co-ordination chemistry. But what exactly is a macrocycle? As far as a co-ordination chemist is concerned, the definition of Melson is probably the most useful. Melson stated that a macrocycle is a cyclic molecule with three or more potential donor atoms in a ring of at least nine atoms. Thus, ethylene oxide (**6.1**), 1,4-dithiane (**6.2**), cyclotetradecane (**6.3**) and cyclooctatetraene (**6.4**) are not commonly thought of as macrocycles (Fig. 6-1), whereas molecules such as cyclam (**6.5**), phthalocyanine (**6.6**), 1,4,7-trithiacyclononane (**6.7**) and dibenzo-18-crown-6 (**6.8**) fit the definition (Fig. 6-2).

This definition is framed to emphasise the metal-binding properties of the ligands and includes most cyclic polydentate ligands that can incorporate a metal ion into the middle of a bonding cavity. It is also worth noting that many macrocyclic ligands have trivial names. This reflects the 'unfriendly' nature of the systematic names; for example, the systematic names of **6.5** and **6.8** are 1,4,8,11-tetraazatetracyclodecane and 2,5,8,15,18,21-hexaoxatricyclo[20.4.0.0^{9,14}]hexacosa-1(22),8,11,13,23,24-hexaene, respectively.

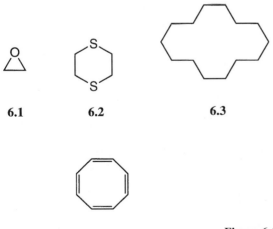

6.1   **6.2**   **6.3**

6.4

Figure 6-1. Some cyclic ligands that are not usually regarded as macrocycles by co-ordination chemists.

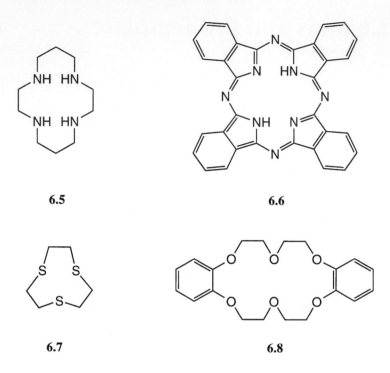

6.5   **6.6**

6.7   **6.8**

Figure 6-2. Some macrocyclic ligands familiar to co-ordination chemists.

In the past thirty years, the study of the co-ordination chemistry of macrocyclic ligands has expanded from the investigation of a chemical curiosity to a major area of research. The origins of this flurry of activity are numerous, but the development of synthetic methodology for the preparation of macrocyclic molecules has played a decisive role. In this chapter and the one following, we will consider one of the primary synthetic methods that is used for the synthesis of macrocyclic ligands and their co-ordination compounds. In doing so, we will see many of the effects that we have discussed in the earlier chapters in operation.

6.1.1 The Synthesis of Large Cyclic Molecules

The preparation of large cyclic compounds (especially those with a ring size of greater than seven atoms) has presented organic chemists with synthetic problems for many years. The origins of the problem may be considered to lie in an unfavourable entropy of reaction, and may be visualised in terms of the low probability of two ends of a chain coming together to form a cyclic molecule. In general, intermolecular processes leading to the formation of oligomers and polymers are favoured over the intramolecular reactions yielding the desired macrocycles (Fig. 6-3).

Intermolecular reaction

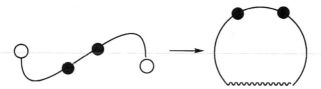

Intramolecular reaction

Figure 6-3. The principal problem associated with the formation of cyclic compounds is in the control of intramolecular as opposed to intermolecular reactions. The open circles represent sites that can react with each other. Intramolecular reaction leads to a cyclic compound, whereas intermolecular reaction leads to oligomers and polymers.

Although some reactions (such as that of furan with acetone to yield the tetradentate cyclic ligand **6.9**, Fig. 6-4) are known to give high yields of macrocyclic products, these are the exceptions rather than the rule. The cyclic product **6.9** is formed in high yields and may be viewed as an oxygen donor analogue of a porphyrin. Alternatively, **6.9** may be viewed as a crown ether and is expected to have a co-ordination chemistry associated with the metal ions of Groups 1 and 2.

In order to circumvent the unwanted intermolecular reactions that are usually observed, syntheses are often performed under conditions of moderate to very great dilution. Under these circumstances, the intramolecular reaction becomes more likely than the intermolecular process. It is more likely for the reactive end of a molecule to encounter the other end of the same molecule as opposed to the reactive site of another molecule. In extreme cases, milligrams of reactants are dissolved in litres of solvent. This places extreme constraints upon the purities of reactants and solvents and also upon the quantities of products which may reasonably be obtained. A further synthetic problem is associated with the observation that many of the reactive precursors are water sensitive; it is very difficult to completely exclude water from a reaction vessel, and many of the solvents used are hygroscopic. Think about a high dilution reaction with 0.1 mmol of a hydrolytically unstable compound in one litre of solvent. It would only require 1.8 mg of water to completely destroy the reactant. This corresponds to 0.0002 % (w/v) of water in the solvent!

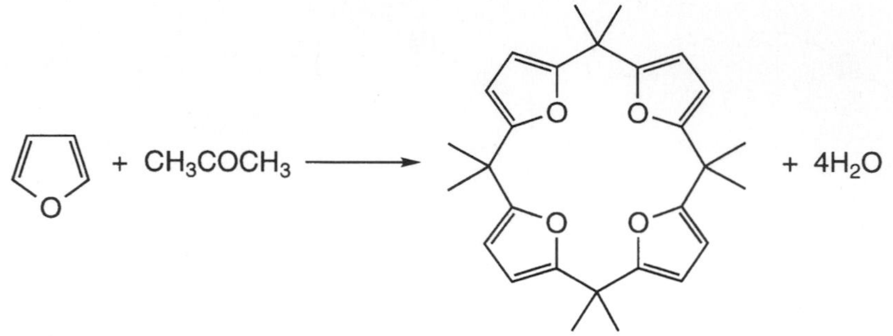

6.9

Figure 6-4. The formation of a macrocycle from the reaction of acetone with furan. This provides a rare example of a high-yield, cyclisation which occurs at 'normal' concentrations of the reactants. In fact, considerable care needs to be taken with this reaction, which may be explosive!

If there are potential donor atoms within the reactant and/or product molecules then it is possible to use a different methodology for the synthesis of cyclic compounds.

6.2 The Template Method

As we considered above, one of the fundamental problems associated with the preparation of macrocyclic ligands is concerned with the orientation of reactive sites such that they give intramolecular (cyclic) rather than intermolecular (acyclic) products. This is associated with the conformation of the reactants and the reactive sites, and so we might expect that judicious location of donor atoms might allow for metal ion control over such a cyclisation process. This is known as a *template synthesis*, and the metal ion may be viewed as a template about which the macrocyclic product is formed. This methodology was first developed in the 1960s, and has been very widely investigated since that time. At the present, template reactions usually prove to be the method of choice for the synthesis of many macrocyclic complexes (with the possible exceptions of those of crown ethers and tetraazaalkanes). When the reactions are successful, they provide an extremely convenient method of synthesis.

The basic strategy is illustrated in Fig. 6-5. This is the same reactant that we saw in Fig. 6-3, but we will now specify that the filled circles represent donor atoms which can bind to a metal ion. When these donor atoms bind in a chelating manner to a single metal centre, the two open circles of the reactive sites are brought close together in space. There is now a greater probability of the intramolecular reaction (giving the macrocyclic product) occurring. Note also, that in this example the cyclic product is obtained as the metal complex.

Template reactions are not limited to those involving a single organic component and a single metal ion. Consider the reaction shown in Fig. 6-6. The two open chain precur-

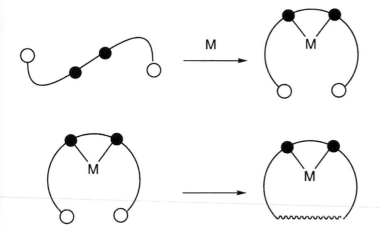

Figure 6-5. A metal-ion-templated cyclisation of a compound containing two donor atoms (filled circles) and two reactive groups (open circles). The co-ordination of the open-chain ligand to the metal ion brings the reactive sites into close proximity and favours the intramolecular reaction.

sors contain mutually reactive functionalities. In the uncontrolled reaction, the normal product is a polymer. However, one of the reactants also contains donor atoms. Co-ordination to a metal ion holds the reactive sites in the correct conformation for reaction with the second organic component. More importantly, after the first reaction has occurred, the intermediate species is correctly oriented for intramolecular reaction.

Naturally, more than two reactant species could be envisaged, as could the introduction of donor atoms into two or more of the reactants. It is not always necessary to isolate the metal complexes of the reactants, and it is often possible to simply mix all of the reactants together in a suitable solvent.

To further exemplify this methodology, let us take a typical example of the application of a template reaction as seen in the synthesis of a mixed N_2S_2 donor macrocyclic ligand **6.11**. This compound is of interest to the co-ordination chemist as it possesses a potentially square–planar array of soft (sulfur) and harder (nitrogen) donor atoms. What sort of co-ordination chemistry is it likely to exhibit? Will the hard or the soft characteristics dominate? The most obvious route for the synthesis of **6.11** would involve the reaction of the dithiol **6.10** with 1,2-bis(bromomethyl)benzene (Fig. 6-7).

Unfortunately, this macrocycle cannot be prepared as a free ligand by this method. The starting diimine **6.10** could apparently be prepared from 2-aminoethanethiol and biacetyl. However, we saw in Fig. 5-79 that the direct reaction of 2-aminoethanethiol with 1,2-dicarbonyls leads to a range of cyclic and acyclic products, rather than to products such as **6.10**. However, we also saw in Fig. 5-78 that the nickel(II) complex (**6.12**) of the **6.12** could be obtained if the reaction was conducted in the presence of an appropriate salt.

The reaction of the nickel(II) complex **6.12** with 1,2-bis(dibromomethyl)benzene occurs smoothly to give the nickel(II) complex of the macrocycle (**6.13**) in respectable yield (Fig. 6-8). If co-ordinating anions are present, these occupy the axial sites to give

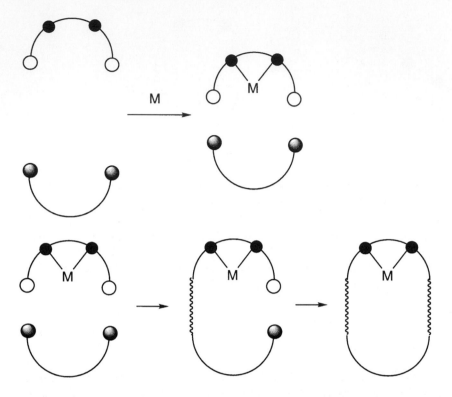

Figure 6-6. A templated reaction between two different open-chain reactants. In this case the two reactants have complementary reactive sites. The shaded circles represent reactive sites which react with the open circles. One of the reactants contains a metal-binding site (filled circles).

a six-co-ordinate complex. Clearly, the metal plays a number of roles in this reaction. The conformation of the 2-aminoethanethiol and of the diiminothiolate are controlled by co-ordination. The complex **6.12** has the sulfur atoms held in the correct orientation for reaction with the electrophile. The metal ion stabilises the imine and modifies the nucleo-

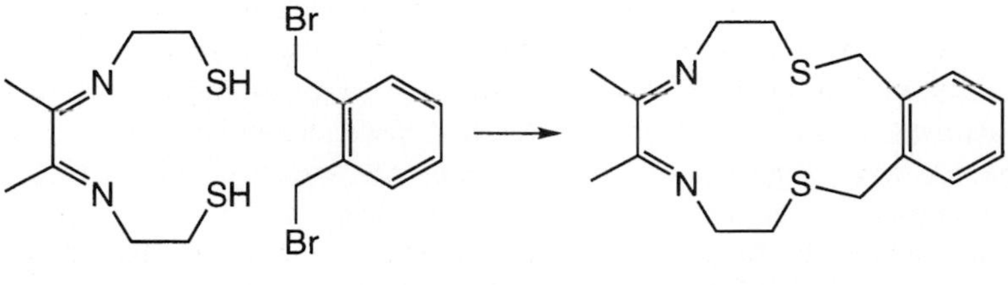

6.10   **6.11**

Figure 6-7. The obvious synthetic route for the preparation of the macrocycle **6.11**.

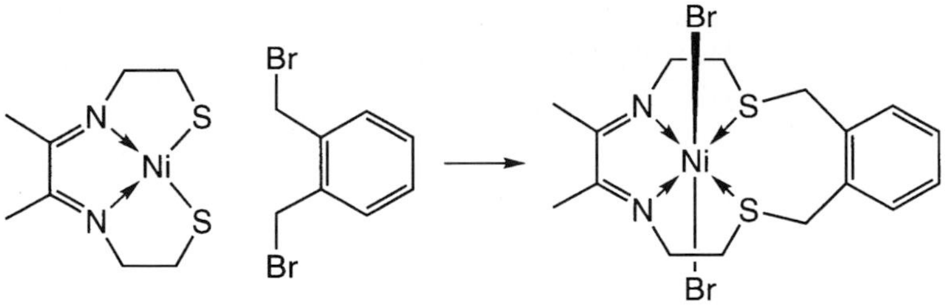

6.12   **6.13**

Figure 6-8. The synthesis of a macrocyclic complex by a template reaction. In this case, the starting organic open-chain compound cannot be obtained metal-free.

philicity of the thiolate. Furthermore, in **6.12** the sulfur atoms are held in such a position that they cannot attack the imine to give the thiazolines discussed in Chapter 5.

Since the first descriptions of template syntheses in the early 1960's innumerable reports of such reactions have appeared. They are frequently easy to perform and high yielding.

6.2.1 Why use Template Reactions?

Template reactions play a very important role in the co-ordination chemistry of macro-cyclic ligands, and it is probably true to say that the huge variety of macrocyclic ligands known reflects the synthetic utility of metal-directed cyclisation reactions. It is somewhat surprising to find that most of these cyclisation reactions rely upon a very few specific types of organic reaction. We will consider these reaction types in Section 6.4. However, before discussing the range of metal-directed chemistry involved in macrocyclic syntheses, it is worth considering the advantages and disadvantages of template reactions. Why do template reactions prove to be so useful?

There are a number of readily appreciated advantages to the use of templated syntheses of macrocycles. The metal complexes of the macrocyclic ligands are usually obtained directly from the templated reactions. There is no need for a metal insertion stage after the ligand has been prepared. As we will see later, macrocyclic ligands which cannot be obtained metal-free are often readily isolated as their complexes. In this case, it is only possible to use templated syntheses for the preparation of such complexes. As we saw in Fig. 6-7, macrocyclic products may be derived from precursor molecules not isolable in the metal-free form. High-dilution methods are not usually needed, and the critical constraints upon solvent and reactant purity are lifted. In ideal cases, the yields may be very good. In many cases, a remarkably high degree of stereo- and regiospecificity over the reaction is achieved.

There are also potential disadvantages associated with the use of template methodology for the formation of macrocyclic ligands. Perhaps the most important is the observation that not all metal ions can act as templates for the specific cyclisation reactions of interest. In many cases it may not be trivial to find an appropriate template ion (if indeed one exists) for the reaction. In general, macrocyclic complexes are relatively non-labile and it may be difficult to remove the metal ion from the complex (which is the primary product of the template reaction) and so obtain the free ligand. If complexes are to be prepared with metal ions which *do not* act as templates, it is necessary to develop demetallation methods, or possibly metal-exchange or *transmetallation* techniques. The kinetic inertness of macrocyclic complexes is also a problem in transmetallation reactions. Even if it is possible to demetallate the complex, the free ligand may not be stable under the demetallation conditions. The products of the template reaction may not always be those which are expected. In many cases, the stoichiometry of the reaction is not as expected or products of other metal-directed reactions may be obtained.

However, it remains a fact that an enormous number of macrocyclic ligands have been, and new examples continue to be, made using template methods.

6.3 Metal-Ion Control of Ring Closure Reactions

The basic reactions which are involved in the synthesis of macrocyclic compounds are very simple and of a very few types. In this section we shall consider the basic organic chemistry of macrocyclic ring-closures and how a metal ion may activate or control these reactions. In most cases, these reactions involve the formation of heteroatom–carbon bonds.

6.3.1 Nucleophilic Displacement

An important reaction which is encountered in macrocycle synthesis involves the nucleophilic displacement of halide or some other good leaving group in the ring-closure step. Typically, the nucleophile is a potential donor atom such as oxygen, sulfur or nitrogen, which might be co-ordinated to a metal ion. Generally, the role of the metal ion in these nucleophilic displacement reactions is predominantly conformational, and it acts to lock the leaving group and the nucleophile in the correct relative orientation for intramolecular reaction. A typical example of macrocycle formation involving nucleophilic ring-closure is seen in the preparation of crown ethers in alkali metal-directed processes. Consider the preparation of 18-crown-6 (Fig. 6-9). The synthesis involves the reaction of triethylene glycol with the ditosylate of triethylene glycol in the presence of a strong base (which serves to deprotonate the glycol). The cyclisation involves sequential displacement of tosylate by alkoxide. If a potassium salt is used as the base, the alkoxide may co-ordinate to the metal ion. The initial stage of the reaction involves the reaction of the potassium complex with the ditosylate. In the key ring-closure step the ligand is still co-ordinated to the metal, and is wrapped around it in such a way that the alkoxide and tosylate groups are held in close proximity. Under these conditions the intramolecular nucleophilic attack leading to cyclisation is favoured over alternative pathways leading to polymerisation. A similar conformational effect is involved in the preparation of the nickel(II) complex shown in Fig. 6-8.

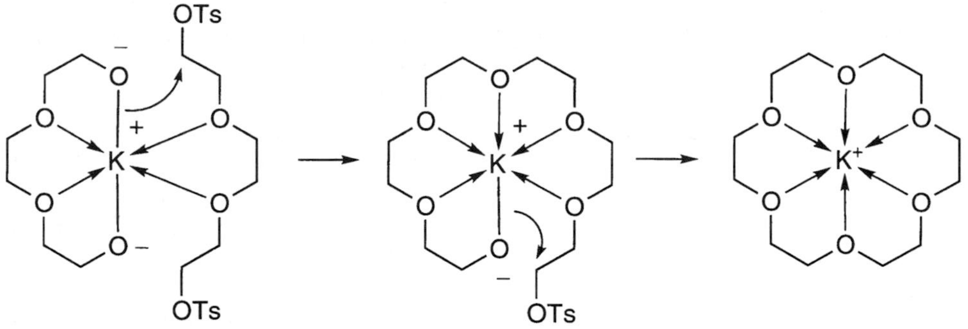

Figure 6-9. The action of a potassium template ion in the formation of 18-crown-6 from open-chain precursors. The important feature is the way in which the metal ion holds the alkoxide and tosylate groups in proximity in the intermediate immediately before the cyclisation.

A second more subtle effect may also be operative in the metal ion control of nucleophilic reactions. When amines, thiolates or alkoxides are used as nucleophiles, they are expected to be highly reactive and hence relatively unselective. However, we saw in Chapter 2 that the proximity of the metal cation to the nucleophile reduces the charge density on the donor atom, and is thus expected to reduce the reactivity. We can use the reduced reactivity, and greater selectivity, of such co-ordinated nucleophiles to direct reaction towards the cyclic products.

Sometimes this deactivation is so great that co-ordinated amines are non-nucleophilic. This is particularly likely when the ligand is co-ordinated to a non-labile metal centre. However, even in these cases, all is not lost. We may also use the enhanced acidity of ligands co-ordinated to a metal centre to generate reactive nucleophiles which would not otherwise be readily accessible. For example, nickel(II) complexes of deprotonated diamines may be prepared, and react with dialkylating agents to yield macrocyclic complexes (Fig. 6-10). To clarify this, consider the reaction in Fig. 6-10 in a little more detail. The amine **6.14** is reactive and unselective, and does not give the desired macrocycle upon reaction with the ditosylate. Deprotonation of the amine under mild conditions is not pos-

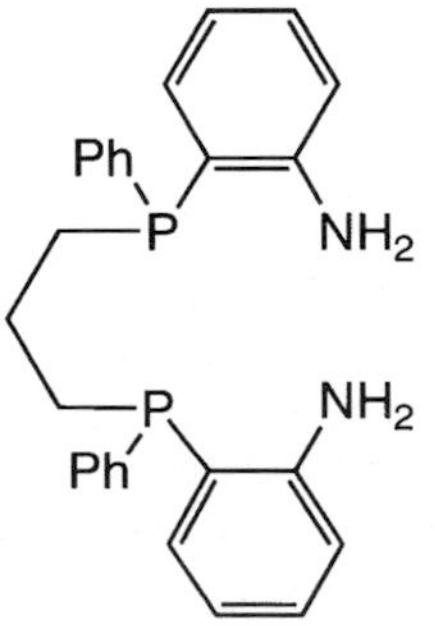

6.14

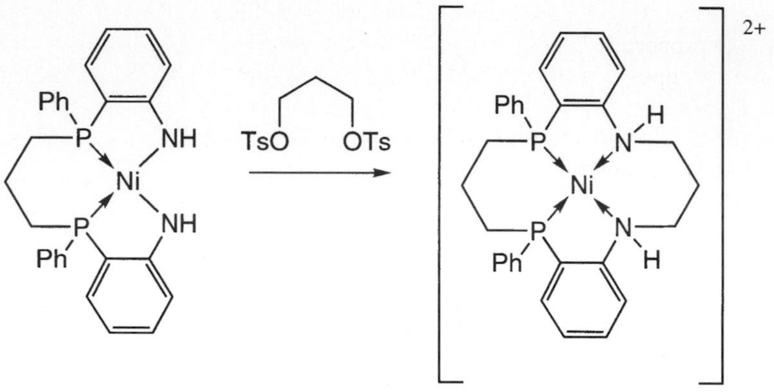

6.15

Figure 6-10. The nickel(II) complex of the deprotonated ligand (H$_2$L = **6.14**) reacts smoothly with the ditosylate to give a nickel(II) macrocyclic complex.

sible (pK_a ≈ 30), and the highly reactive anion would be extremely unselective in its reactions. Co-ordination of **6.14** to a nickel(II) centre results in polarisation of the N–H bonds, as discussed in Chapter 2, and a lowering of the pK_a. Accordingly, the complex can be deprotonated to give the neutral species **6.15**. This is now correctly oriented for reaction with the tosylate to give the macrocycle and is also moderated in its reactivity by the co-ordination of the nucleophilic sites to the metal.

Another elegant example of a nucleophilic cyclisation is seen in the reaction of the nickel(II) complex of dimethylglyoxime with BF$_3$. The starting nickel(II) complex **6.16** is the red insoluble material familiar to anyone who has performed a gravimetric analysis for nickel. Each of the two dimethylglyoxime ligands has been deprotonated once. The two ligands are then hydrogen bonded to give the square-planar complex **6.16**. This is another

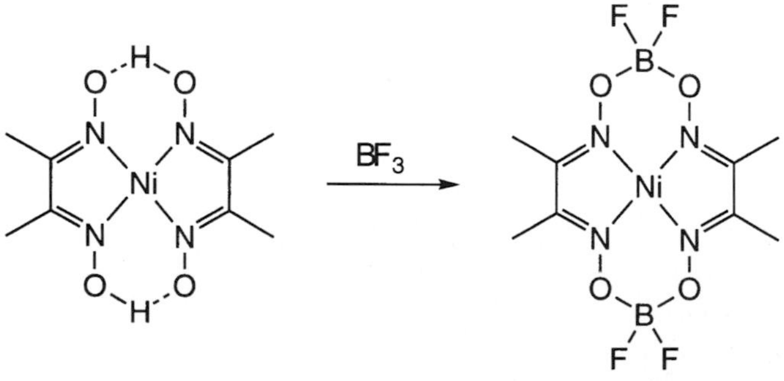

6.16

Figure 6-11. The reaction of **6.16** with BF$_3$ gives the nickel(II) complex of a new macrocyclic ligand. The ligand is dianionic, with the charge formally localised upon the tetrahedral boron atoms.

example of a pseudo-macrocyclic complex, in which hydrogen bonds assemble a structure with many of the properties of a macrocycle, as we saw in Fig. 5-13. However, the hydroxy and alkoxide groups of the dimethylglyoximato ligands are still nucleophilic, and reaction of **6.16** with boron trifluoride gives a nickel(II) complex of a dianionic macrocyclic ligand (Fig. 6-11). We will encounter further examples of nucleophilic cyclisation later in this chapter and in Chapter 7.

6.3.2 Imine Formation

Probably one of the commonest reactions encountered in the template synthesis of macrocycles is the formation of imine C=N bonds from amines and carbonyl compounds. We have seen in the preceding chapters that co-ordination to a metal ion may be used to control the reactivity of the amine, the carbonyl or the imine. If we now consider that the metal ion may also play a conformational role in arranging the reactants in the correct orientation for cyclisation, it is clear that a limitless range of ligands can be prepared by metal-directed reactions of dicarbonyls with diamines. The π-acceptor imine functionality is also attractive to the co-ordination chemist as it gives rise to strong-field ligands which may have novel properties. All of the above renders imine formation a particularly useful tool in the arsenal of preparative co-ordination chemists. Some typical examples of the templated formation of imine macrocycles are presented in Fig. 6-12.

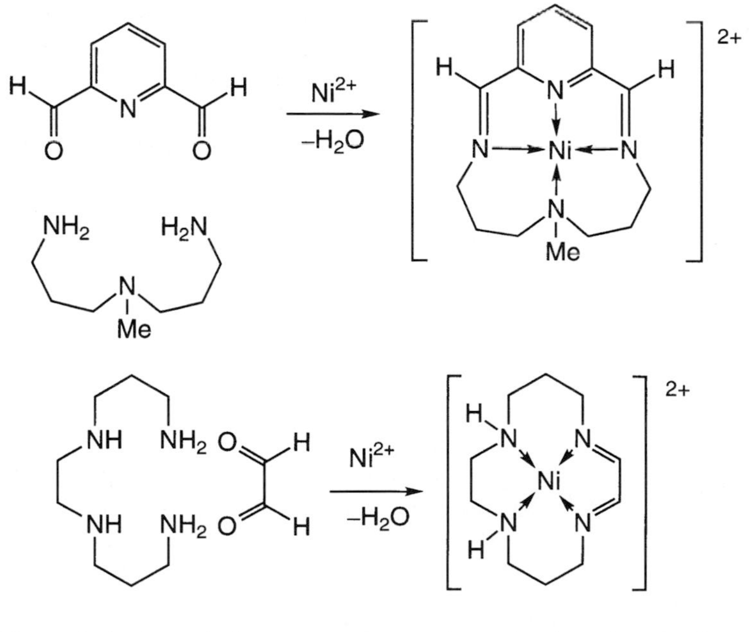

Figure 6-12. Two typical examples of nickel(II) templated reactions leading to macrocyclic ligands containing imine donor groups.

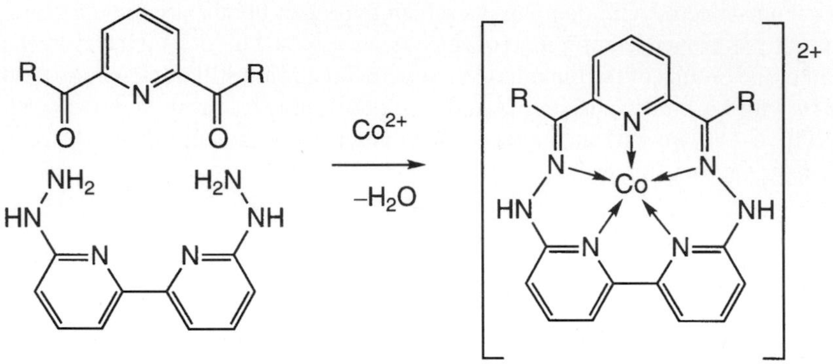

Figure 6-13. The templated formation of a hydrazone macrocycle. This ligand contains five nitro-gen donor atoms and was designed to investigate the properties of metal complexes containing a pen-tagonal planar donor set. In practice, the complexes of these ligands usually acquire one or two axial ligands to give six- or seven-co-ordinate complexes.

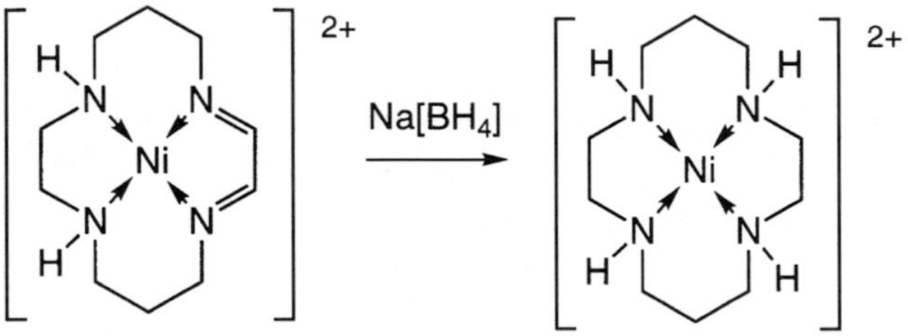

Figure 6-14. The reduction of co-ordinated macrocyclic imines provides a method for the prepara-tion of macrocyclic amines. The reaction above illustrates one of the standard methods for the pre-paration of cyclam. The metal ion may be removed from the nickel(II) complex by prolonged reac-tion with cyanide.

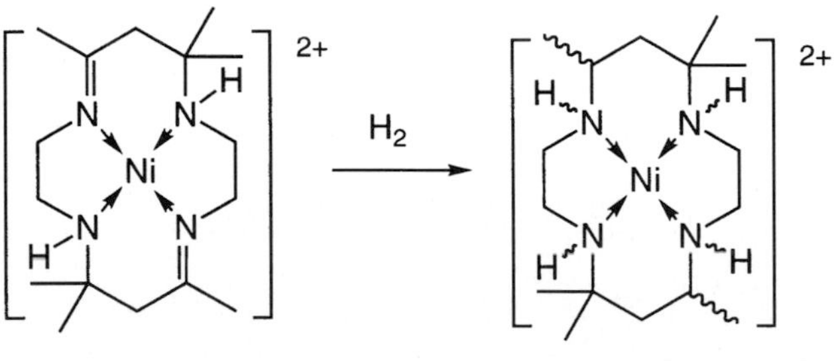

6.17   6.18

Figure 6-15. The reduction of a co-ordinated imine macrocycle may lead to a number of different diastereomers of the complex of the saturated macrocyclic ligand

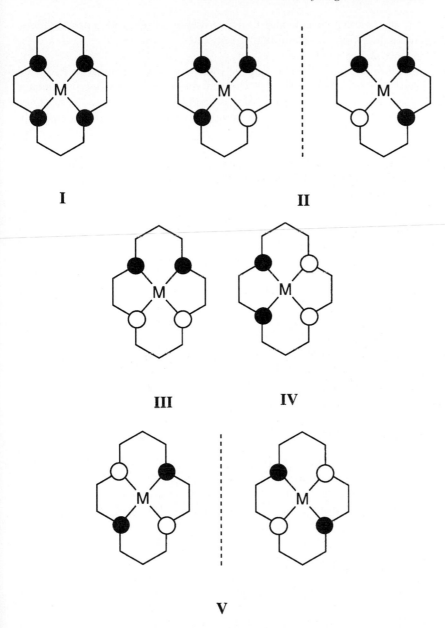

Figure 6-16. Co-ordination of the four nitrogen atoms of a tetraaza macrocyclic ligand to a metal results in a restricted inversion at each nitrogen. The filled circles represent a substituent lying above the plane of the paper, and the open circles one lying below the plane. In the case of cyclam, with hydrogen substituents, the barriers to inversion are relatively low, but with bulkier substituents on the nitrogen the different diastereomers are readily isolable. Note also that the isomers labelled **II** and **V** are chiral and will exist in two enantiomeric forms.

The template method may be extended to derivatives of imines, and hence to the synthesis of cyclic hydrazones. An example of a templated cyclisation leading to a cyclic hydrazone is shown in Fig. 6-13.

The co-ordination chemistry of macrocyclic ligands containing imine or hydrazone groups has been widely studied and, as expected, the presence of the imine functionality in the ring confers unusual redox properties to the complexes.

The synthetic method may be seen to be complementary to direct nucleophilic displacement. Whereas amines often react relatively sluggishly in metal-mediated nucleophilic displacements, they usually undergo facile reaction with carbonyls to form imines. The reduction of the imines (free or co-ordinated) may then be achieved by reduction with Na[BH$_4$] or (less conveniently) by direct hydrogenation. This provides a very convenient method for the preparation of cyclic *amines* (Fig. 6-14).

The conversion of a co-ordinated imine to a co-ordinated amine results in a change in geometry about the nitrogen atom from trigonal-planar (sp^2) to tetrahedral (sp^3), with the formation of a potential chiral centre. In many cases, inversion of the *co-ordinated* amine is a high energy process, and the reduction of the more developed macrocyclic systems may result in the formation of a number of different diastereomers. A typical example is shown in Fig. 6-15. The nickel(II) complex **6.17** is obtained from a template condensation of 1,2-diaminoethane with acetone (see Section 6.3.3) and may be reduced by dihydrogen to a saturated macrocycle containing only amine donors. However, inversion at the co-ordinated amine centres is restricted, and a number of diastereomers can result, depending upon the relative orientation of the hydrogen atoms. This is illustrated for a generic cyclam complex in Fig. 6-16. In the case of the derivative **6.18**, the orientation of the methyl groups also needs to be taken into consideration.

6.3.3 C–C Bond Formation

The final type of reaction generally encountered in the templated preparation of macrocyclic ligands involves the direct formation of C–C bonds. The simplest reactions are of the Claisen or aza-Claisen type, in which an enol, enolate or enamine reacts with a carbonyl or an imine (Fig. 6-17). Although the reaction shown in Fig. 6-17 appears to be rela-

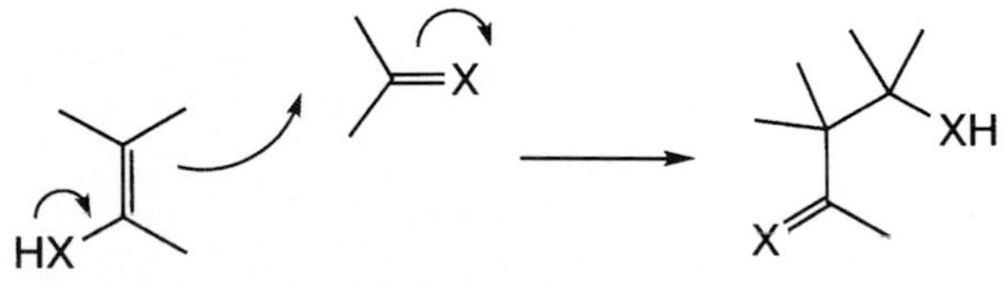

X = O or NH

Figure 6-17. The formation of a C–C bond using a Claisen-type condensation. A nucleophilic enol, enolate or enamine reacts with the electrophilic carbon of a carbonyl compound or an imine.

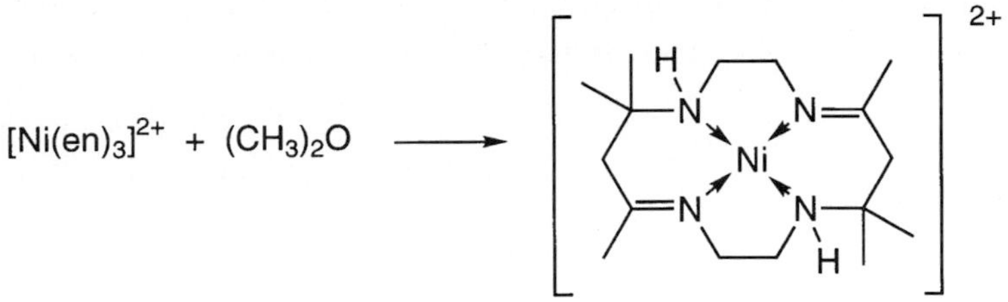

$[Ni(en)_3]^{2+}$ + $(CH_3)_2O$ $\longrightarrow$

Figure 6-18. The condensation of the $[Ni(en)_3]^{2+}$ salts with acetone yields nickel(II) complexes of a new tetraaza macrocyclic ligand.

tively complex, it is not usually necessary to isolate reactive precursors, since they are formed in the course of the template reaction.

The role of the metal ion may be purely conformational, acting to place the reactants in the correct spatial arrangement for cyclisation to occur, or it may play a more active role in stabilising the enol, enolate, imine or enamine intermediates. The prototypical example of such a reaction is shown in Fig. 6-18. The nickel(II) complex of a tetradentate macrocyclic ligand is the unexpected product of the reaction of $[Ni(en)_3]^{2+}$ with acetone. There are numerous possible mechanisms for the formation of the tetradentate macro-cyclic ligand and the exact mechanism is not known with any certainty.

Notice that, like all of the other template condensations considered so far, this is essen-tially a two-dimensional process. The three-dimensional structure of the $[Ni(en)_3]^{2+}$ cation is not made use of and only two of the diamine ligands are incorporated into the product. The overall stoichiometry of the reaction involves two diamine ligands and four acetone molecules. We will elaborate upon this observation in the next section.

Let us close this section with a few other observations about the reaction shown in Fig. 6-18. It is possible to eliminate the C–C bond formation in the template reaction by using reactants in which it has already been achieved. The Claisen condensation of two acetone molecules can give either diacetone alcohol (**6.19**) or mesityl oxide (**6.20**). Use of either of these in the reaction with $[Ni(en)_3]^{2+}$ salts gives the same macrocyclic product as in Fig. 6-18. In fact, it is not necessary to use a templated cyclisation for the preparation of this macrocycle at all. The reaction of acetone or **6.19** or **6.20** with the hydrochloride salt of 1,2-diaminoethane gives good yields of the chloride salt of the protonated macrocycle. This could, of course, be regarded as a proton-templated cyclisation. In a strictly terminological sense this is a *specific* acid template reaction whereas those invol-ving metal ions are *general* acid templated. This does not seem to be a useful way to think

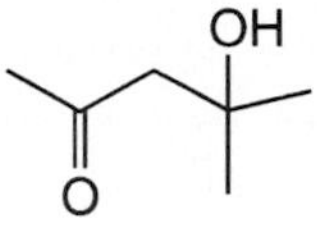

6.19

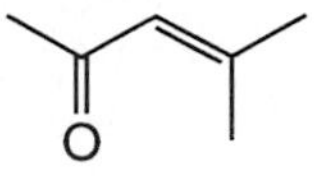

6.20

about these reactions, however. Finally, we should note that there is no reason why the two double bonds should be arranged *trans* about the metal. In reality, variable amounts of the *cis* and *trans* isomers are formed.

6.4 Stoichiometry – [1+1], [2+2] or Other?

The formation of cyclic ligands by template reactions has another associated complexity which we encountered in Section 6.3.3. This concerns the number of reacting molecules involved in the formation of the cyclic products and the overall stoichiometry of the reaction. We have not yet considered the control of stoichiometry of the cyclisation reactions in any great detail.

The formation of macrocyclic ligands by template reactions frequently involves the reaction of two difunctionalised precursors, and we have tacitly assumed that they react in a 1:1 stoichiometry to form cyclic products, or other stoichiometries to yield polymeric open-chain products. This is certainly the case in the reactions that we have presented in Figs 6-8, 6-9, 6-10, 6-12 and 6-13. However, it is also possible for the difunctionalised species to react in other stoichiometries to yield discrete cyclic products, and it is not necessary to limit the cyclisation to the formal reaction of just one or two components. This is represented schematically in Fig. 6-19 and we have already observed chemical examples in Figs 6-4, 6-11 and 6-18. We have already noted the condensation of two molecules of 1,2-diaminoethane with four molecules of acetone in the presence of nickel(II) to give a tetraaza-macrocycle. Why does this particular combination of reagents work? Again, why are cyclic products obtained in relatively good yield from these multi-component reactions, rather than the (perhaps) expected acyclic complexes? We will try to answer these questions shortly.

Firstly, let us introduce a shorthand way of describing these reactions. We may describe a reaction involving one molecule of each of two reactants as being of [1+1] stoichiometry. The reactions in Figs. 6-8, 6-9, 6-10, 6-12 and 6-13 are of this type. This shorthand only describes the stoichiometry in the organic products of the reaction, and does not specifically tell us how many metal ions are involved. A reaction involving two molecules of each of two reactants is of [2+2] stoichiometry. The reactions shown in Figs. 6-4, 6-11 and 6-17 could be described as being [4+4], [2+2] and [2+4], respectively.

Many examples are known in which multiple components are brought together about a metal ion to form macrocyclic complexes. Typical examples include the formation of *meso*-tetraphenylporphyrin (**6.21**) from benzaldehyde and pyrrole (Fig. 6-20, [4+4]), or phthalocyanine (**6.6**) from phthalonitrile (Fig. 6-21). The formation of the tetraphenyl-porphyrin is catalysed by a range of Lewis acids, and the facile preparation from aldehydes and pyrroles has obvious implications for the bioevolution of porphyrin pigments. Virtually any benzene derivative with *ortho* carbon-bearing substituents can be converted to a phthalocyanine complex on heating with a metal or metal salt in the presence of ammonia or some other nitrogen source.

Another spectacular example is seen in the formation of **6.22** from the [4+4] condensation of 2,6-diacetylpyridine with 2-hydroxy-1,3-diaminopropane in the presence of a manganese(II) template (Fig. 6-22). The cavity of this macrocycle is very large, and the final complex obtained contains a total of four manganese centres connected by the four deprotonated hydroxy oxygen atoms in a cubic arrangement.

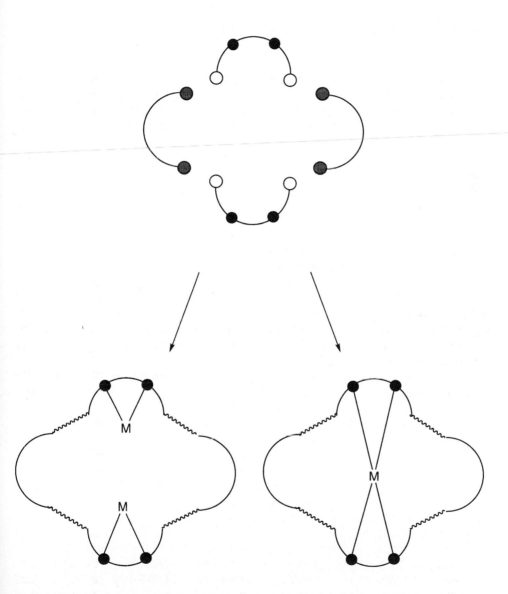

Figure 6-19. The formation of a [2+2] macrocyclic product from the cyclisation of two molecules of each of two components. The final ring size will dictate the number and geometry of metal ions incorporated into the product.

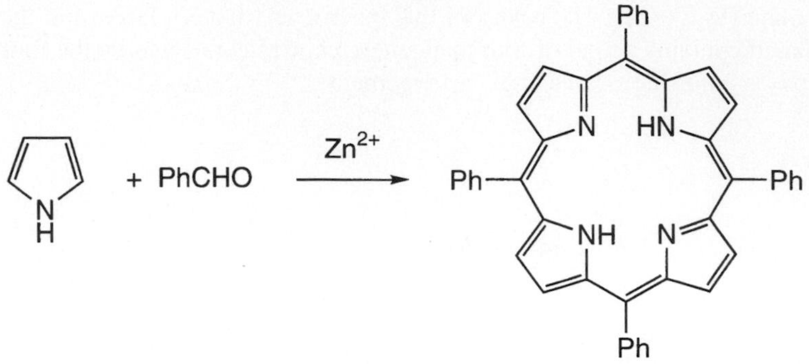

6.21

Figure 6-20. The [4+4] condensation of benzaldehyde and pyrrole in the presence of Lewis acids yields *meso*-tetraphenylporphyrin.

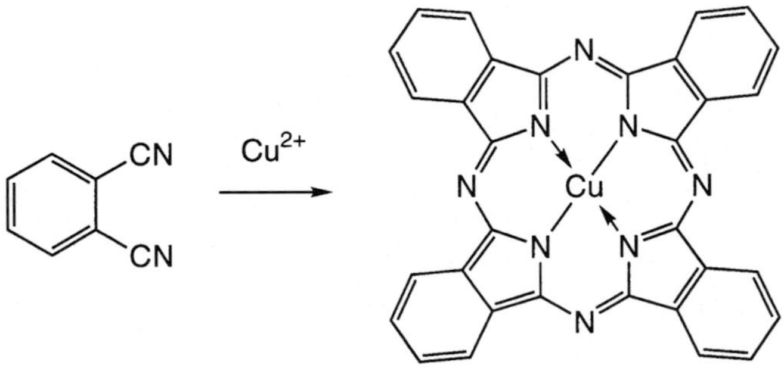

Figure 6.21. Copper phthalocyanine is formed as a beautiful blue compound when phthalonitrile is heated with copper(II) salts. The compounds are so stable that they can be formed from heating a huge range of substituted benzene derivatives with a source of nitrogen and a metal salt. The highly coloured compounds are used as pigments.

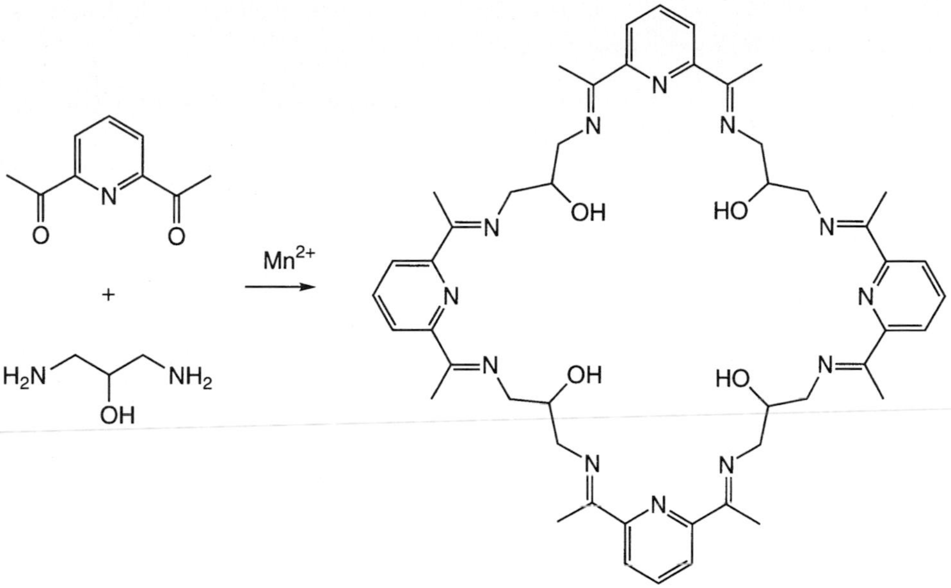

6.22

Figure 6-22. The [4+4] condensation of a diamine with 2,6-diacetylpyridine in the presence of manganese(II) template ions. The final macrocyclic ligand binds four manganese ions.

6.5   The Macrocyclic Effect

Why is it that these metal-directed reactions are so dominant in the preparative chemistry of macrocyclic ligands? The answer to this lies in part in the great stability that is often associated with macrocyclic complexes. Very often, the complexes exhibit high kinetic and thermodynamic stabilities.

It is well known that complexes of polydentate ligands are found to be more thermo-dynamically stable than those containing an equivalent number of monodentate ligands; this is the *chelate* effect. In a similar manner, complexes of macrocyclic ligands are gene-rally more stable than those of equivalent open-chain ligands, and this is known as the *macrocyclic* effect. Specifically, it is usually observed that *for a particular metal ion* complexes with macrocyclic ligands are more stable than those with polydentate open-chain ligands containing *an equal number of equivalent donor atoms*. This is illustrated in Table 6-1 for a series of copper complexes containing four nitrogen donor atoms. Note that the ligands chosen contain similar donor atoms arranged in roughly the same manner. By choosing a large series of ligands, it is possible to probe the various features relating to chelate ring size and substitution patterns on the carbon and nitrogen atoms.

Why are complexes with macrocyclic ligands particularly stable? This question has generated considerable controversy over the years, and it is now clear that the factors

involved are remarkably subtle and depend critically upon the medium in which the measurements are made. We will discuss this briefly in the next section.

Table 6-1. Thermodynamic stability data for copper(II) complexes of a series of nitrogen donor
ligands, illustrating the macrocyclic effect.

Ligand	log K
$H_2N(CH_2)_2NH_2$	10.7
$H_2N(CH_2)_3NH_2$	9.8
$H_2N(CH_2)_2NH(CH_2)_2NH_2$	16.1
$H_2N(CH_2)_2NH(CH_2)_2NH(CH_2)_2NH_2$	20.1
$H_2N(CH_2)_2NH(CH_2)_3NH(CH_2)_2NH_2$	23.9

	28.0

6.5.1 Origin of the Macrocyclic Effect

It is relatively easy to make the measurements that show that macrocyclic complexes are
thermodynamically more stable than those with the open-chain ligands for a particular
metal ion. It is a much more subtle and vexed question to ask what the thermodynamic
origin of this macrocyclic effect is. We know that the thermodynamic stability constant K
for a reaction is related to the change in free energy $\Delta G^\ominus$ as shown in Eq. (6.1), and that
the free energy change may be expressed in terms of the enthalpy change ($\Delta H^\ominus$) and entropy change ($\Delta S^\ominus$), Eq. 6.2. The combination of Eqs. (6.1) and (6.2) gives Eq. (6.3), which
relates the stability constant K to the enthalpy and entropy terms for the reaction.

$$\Delta G^\ominus = -RT \ln K \tag{6.1}$$

$$\Delta G^\ominus = \Delta H^\ominus - T\Delta S^\ominus \tag{6.2}$$

$$RT \ln K = T\Delta S^\ominus - \Delta H^\ominus \tag{6.3}$$

It should now be possible to determine whether the macrocyclic effect is entropic or
enthalpic in origin. Initial investigations were made on transition metal complexes and
most workers had a prejudice towards an entropic origin, similar to that of the chelate
effect. More recently, it has become apparent that there is no single cause to which the

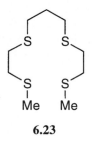

6.23

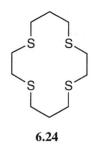

6.24

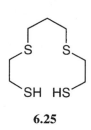

6.25

macrocyclic effect may be attributed, and indeed it is increasingly looking as though the precise origin varies from case to case.

One of the problems with studying both the chelate effect and the macrocyclic effect is to try to separate *local* from *global* energy terms. The local terms are those relating to the metal–ligand bonding that co-ordination chemists immediately think of. However, it is also apparent that the process of binding the ligand to the metal has some global conse-quences relating to the interactions of the ligand, the metal ion and the new macrocyclic complex with the solvent. These latter terms are much more difficult to assess. Let us start by considering the copper(II) complexes of the thioether ligands **6.23** and **6.24**. These have been chosen to be as similar as possible. It would be unreasonable to compare the dithiol ligand **6.25** with the macrocycle **6.24**, because the hydrogen bonding interactions with solvent or other molecules would be very different. Ligands such as **6.23** and **6.24** are of importance as they are not expected to be particularly strongly solvated in protic solvents. This allows us to probe, primarily, the local effects associated with the metal–ligand bin-ding.

The value of log K for the copper complex of **6.24** is 4.3, whilst for that of **6.23** it is 1.97. The macrocyclic complex is thus about 100 times more stable than the open-chain complex, and this is presumably due to the macrocyclic effect. In this case, thermo-dynamic measurements have shown that $\Delta H^{\ominus}_f$ for the macrocyclic and open-chain com-plexes are almost identical, and so the macrocyclic effect is due almost entirely to the entropy term. However, even with these ligands the involvement of solvation may not be neglected entirely. The stability values given above are for the complexes in aqueous solu-tion; if the measurements are repeated in 80 % aqueous methanol, the value of log K for the formation of the macrocyclic complex is only 3.5. A hole-size effect (section 6.6) is also apparent if we move to the larger thioether macrocycle **6.26**. For the formation of the copper complex of **6.26** (again in 80 % aqueous methanol) log K is found to be 0.95.

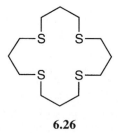

6.26

Unfortunately, the majority of macrocyclic ligands of interest to co-ordination chemists contain donor atoms such as nitrogen or oxygen which can interact strongly (usually by hydrogen-bonding) with solvent. What happens if we now turn to study the macrocyclic effect with these ligands. As indicated earlier, initial studies with these ligands were
contradictory. This was due, in part, to variations in the 'standard' conditions used for the
stability constant determinations. In particular, small variations in the solvents and ionic
strengths, and the use of non-rigorous thermodynamic derivations, were found to have
profound effects on the results. This is not to say that this early work was careless, merely that approximations which had been acceptable in related studies were not valid for
macrocyclic complexes.

However, enough compatible data are available, and we are now in a position to make
a few general conclusions. Perhaps the most important feature is that in comparison of
macrocyclic complexes with complexes of analogous open-chain ligands, $\Delta G^{\ominus}$ always
favours formation of the macrocyclic complex. The entropy term, also, almost always
favours the formation of the macrocyclic complex. This latter observation is as might be
expected by analogy to the chelate effect. The enthalpy term may favour *either* the
macrocyclic *or* the open-chain complex. With macrocyclic tetraamine ligands, the entropy term always favours the macrocyclic complex and the enthalpy term is variable. With
the tetraamine ligands, the differences have been shown to arise from variations in the solvation of the product and reactant solution species, and in particular from differences between the solvent hydrogen bonding interactions shown by the free, open-chain and cyclic
amines. The important feature is, perhaps, not the precise origin of the effect, but the
observation that the macrocyclic complexes *are* thermodynamically more stable than
those of the open-chain ligands. To stress this, consider Fig. 6-23. The greater stability of
the macrocyclic complex means that log K for the reaction is 5.2. This corresponds to a
free energy change of -30 kJ mol^{-1} at 300 K. Remember also, that this only tells us about
the position of the equilibrium, and nothing about the rate at which it is attained.

The thermodynamic stability of the macrocyclic complex provides one of the driving
forces for cyclisation in template reactions. In a way, co-ordination of the macrocycle to
the metal ion provides a thermodynamic sink into which the reaction product can fall. This
is clearly of importance when we consider the reactions such as the formation of metal

$$\log K = 5.2$$

Figure 6-23. The greater stability of the macrocyclic complex means that when the copper(II) complex of an open-chain tetraamine ligand is reacted with cyclam, the copper(II) complex of cyclam is
formed.

phthalocyanine complexes (Fig. 6-21); the thermodynamic stability of the metal complexes is such that some may be sublimed at several hundred degrees centigrade without decomposition. Metal complexes of phthalocyanines have found very widespread applications as colouring agents and pigments, a good testament to their extreme air, light and water stability!

It is tempting to suggest that the origins of these favourable thermodynamic terms are associated with the changes in conformational entropy between free and co-ordinated ligands, but this is probably over-naïve. Although detailed studies of the thermodynamic changes involved in the formation of macrocyclic complexes have been made, the results are not all in accord with each other. The relative importance of the $\Delta H^{\ominus}$ and $T\Delta S^{\ominus}$ terms appears to vary with the metal and the ligand. What is clear, however, is that very often the stability of the macrocyclic complex may be traced to differences in the solvation of the free ligands and the metal complexes.

In addition to their thermodynamic stability, complexes of macrocyclic ligands are also kinetically stable with respect to the loss of metal ion. It is often very difficult (if not impossible) to remove a metal from a macrocyclic complex. Conversely, the principle of microscopic reversibility means that it is equally difficult to *form* the macrocyclic complexes from a metal ion and the free macrocycle. We saw earlier that it was possible to reduce co-ordinated imine macrocycles to amine macrocyclic complexes; in order to remove the nickel from the cyclam complex that is formed, prolonged reaction with hot potassium cyanide solution is needed (Fig. 6-24).

This kinetic stability follows directly from the mechanism of complexation. Normally, the metal-ligand bonds in a complex are formed sequentially. Thus, for $[Ni(NH_3)_6]^{2+}$, there are six separate stages involved in its formation from $[Ni(H_2O)_6]^{2+}$, in which the water ligands are replaced in a stepwise manner by ammonia ligands. Similarly, the formation of a complex with $H_2NCH_2CH_2NHCH_2CH_2NHCH_2CH_2NH_2$ will also be a stepwise process, with the sequential formation of four metal-nitrogen bonds. In contrast, in a macrocyclic ligand such as cyclam, the donor atoms are constrained such that it is difficult to distort the ligand for an initial step with only one amine co-ordinated. The result is a process which, essentially, involves the simultaneous formation of all four metal-nitrogen bonds. This has a high activation energy as it also involves the simultaneous *breaking*

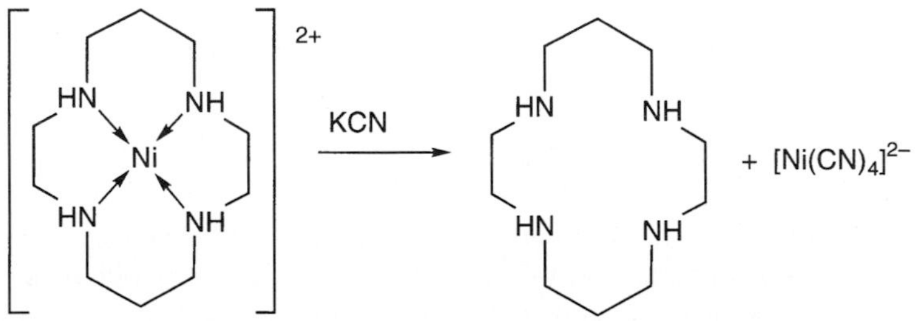

Figure 6-24. The nickel(II) complex of cyclam is extremely stable. To remove the metal it is necessary to react the complex with the strong field ligand, cyanide. In this case, the thermodynamic driving force for demetallation comes from the very high stability of the $[Ni(CN)_4]^{2-}$ ion.

of four metal–water bonds. Similarly, demetallation requires the simultaneous breaking of all of the metal–macrocyclic ligand bonds. These effects will be even more pronounced in rigid ligands. The net effect is that macrocyclic complexes may be slow to form, but once they have been made a combination of kinetic and thermodynamic stability makes metal ion loss unlikely.

In terms of template reactions, this combination of kinetic and thermodynamic stability usually means that the metal ion remains co-ordinated to the macrocyclic ligand and the isolation of the metal complex of the macrocycle provides strong circumstantial evidence for the existence of a metal-directed process. This is particularly easy to establish if the incorporation of the metal ion into the macrocyclic ligand can be shown to be slower than the metal-directed formation reaction.

6.6 Metal Ion Dependence

In this section we will consider the choice of particular metal ions for particular template syntheses. We noted earlier that not all metal ions could act as templates for a particular reaction. What criteria can we use to match a potential template ion to a given macrocyclic product? To a certain extent, the choice of a template ion is dictated by experience, intuition and prejudice. In reality, macrocyclic chemists have their own favourite metal ions that they tend to try first of all! Very often, the first choice of a template ion is nickel(II), and this probably partly explains the vast number of nickel(II) macrocyclic complexes which have been prepared.

In general, we would expect to use soft metal ions in conjunction with soft donor atoms, whilst harder metal ions would be used with harder donors. This rule of thumb usually provides a good working basis for the selection of template ions. Thus, alkali metal ions tend to be used for the preparation of crown ethers and other ligands containing hard oxygen donors, whilst the softer transition metal ions tend to be used for the preparation of complexes with nitrogen or sulfur donor ligands. However, it is usually more a matter of luck than judgement in finding a successful template ion.

On a number of occasions, we have referred to the size of the bonding cavity within a macrocyclic ligand, and this has proved to be a good criterion for the selection of a template ion. How can we quantify this concept of the size of the bonding cavity?

6.6.1 Macrocyclic Hole Size – An Abused Concept

In our discussion concerning open-chain ligands, we tacitly assumed that there was sufficient flexibility within the system to allow the metal–ligand distance to be optimised. This is actually the reason why we get such good comparisons between the bond energies of metal–ligand bonds in complexes with monodentate and chelating ligands. For a given metal ion, the bond lengths in the chelated complexes are very similar to those in the complexes with the monodentate ligands. This distance is simply defined as the sum of the ionic radius of the metal ion and the covalent radius of the donor atom. We assume that

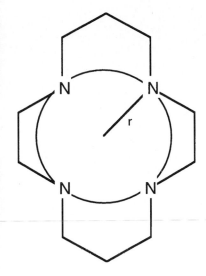

Figure 6-25. The hole size of a macrocyclic ligand defines the available bonding cavity for a metal ion. As a first step, a best-fit circle is drawn through the donor atoms, and the radius of this circle measured.

conformational changes occur in the ligand to allow the metal–ligand distance to be optimised. However, macrocyclic ligands tend to be less flexible, and a more limited range of metal-donor atom distances can be optimally accommodated within a given ligand.

This allows us to introduce the concept of the *hole size* of a macrocyclic ligand. The hole size of a macrocyclic ligand is defined as the radius of the *available* bonding cavity within the ligand. It follows immediately that *the hole size of a macrocyclic ligand may vary with conformational changes within the ligand.* However, even for very flexible macrocycles, we may usefully define the hole size of the available bonding cavity of the ligand in the lowest energy conformation.

Consider a planar tetradentate macrocycle, and draw a best-fit circle through the centres of the four donor atoms (Fig. 6-25). The radius of this circle (from the centre of the circle to the centre of the donor atoms) may be considered to be dictated by two

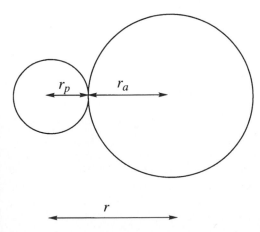

Figure 6-26. The radius of the best-fit circle through the donor atoms is composed of two components. The first is the radius of the donor atoms, r_a, and the second is the hole size, r_p.

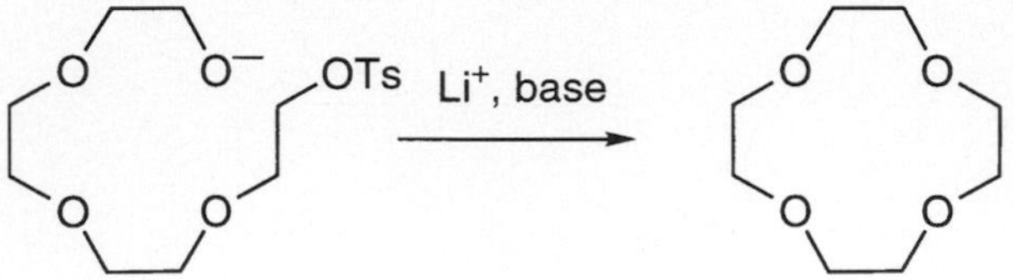

6.27

Figure 6-27. Lithium cations are the correct size for the bonding cavity of **6.27** and are therefore the template ion of choice for its synthesis.

parameters. The first component represents the effective covalent radius of the donor atom (r_a) and the other the effective radius of the available bonding cavity (r_p). It is this last parameter r_p that relates to the size of a metal ion that can be bound in the cavity and defines the hole size of the macrocycle (Fig. 6-26). Obviously, the method described works best for macrocycles with a planar or near-planar set of donor atoms that are arranged regularly. More sophisticated methods exist for determining the hole size of macrocycles containing a variety of donor atoms arranged irregularly or in a non-planar manner.

We can now make the reasonable assumption that the closer the radius of the metal ion is to the hole size of the ligand, the more stable the complex is likely to be. This follows from an optimal metal-donor atom distance. This also allows us to make some further discrimination beyond hardness and softness in the selection of metal ions for use in template reactions. The small crown ether 12-crown-4 (**6.27**) has a hole size of about 0.8 Å. The presence of the relatively hard oxygen donors in **6.27** suggests that we should use a Group 1 or Group 2 metal ion as a template for its preparation. The ionic radii of the Group 1 metal ions increase as we descend the group (Li⁺, 0.76 Å; Na⁺, 1.02 Å; K⁺, 1.38 Å; Rb⁺, 1.52 Å; Cs⁺, 1.67 Å). We would thus predict that Li⁺ will be the ion best suited for the bonding cavity, and hence the best template ion to use in the synthesis. This is indeed

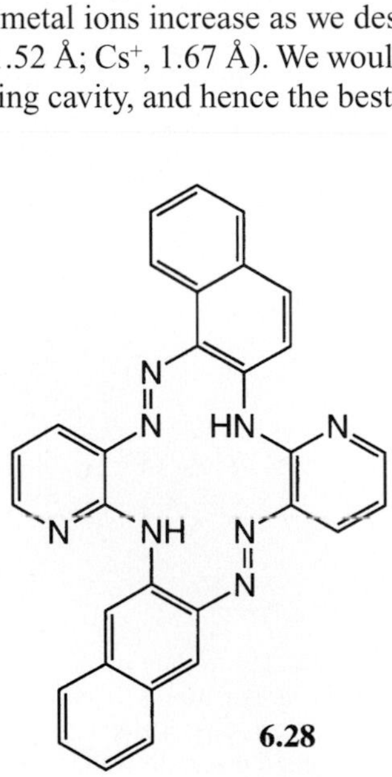

6.28

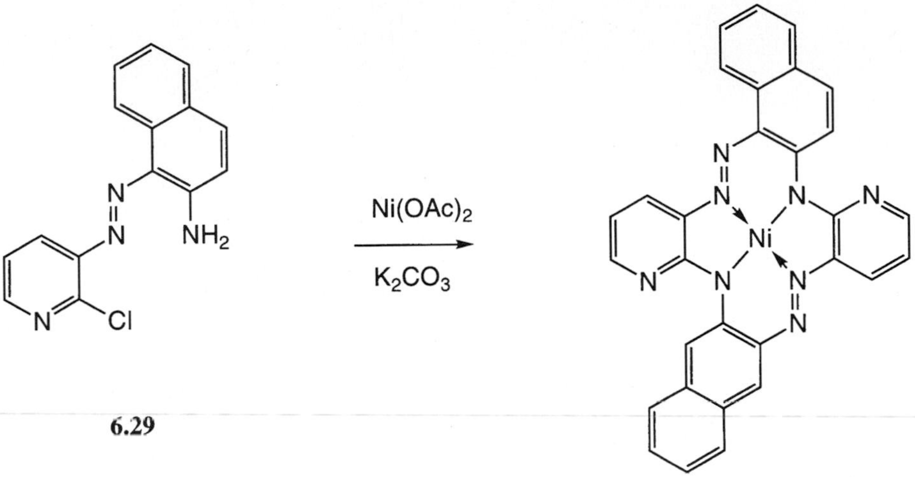

6.29

Figure 6-28. The template formation of the nickel(II) complex of **6.28**. The nickel(II) ion is the correct size for the cavity of the macrocycle. The macrocycle is doubly deprotonated to give a neutral complex.

the case. A partial confirmation comes from a study of the stability constants for the Group 1 complexes with **6.27**; the lithium complex is the most stable. The formation of **6.27** in a process templated by lithium ions is shown in Fig. 6-27. It is pertinent at this point to warn the reader that the ionic radius of a given metal ion is probably as variable a concept as the hole size of a ligand; the apparent radius of a metal ion depends upon both the co-ordination number and the geometry.

In a similar manner, nickel(II) has the correct ionic radius for the bonding cavity of the fourteen–membered ring, tetraazamacrocycle **6.28**. The reaction of **6.29** with nickel(II) acetate in the presence of base gives the nickel(II) complex of **6.28** (Fig. 6-28). This is an example of a template reaction that involves a nucleophilic displacement as the ring-formation process.

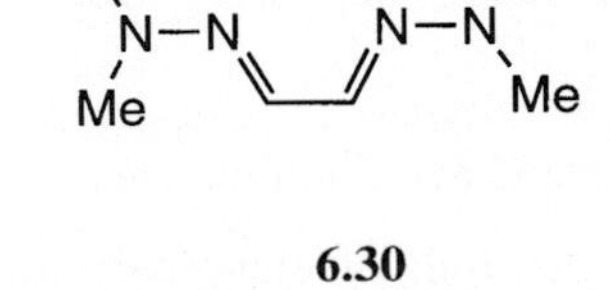

6.30

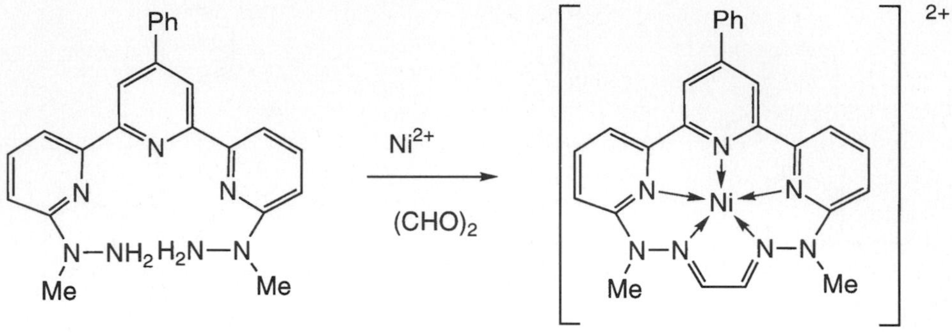

6.31

Figure 6-29. The condensation of **6.31** with glyoxal in the presence of nickel(II) acetate gives the nickel(II) complex of **6.30**. In the product, the nickel(II) ion is actually seven co-ordinate with two additional axial solvent molecules which are not shown.

The pentadentate nitrogen donor ligand **6.30** has a hole size of about 0.7 Å and we predict that we could use a first–row transition metal ion as the template for its synthesis. The macrocycle is best prepared by the condensation of **6.31** with glyoxal about a nickel(II) template. In this condensation, most other metal ions are ineffective as templates.

6.7 Consequences of Metal Ion/Hole Size Mismatch

In the preceding section, we used the correct matching of the size of the metal ion to the size of the macrocyclic cavity to select metal ions that would be effective as templates. We should now consider what the consequences will be if there is a gross discrepancy between the two parameters. One can envisage a number of possible consequences of template reactions if the metal ion used as the template is not the correct size for the desired macrocyclic product:

1. Perhaps the simplest would be that the metal ion does not act as a template, and no reaction occurs;
2. If the metal ion is too large or too small for the cavity, a macrocycle with a different ring size to that expected might be formed;
3. Perhaps the metal ion will act as a template, but the ligand then undergoes a conformational change to optimise metal–ligand interactions;
4. The expected macrocycle could be formed, but the mismatch in sizes makes the metal ion labile;
5. The initially formed macrocycle undergoes a metal-directed rearrangement to produce a cavity that is more suitable for the metal ion involved.

Let us now consider each of these possibilities in more detail.

6.7.1 The Metal Ion does not act as a Template

This first case is the least interesting, although it is by far the most common result in putative template reactions, even when the metal ion is apparently the right size for the desired product! A good example is found in the reaction presented in Fig. 6-29. All of the dipositive first-row transition metal ions have similar sizes, but only nickel(II) is effective for the formation of complexes of **6.30** in a template condensation. The other metals either give polymers, complexes of **6.31**, or in the case of cobalt(II), compounds that are half-way to the desired product containing **6.32**.

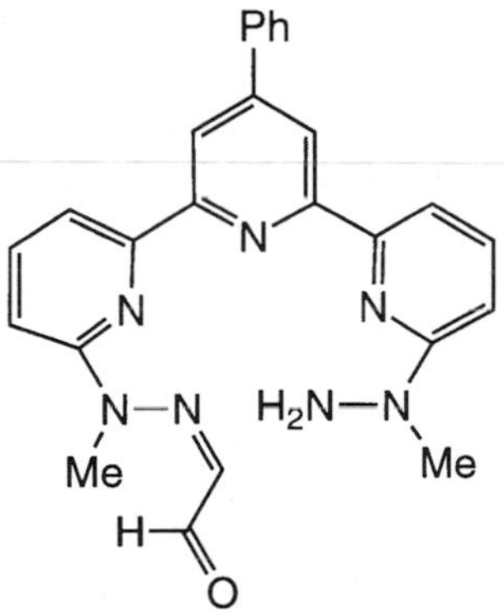

6.32

6.7.2 Control Over the Ring Size

The operation of the mismatch effects may be seen to best advantage when a range of products is possible from a single reactant or set of reactants. The reaction of ethylene oxide with metal salts results in the formation of crown ethers (Fig. 6-30). Obviously, a whole range of different cyclic oligomers and acyclic polymers could be formed from ethylene oxide. If we specifically wanted to obtain 18-crown-6, with a hole size of about 1.4 Å, we would expect to use a potassium ion as template ($r = 1.38$ Å). In fact, 18-crown-6 *is* obtained in good yield from the reaction of ethylene oxide with potassium tetrafluoroborate. In contrast, if we wanted 12-crown-4, with a hole size of about 0.8 Å, it

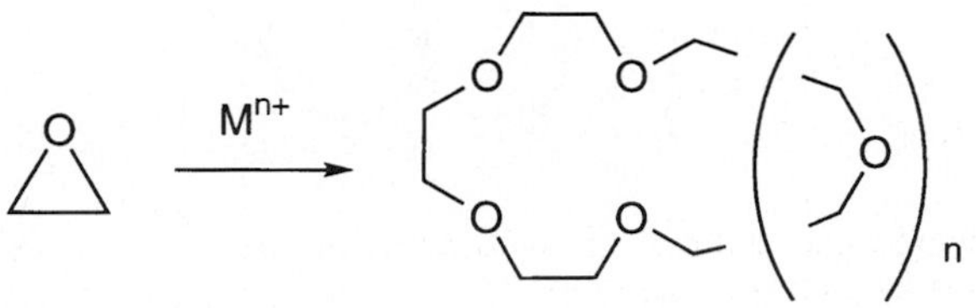

Figure 6-30. The reaction of ethylene oxide with metal salts gives crown ethers of various sizes.

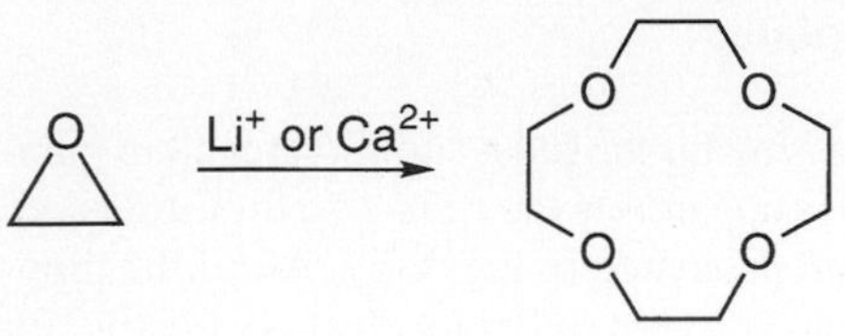

Figure 6-31. The reaction of ethylene oxide with lithium or calcium salts gives 12-crown-4, as expected from hole size arguments.

would be more appropriate to use a lithium ($r = 0.76$ Å) or calcium ($r = 0.94$ Å) salt as template (Fig. 6-31). Once again, these arguments appear to be reasonable and indeed 12-crown-4 is obtained on reaction with calcium salts.

In many cases it is possible to utilise the hole size effects for the synthesis of specific types of macrocycle. Thus, a tetradentate macrocycle (**6.33**) is expected to be obtained from a template condensation of 2,6-diacetylpyridine with 1,5,9-triazanonane in the presence of small, first-row transition metal dications. The hole size of **6.33** closely matches the size of these metal ions. This is indeed what happens when Ni^{2+} ($r = 0.8$ Å) is used as a template for the condensation and the nickel(II) complex of **6.33** is obtained in good yield (Fig. 6-32). However, when Ag^+ ($r = 1.0$ Å) is used as a template, the metal ion

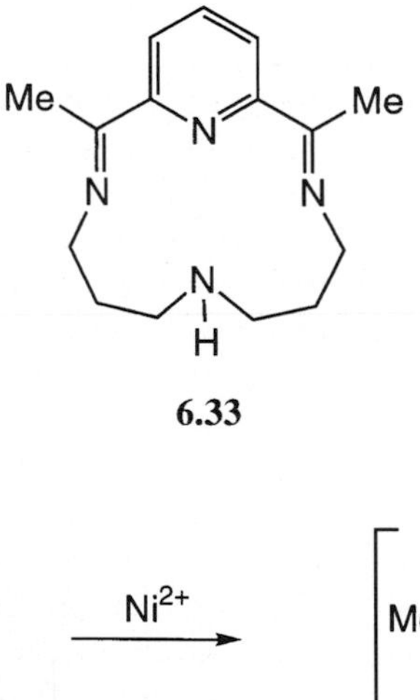

6.33

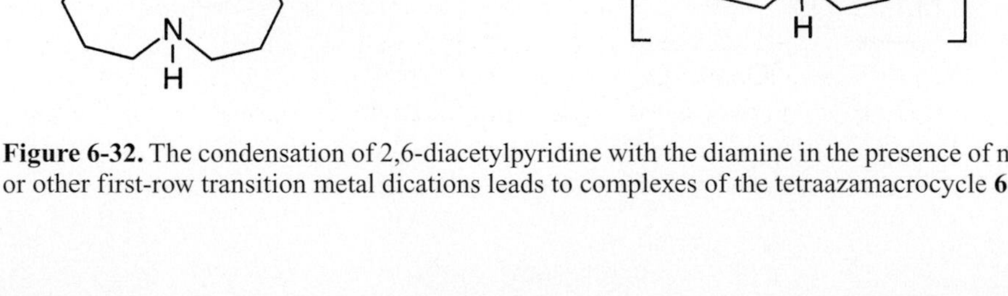

Figure 6-32. The condensation of 2,6-diacetylpyridine with the diamine in the presence of nickel(II) or other first-row transition metal dications leads to complexes of the tetraazamacrocycle **6.33**.

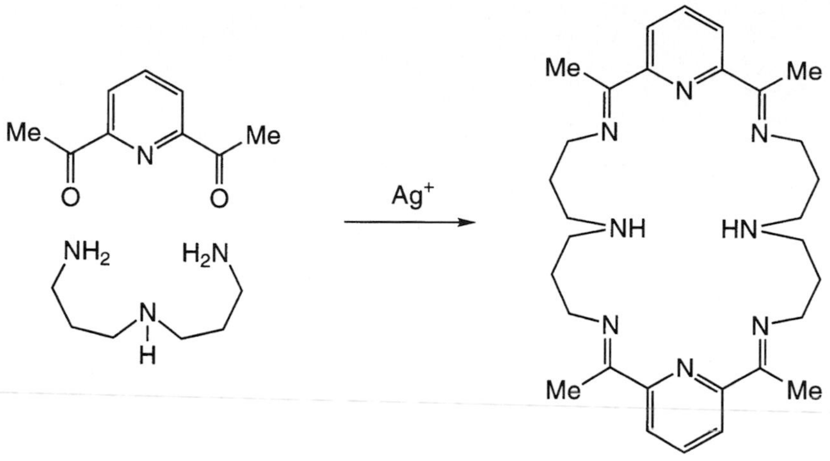

6.34

Figure 6-33. When a larger metal ion such as silver(I) is used as a template, the condensation of 2,6-diacetylpyridine with the diamine gives a complex of the larger [2+2] macrocycle **6.34**. Actually, the macrocycle **6.34** is isolated as a disilver complex.

is too large for the cavity of **6.33** and a complex of the new [2+2] macrocycle **6.34** is obtained (Fig. 6-33). Exactly why silver(I) gives the [2+2] product is not quite clear, because, in reality, a disilver complex is obtained, suggesting that the cavity is too great for a single silver(I) ion. This suggests that the preferred co-ordination geometry of the metal ion is probably as important as the actual ionic size.

Clearly it is possible to exert relatively subtle control over the direction of cyclisation reactions by a judicious choice of template ion, even if the effects may not be fully quantified or understood.

6.7.3 Conformational Change May Alter the Apparent Hole Size

We saw in Fig. 6-30 the conversion of ethylene oxide to crown ethers upon reaction with appropriate metal salts, and demonstrated that the hole sizes of the products corresponded to the ionic radius of the template ion. However, lest we become over-confident, it should be pointed out that the major product from the reaction of ethylene oxide with caesium salts ($r = 1.67$ Å) is not the expected 21-crown-7 with a hole size of about 1.7 Å) but 18-crown-6 (hole size, 1.4 Å) (Fig. 6-34). The reason for this lies in the structure of the complex formed. We have always assumed that the metal ion will try to lie in the middle of the bonding cavity of the macrocycle. There is no real reason why this should be. Caesium could form a complex with 21-crown-7 in which all of the oxygen atoms lie approximately planar with the metal in the centre of the cavity. It is also apparent that caesium could not occupy the middle of the cavity in 18-crown-6. However, a different type of complex can be formed with 18-crown-6, in which a caesium ion is sandwiched bet-

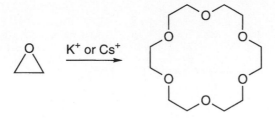

Figure 6-34. The reaction of ethylene oxide with caesium salts gives 18-crown-6 rather than the 21-crown-7 expected on the basis of hole sizes

ween two of the macrocyclic ligands (Fig. 6-35). This complex has a total of twelve Cs–O bonds and is favoured over the 1:1 complex with 21-crown-7.

When macrocycles containing flexible chains are considered, it is possible for the ligand to undergo a conformational change to optimise the metal–ligand bonding in cases where there is an apparent mismatch. We have seen this in the formation of the various crown ethers above and another example is to be found in the transition metal mediated formation of the 12-membered ring macrocycle **6.35**. The cavity in the 12-membered ring of **6.35** is apparently too small to accommodate a nickel(II) ion. It is surprising, therefore, that the nickel(II) complex of **6.35** is obtained in excellent yield from a template reaction **6.36** and **6.37** with Ni^{2+} (Fig. 6-36).

However, the problem arises from our measurement of the hole size of **6.35**. Certainly the hole size of the ligand in a *planar* configuration is too small for the nickel(II), but it is possible for the ligand to fold to adopt a *cisoid* configuration. It is in this form that the ligand is co-ordinated to the nickel; the metal is octahedral, with the two remaining co-ordination sites occupied by bromide ions and with near optimal Ni–N and Ni–S distances (Fig. 6-37).

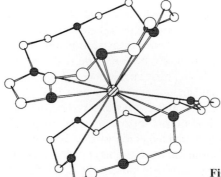

Figure 6-35. The structure of the 2:1 complex cation formed from 18-crown-6 and caesium ions

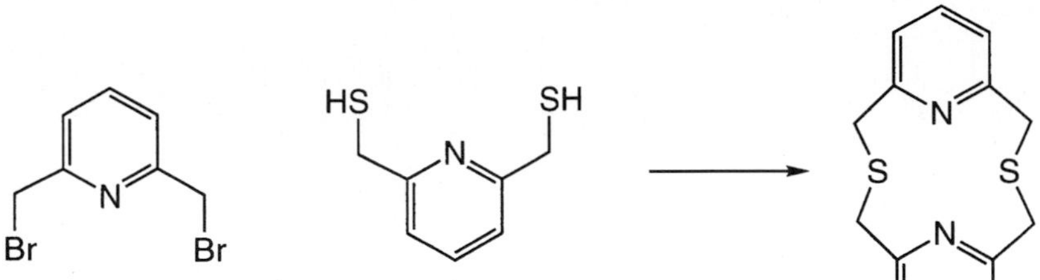

6.36   **6.37**   **6.35**

Figure 6-36. The 12-membered ring macrocycle is obtained as its nickel complex from a template condensation of **6.36** with **6.37** in the presence of nickel(II). The cavity of **6.35** is formally too small for a nickel(II) ion!

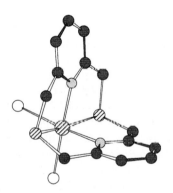

Figure 6-37. The structure of the nickel(II) complex of **6.35** obtained from the template condensation in Fig. 6-36, showing the folded conformation of the ligand.

6.7.4 Mismatch in Sizes Makes the Metal Ion Labile

One of the more interesting hole size effects arises when the metal ion successfully acts as a template, but is labilised in the macrocyclic complex that is formed. The consequence of this is that the metal ion acts as a *transient template*. The metal ion may be viewed as pre-organising the reactants to form the macrocyclic products, but then finding itself in an unfavourable environment after the cyclisation. The effect is best observed when a small metal ion is used as a template for a reaction that can only give one product (or at least, only one *likely* product). What happens to the metal ion when it finds itself in an environment that does not match up to its co-ordination requirements? The most useful consequence would be labilisation of the metal ion, with resultant demetallation and formation of the metal-free macrocycle. This would overcome one of the major disad-

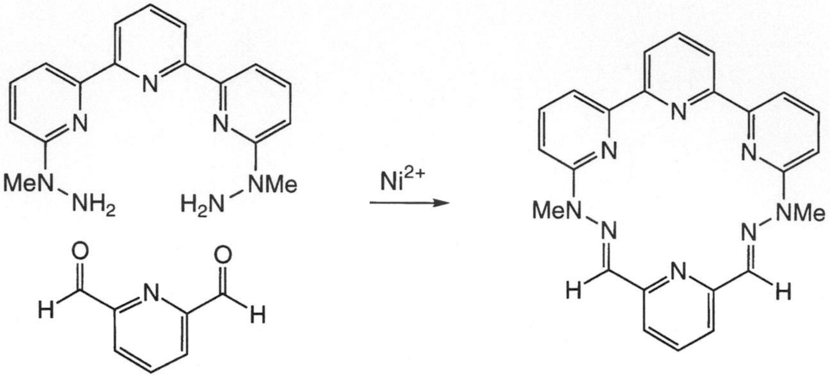

6.38

Figure 6-38. The transient template preparation of **6.38**. The nickel(II) ion is too small for the cavity of the macrocycle **6.39**.

vantages of template reactions – that of removing the metal ion from the complexes usually obtained.

An example of a transient template reaction is seen in the preparation of the hexadentate ligand **6.38**. The open-chain pentadentate bishydrazine ligand **6.31** has an effective bonding cavity of 0.7 Å and forms a stable nickel(II) complex with optimal Ni–N bonding distances. Reaction with 2,6-diformylpyridine gives the nickel(II) complex of macrocycle **6.38** that has a hole size of about 1.1 Å. The nickel(II) is thus too small for the macrocycle that it has formed about itself and in the presence of chloride ion is lost as [NiCl$_4$]$^{2-}$, leaving the free macrocycle in solution (Fig. 6-38). This reaction illustrates a number of features of transient template reactions; the metal ion matches well with the precursors but not with the products and only a single product is expected. The reader is left to verify that [2+2] products are not expected from this reaction.

6.7.5 Metal-Directed Rearrangement to Produce a More Suitable Cavity

We have now seen that the relative sizes of the metal ion and the bonding cavity of the macrocyclic ligand may have profound effects upon the course of template syntheses of macrocycles. These have been expressed in the stoichiometry of the reaction and in the conformation of the macrocyclic products. However, it is also possible to observe changes in the reactivity of the macrocyclic ligand itself as a result of interaction with metals of the 'wrong' size.

Let us consider the [2+2] macrocyclic ligand **6.39**, which is prepared by the non-template condensation of 1,2-diaminobenzene with 2,6-diformylpyridine. The hole size of this ligand is about 1.3 Å, so we would expect it to be too large to bind first-row transition metal dications. As a matter of interest, the ligand binds K$^+$, with an ionic radius of 1.38 Å, more strongly than does the crown ether, 18-crown-6. In the absence of any che-

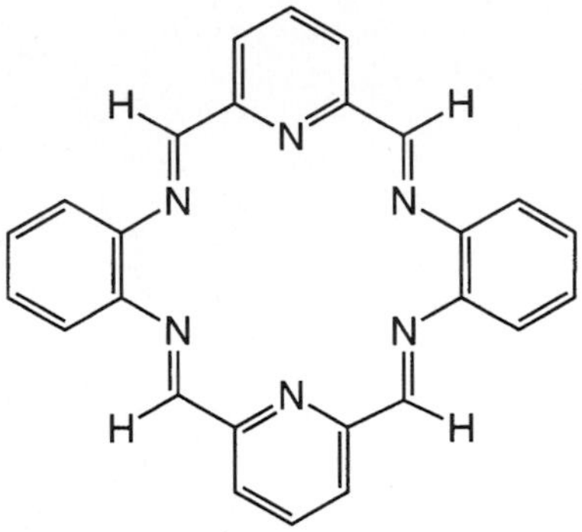

6.39

mical transformation, we would expect the small first-row transition metal ions to 'rattle' in the cavity of the ligand. We saw earlier that this is expected to lead to a distortion of the macrocycle, or loss of the metal ion. In this case, the ligand is rigidly planar and so cannot undergo any conformational change to relieve the mismatch. It is, thus, surprising that apparently stable non-labile complexes with first row transition metal ions are formed. What is happening?

In earlier chapters we noted that metal ions could *either* activate *or* deactivate an imine with respect to addition of a nucleophile. We will now see an example of metal-ion activation in action. In fact, the complexes that are formed from **6.39** arise as a result of metal-initiated nucleophilic attack at the imine groups. The reaction of the free ligand **6.39** with methanolic cobalt(II) acetate results in the attack of methanol upon one of the imine bonds of the initially formed complex (Fig. 6-39).

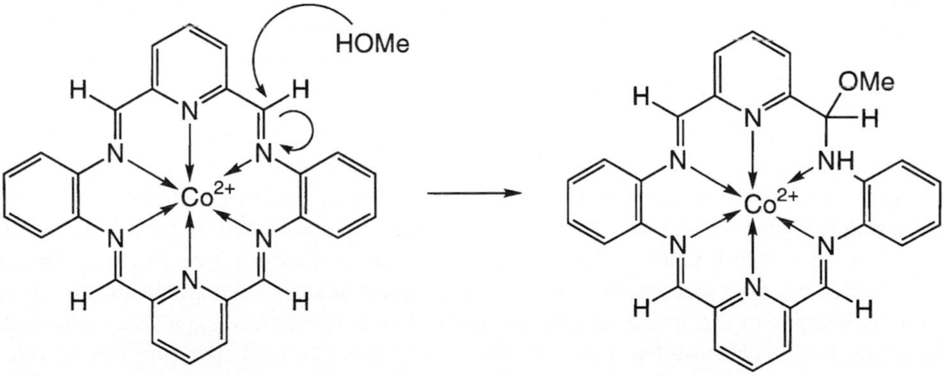

6.40   **6.41**

Figure 6-39. The macrocyclic complex **6.40** is too large for a cobalt(II) ion to bind effectively. The imine bonds are activated towards attack by a methanol molecule. This generates a slightly smaller cavity in the complex **6.41**.

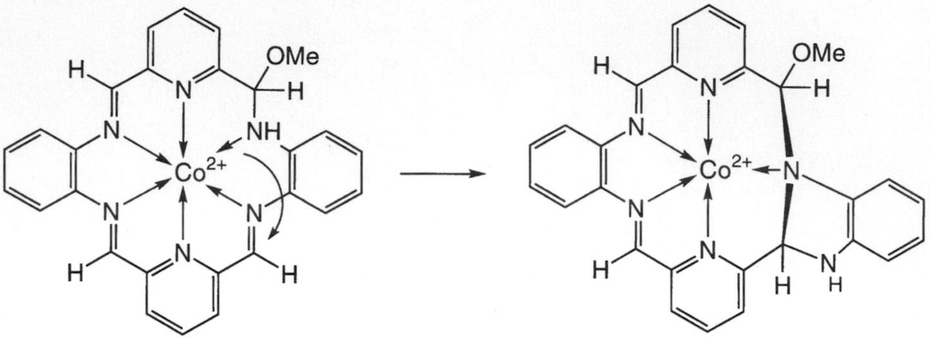

6.41

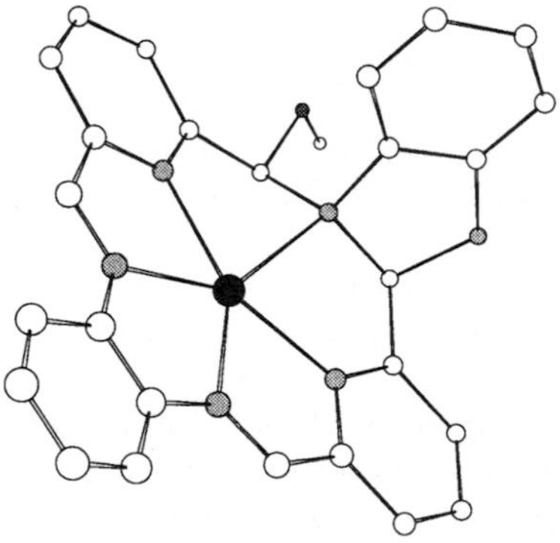

Figure 6-40. The complex **6.41** undergoes an intramolecular nucleophilic attack at a second imine to generate a new macrocycle with the correct cavity size for cobalt(II). The lower structure shows the complex cation as it is found in the solid state. The cobalt ion is actually seven-co-ordinate, with axial water and methanol ligands (omitted for clarity).

The complex **6.41** now contains newly generated sp^3 hybridised nitrogen and carbon atoms and the resultant changes in bond angles result in a puckering of the macrocycle and a reduction in the hole size. However, this reduction in hole size is still not sufficient to ideally accommodate the cobalt(II) ion and a second reaction of an imine occurs to reduce the hole size still further. In this reaction, the newly formed, and nucleophilic, secondary amine of the aminol attacks an adjacent imine group to form a five-membered ring (Fig. 6-40). The result is a change from the planar 18-membered ring macrocycle, which is too large for the cobalt(II) ion, to a non-planar, 16-membered ring macrocycle which possesses a more suitable bonding cavity.

A similar complexity of rearrangements is observed in the related compound **6.42**, which is derived from the metal-free reaction of 2,6-diacetylpyridine with 1,2-diamino-benzene (Fig. 6-41). In this case, the steric interactions between the methyl groups and the

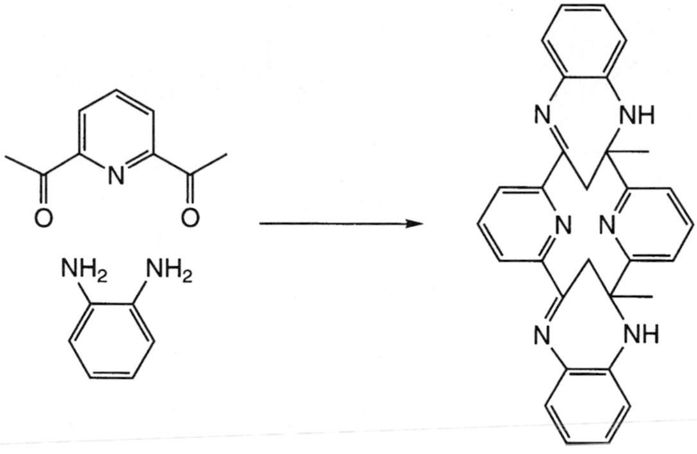

6.42

Figure 6-41. The reaction of 2,6-diacetylpyridine with 1,2-diaminobenzene gives a complex product **6.42**. Steric interactions prevent **6.43** from being formed.

phenyl rings initiate a rearrangement in the free ligand itself. The expected product was the macrocycle **6.43**.

When **6.42** is treated with copper(II) salts a complicated rearrangement occurs to give **6.45**. It can be seen that the ligand in **6.45** is derived from **6.43**. However, **6.43** is sterically strained because of the interactions between the methyl groups and the phenylene protons and the cavity is too large for the small copper(II) ion. A formal mechanism is presented in Fig. 6-42. It is most unlikely that this is the precise mechanism, or that complexes of **6.43** are actually involved, but it does serve to illustrate the way in which sequential simple metal-directed reactions can give a complex rearrangement. The steps

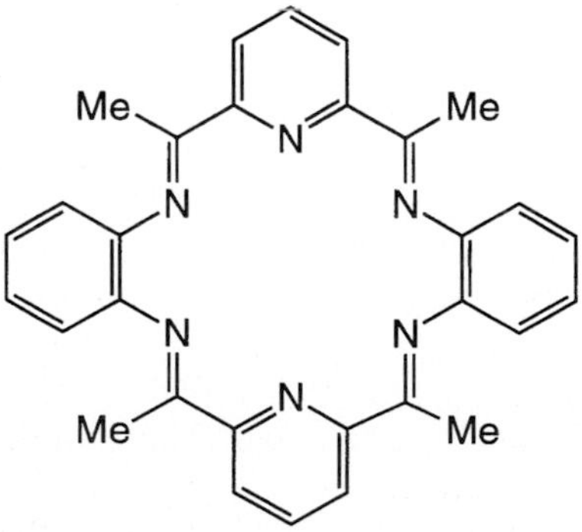

6.43

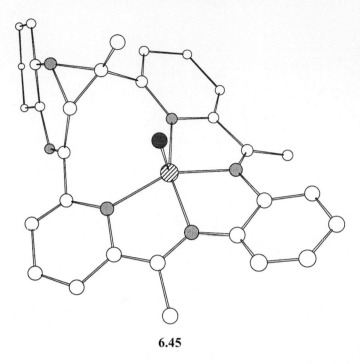

6.45

involved include nucleophilic attack at an imine, tautomerisation of an imine to a vinyl-amine, and aza-Claisen reaction, *and* oxidation! A view of the cation as observed in the crystal (**6.45**) reveals that the ligand is four co-ordinate and greatly distorted from pla-narity. The copper(II) centre is five co-ordinate with an axial water ligand. The overall effect of the rearrangement is to create a smaller 15-membered ring cavity for the metal ion.

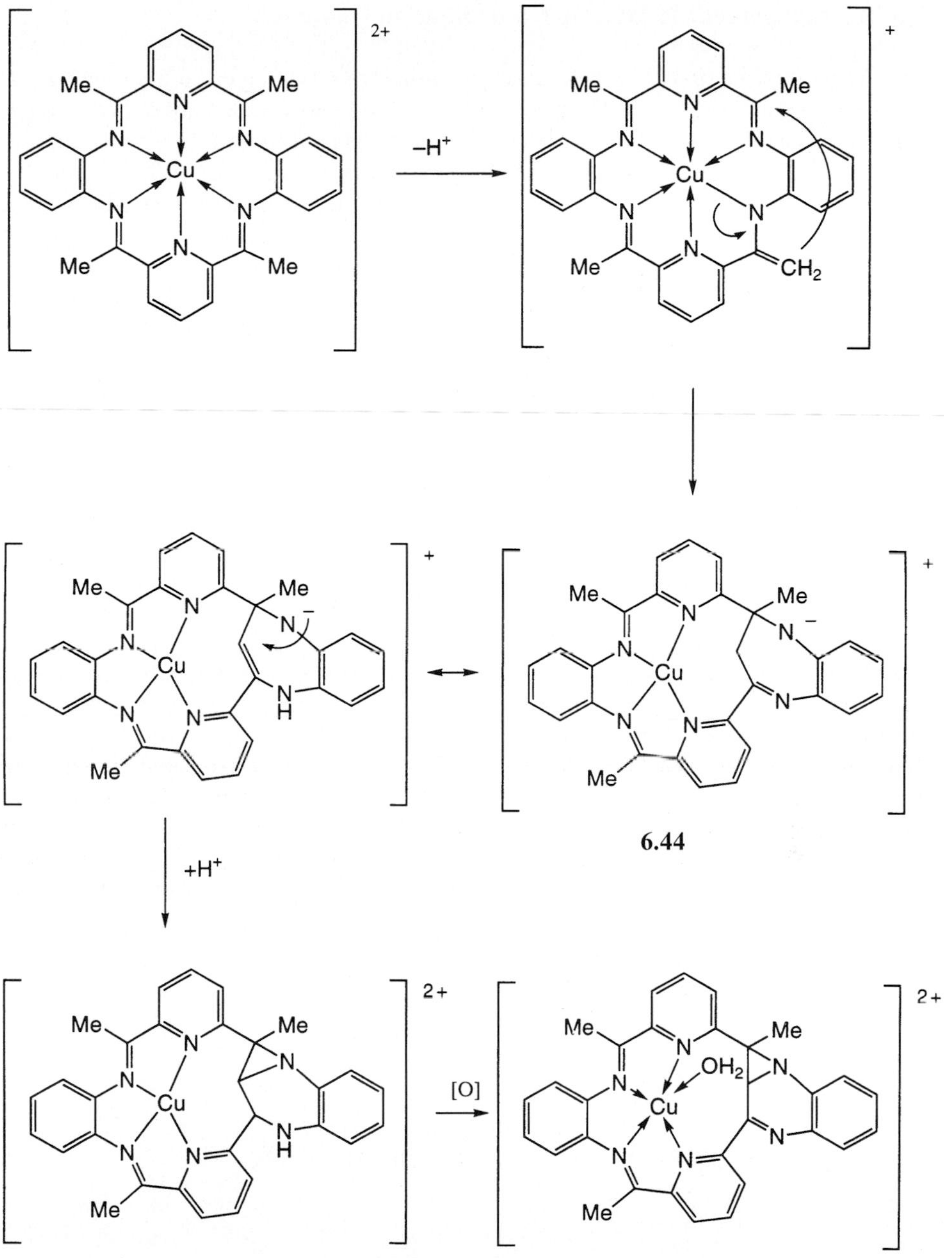

Figure 6-42. A formal mechanism by which the product **6.45** could arise from **6.42** on treatment with copper(II) salts.

6.7.6 Rearrangement Related to Co-ordination Geometry

So far, we have concentrated upon reactions resulting from discrepancies between the size of the metal ion and the size of the macrocyclic cavity. However, it is not only the size of the metal ion that may result in a mismatch; what happens if the favoured conformation of the ligand does have an arrangement of donor atoms that matches with the preferred co-ordination geometry of the metal ion? This is exactly the situation that we observe with metal complexes of some pentadentate macrocycles. We have previously observed the formation of tetraaza macrocycles from the template condensation of 2,6-diacetylpyridine with diamines in the presence of a transition metal ions. We also noted that if the size of the metal ion were incorrect, it was possible to get [2+2] or other products. Now let us look at this topic in a little more detail.

Macrocyclic ligands containing five donor atoms are of interest as they provide the co-ordination chemist with a way of studying compounds with unusual co-ordination numbers and geometries under closely defined conditions. Although nickel(II) might be expected to act as an effective template on the basis of the hole size for the condensation of 2,6-diacetylpyridine with 1,4,8,11-tetraazaundecane to give the macrocycle **6.46**, it is found to be inactive. This is thought to be due to the preference of the nickel(II) ion for an octahedral geometry, rather than a five- or seven-co-ordinate one. The ligand **6.46** is not flexible enough to fit any of the preferred geometries in a strain-free manner.

The origin of this preference is partly in the crystal field stabilisation energy associated with the d^8 configuration, which strongly favours the octahedral geometry. It is reassuring that non-transition metals, which have no electronic preference for any particular geometry, act as effective templates. If we attempt to use a large metal ion such as silver(I) as template, a complex of the [2+2] product **6.47** is obtained. What happens if we take this silver(I) complex of the [2+2] ligand and attempt to remove or change the metal? If we

6.46   6.47

attempt to remove the silver ion in water, the free ligand is hydrolysed, but treatment with methanolic nickel(II) salts gives a derivative of the [1+1] macrocycle, **6.48**. In **6.48**, a methanol molecule has added to an imine bond of **6.46**. In this compound, the five nitrogen donor atoms are arranged such that they may occupy five of the six co-ordination sites of an octahedral nickel(II) centre. This is of particular interest as this product is not obtained directly from the template reaction with nickel(II); in other words, we are observing the product of some crystal-field dictated intramolecular reaction upon transmetallation.

6.48

6.8 Multicomponent Assembly of Macrocycles

In this chapter we have seen a variety of reactions in which one or two reactants have undergone metal-controlled reactions to give macrocyclic ligands. We have seen how very precise control over the reaction may be achieved by a consideration of the sizes and bonding character of the metal ion and the ligands. If this control is so precise, we can now ask whether it is possible to control the assembly of a variety of different components in a sequential but programmed manner to give a single product. It is this topic that we shall consider in this penultimate section of the chapter.

Let us start by considering the reaction of the copper(II) complex **6.49** with formaldehyde. Initially we might expect the diimine **6.50** to be formed, but this ignores the nature of the intermediates. As we saw earlier, the reaction of an amine with an aldehyde initially produces an aminol. Consider the addition of the second molecule of formaldehyde to **6.49**. The product will be **6.51**, which contains an imine and an aminol (Fig. 6-43). The imine is co-ordinated to a metal ion, and the polarisation effect is likely to increase the electrophilic character of the carbon. The hydroxy group of the aminol is nucleophilic and it is correctly oriented for an intramolecular attack upon the co-ordinated imine. The result is the formation of the copper(II) macrocyclic complex **6.52**.

The important feature is that the imine that is initially formed is susceptible to attack by a nucleophile. What would happen if we were to have another nucleophile present in the solution? Conceptually this is exactly the situation when we react **6.49** with a mixture

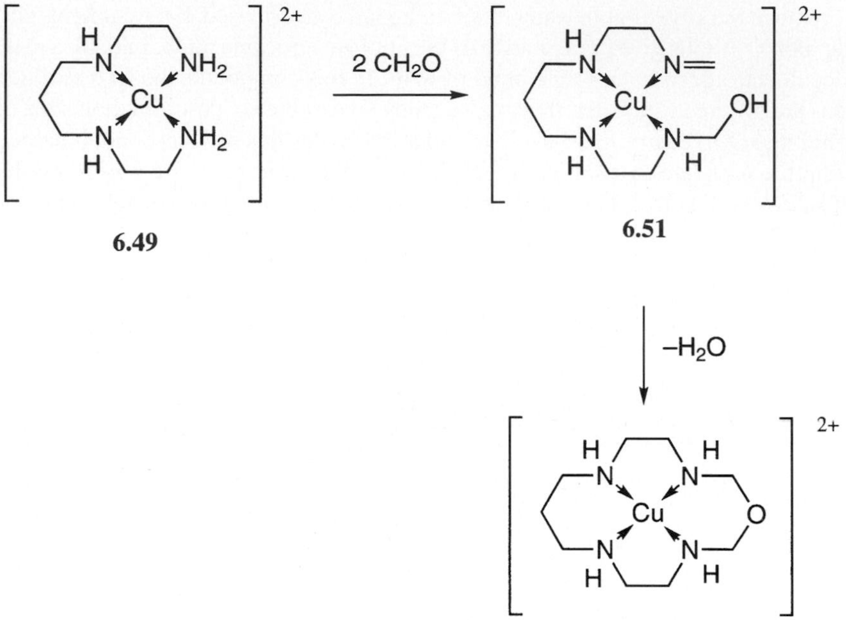

6.49

2 CH₂O

6.51

−H₂O

6.52

Figure 6-43. The reaction of **6.49** with formaldehyde gives the macrocyclic complex **6.52** rather than **6.50**. The reaction proceeds through the intermediate **6.51**.

of formaldehyde and nitroethane under basic conditions. Nitroethane generates a nucleophilic anion, which can react with intermediates such as **6.50**. Note that the precise sequence of events is not known for certain, but almost certainly does not involve **6.50**. A possible sequence which illustrates the various metal directed processes is shown in Fig. 6-44. In this sequence the initially formed imine is attacked by the anion from nitromethane to give an intermediate **6.53**. This is, in turn, attacked by formaldehyde to give a new imine, which undergoes an intramolecular nucleophilic attack to give the macrocycle **6.54**. Of course, in reality, the reaction of the nitroalkane with the aldehyde could occur

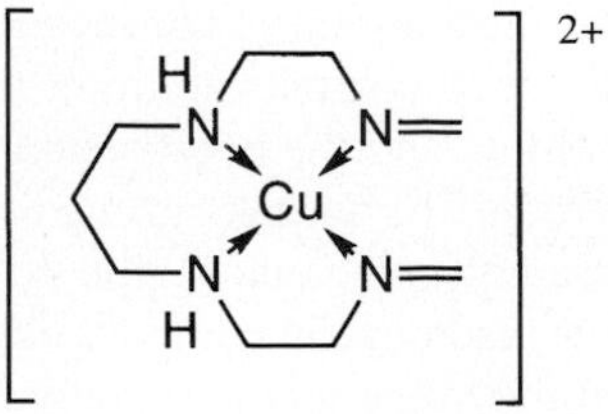

6.50

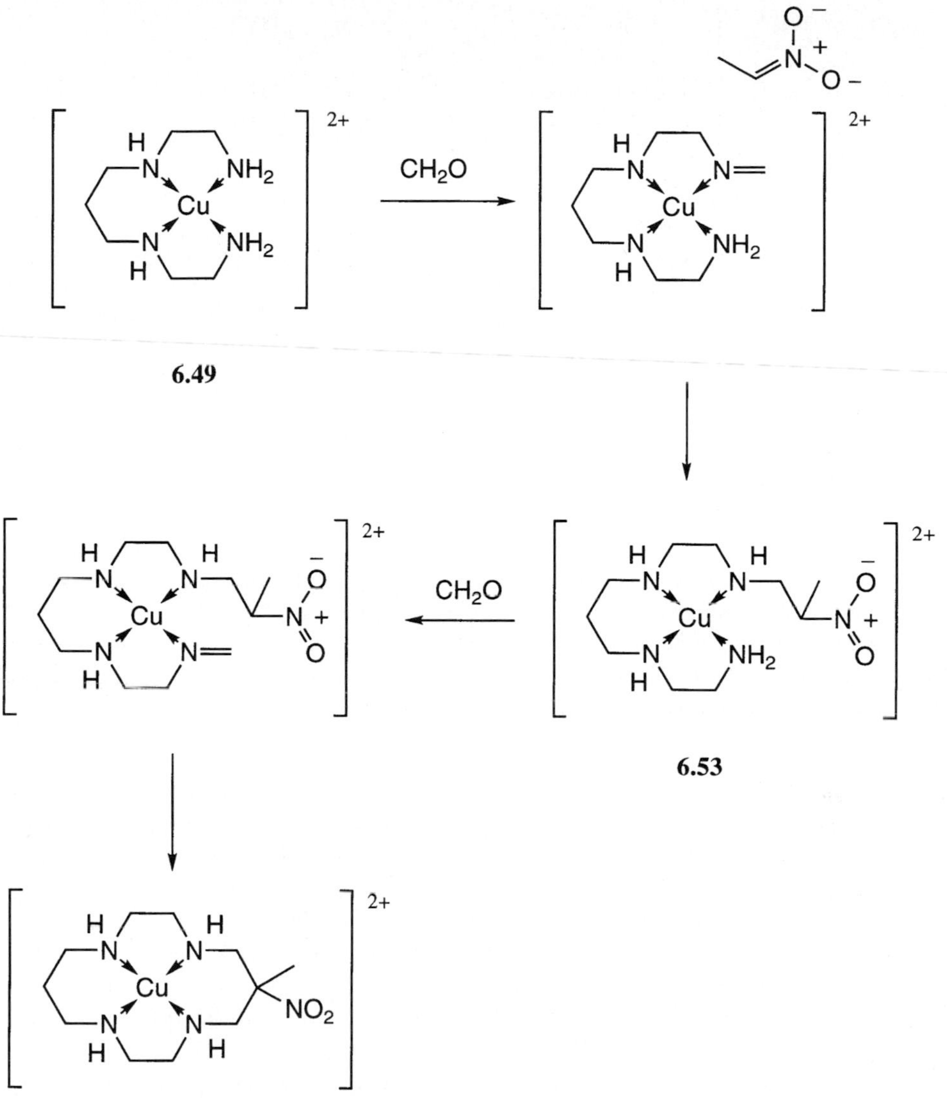

Figure 6-44. The formation of a copper(II) macrocyclic complex from a three component reaction involving **6.49**, formaldehyde and nitroethane. The product **6.54** arises from the reaction of one equivalent of **6.49** with two equivalents of formaldehyde and one of nitroethane.

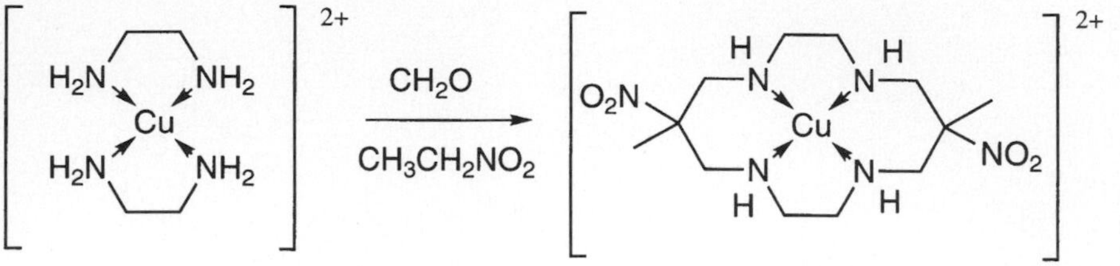

6.55

Figure 6-45. The formation of a copper(II) macrocyclic complex from a three component reaction involving [Cu(en)$_2$]$^{2+}$, formaldehyde and nitroethane. The product **6.55** arises from the reaction of one equivalent of [Cu(en)$_2$]$^{2+}$ with four equivalents of formaldehyde and two of nitroethane.

prior to reaction with the co-ordinated amine. The metal ion is templating the formation of the planar tetraaza macrocyclic ligand.

The remarkable feature is that the mixture of these three reactants follows a precisely coded reaction sequence to give the macrocyclic product. A series of electrophiles and nucleophiles of different relative strengths are involved. The reaction may be extended a stage further, as shown in Fig. 6-45. The reaction of [Cu(en)$_2$]$^{2+}$ with formaldehyde and nitroethane results in the formation of the macrocyclic complex **6.55**.

The nature of the additional nucleophile may be varied. For example, the reaction of the nickel(II) complex **6.56** with formaldehyde and methylamine gives the macrocyclic complex **6.57** (Fig. 6-46). Again, it is not clear whether the first steps of the reaction involve reaction with formaldehyde, followed by attack of amine upon the imine, or initial formation of an electrophile such as H$_2$C=NMe, which attacks **6.56**.

As a final example, we illustrate the reaction of [Cu(en)$_2$]$^{2+}$ with formaldehyde and methylamine to give the macrocyclic complex **6.58** (Fig. 6-47).

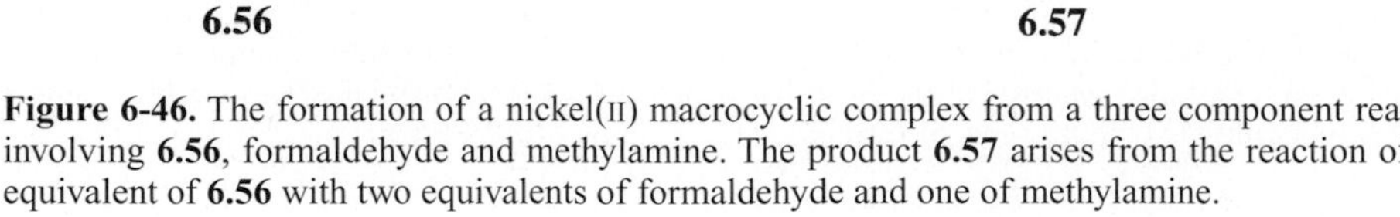

6.56   **6.57**

Figure 6-46. The formation of a nickel(II) macrocyclic complex from a three component reaction involving **6.56**, formaldehyde and methylamine. The product **6.57** arises from the reaction of one equivalent of **6.56** with two equivalents of formaldehyde and one of methylamine.

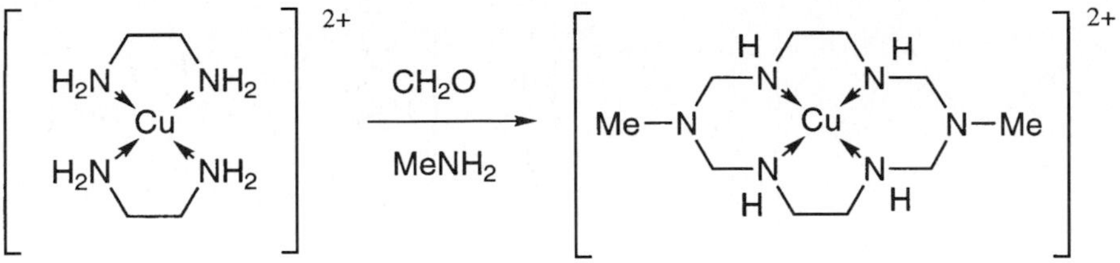

6.58

Figure 6-47. The formation of a copper(II) macrocyclic complex from a three component reaction involving [Cu(en)₂]²⁺, formaldehyde and methylamine The product **6.58** arises from the reaction of one equivalent of [Cu(en)₂]²⁺ with four equivalents of formaldehyde and two of methylamine

6.9 Into the Third Dimension

It has probably not escaped the readers notice that almost all of the macrocycles that we have discussed to date are topographically two-dimensional. This is not to say that the molecules are all rigidly planar, but that they can be represented on the page in two dimensions without the crossing of any bonds. Where three-dimensional structure is imposed by a metal ion, it has been limited to axial ligands not involved in the macrocyclic system. Even when we have used octahedral metal ions, such as in the condensation of [Ni(en)₃]²⁺ with acetone, the products contained a planar tetraaza macrocycle. We shall now begin to venture into the third dimension, by showing that it is possible to introduce three-dimensional order in the ligand utilising a *planar* metal centre, and will elucidate this idea in the next chapter by developing three-dimensional structure from the *metal* centre.

The reactions that we considered in Section 6.8 all gave planar macrocyclic systems. It is also possible to introduce a little more three dimensional structure into the macrocyclic ligands by the use of suitable structured diamines in these reactions. For example, the reaction of **6.59** with formaldehyde and ammonia in the presence of nickel(II) salts gives the complex **6.60**. Notice that the overall stoichiometry of the reaction involves one equi-

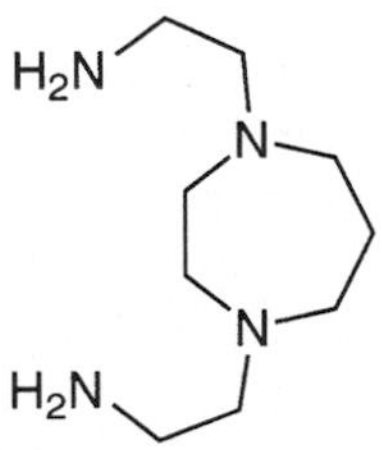

6.59

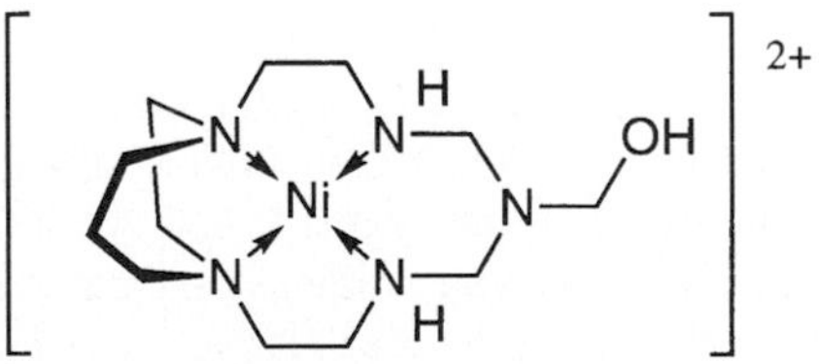

6.60

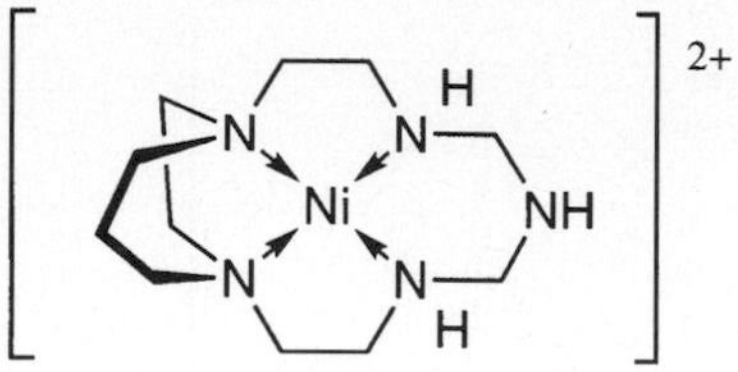

6.61

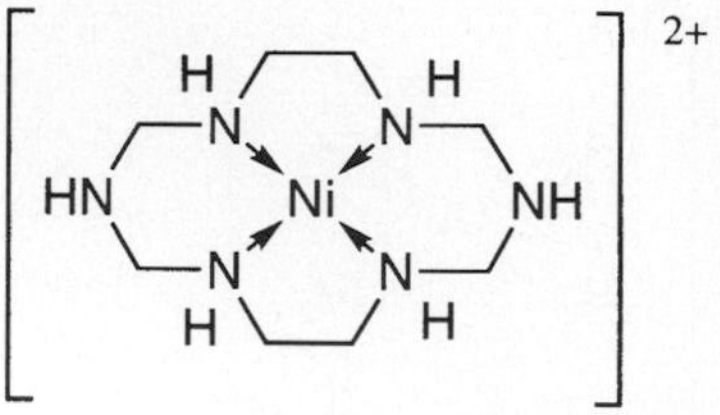

6.62

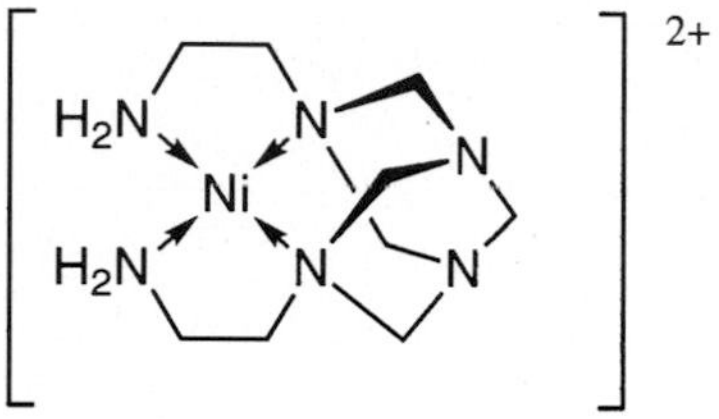

6.63

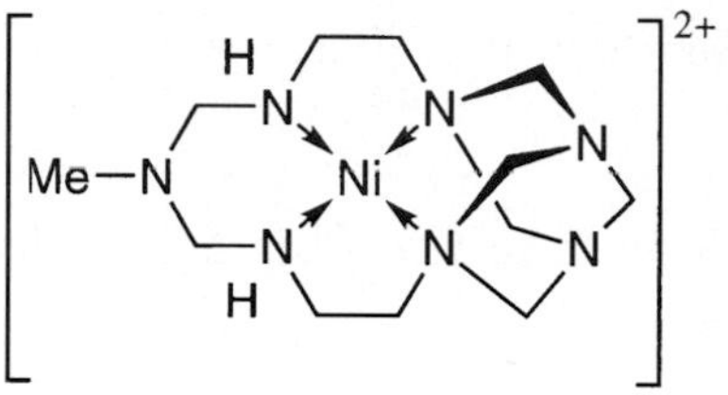

6.64

valent of **6.59**, one ammonia and three formaldehyde molecules. The third formaldehyde formally adds to the intermediate **6.61**, although the precise sequence of events in this reaction, like those discussed above, is not known with any certainty.

As a final example of the introduction of three-dimensional structure into the ligand, we consider the reaction of 1,2-diaminoethane with formaldehyde and ammonia in the presence of nickel(II). The optimistic researcher might expect (hope?) to obtain **6.62** from this reaction. The actual product is **6.63**. Only one side of each 1,2-diaminoethane ligand has reacted, and the overall stoichiometry is two 1,2-diaminoethane molecules, five formaldehyde molecules and two ammonia molecules. The reaction of **6.63** with formaldehyde and methylamine gives **6.64**!

6.10 Conclusions

In this chapter we have covered a great deal of material relating to the preparation of macrocyclic complexes. The basic reactions that we have introduced in earlier chapters have now found a synthetic use. At the very end of the chapter we began to ponder ways of introducing three dimensional structure into macrocyclic systems. This is the topic that we consider in the next chapter.

Suggestions for further reading

1. L.F. Lindoy, *The Chemistry of Macrocyclic Ligand Complexes*, Cambridge University Press, Cambridge, **1989**.
   – This is an easy-to-read text which provides a general introduction to the area of macrocyclic chemistry and has much to say about template reactions.
2. *Co-ordination Chemistry of Macrocyclic Compounds*, (ed. G.A. Melson),Plenum, New York, **1979**.
   – Slightly dated now, but in its time it was *the* standard reference book. Still worth a look at to see the diversity of macrocyclic ligands which can be prepared.
3. *Stereochemical and Stereophysical Behaviour of Macrocycles*, (ed. I. Bernal), Elsevier, Amsterdam, **1987**.
   – This book consists of a collection of articles discussing the conformational properties of macrocyclic ligands. It is an excellent starting point for a better understanding of this topic.
4. N.F. Curtis, *Coord. Chem. Rev.* **1968**, *3*, 3.
   – An early review by one of the pioneers of macrocyclic chemistry.
5. D.St.C. Black, A.J. Hartshorn, *Coord. Chem. Rev.* **1972–1973**, *9*, 219.
   – An interesting review on the design of macrocyclic ligands.
6. D.E. Fenton, *Pure Appl. Chem.* **1986**, *58*, 1437.
   – A good review of macrocyclic tetraimine macrocycles.
7. M. de Sousa Healy, A.J. Rest, *Adv. Inorg. Chem. Radiochem.* **1978**, *21*, 1.
   – A key review of metal templated reactions in macrocyclic chemistry. Although it is a little old, it contains a great deal of interesting material.
8. D.St.C. Black in *Comprehensive Co-ordination Chemistry*, *Vol. 6*, Chapt. 61.1, (eds. G. Wilkinson, R.D. Gillard, J.A. McCleverty), Pergamon, Oxford, **1987**.
   – This chapter contains much interesting material relating to the template preparation of macrocycles.
9. D.St.C. Black in *Comprehensive Co-ordination Chemistry*, *Vol. 1*, Chapt. 7.4, (eds. G. Wilkinson, R.D. Gillard, J.A. McCleverty), Pergamon, Oxford, **1987**.
   – This chapter also contains material relating to the template preparation of macrocycles.
10. T. Mashiko, D. Dolphin in *Comprehensive Co-ordination Chemistry*, *Vol. 2*, Chapt. 21.1, (eds. G. Wilkinson, R.D. Gillard, J.A. McCleverty), Pergamon, Oxford, **1987**.
    – This chapter describes porphyrins and related macrocycles and deals with a range of interesting metal-mediated syntheses and reactions.
11. N.F. Curtis in *Comprehensive Co-ordination Chemistry*, *Vol. 2*, Chapt. 21.2, (eds. G. Wilkinson, R.D. Gillard, J.A. McCleverty), Pergamon, Oxford, **1987**.
    – A review article about polyaza macrocycles including material relating to templated synthesis.
12. S.H. Laurie in *Comprehensive Co-ordination Chemistry*, *Vol. 2*, Chapt. 22, (eds. G. Wilkinson, R.D. Gillard, J.A. McCleverty), Pergamon, Oxford, **1987**.
    – Something a little different. An introduction to the world of macrocycles in life processes.

13. C.A. McAuliffe in *Comprehensive Co-ordination Chemistry*, *Vol. 2*, Chapt. 14, (eds. G. Wilkinson, R.D. Gillard, J.A. McCleverty), Pergamon, Oxford, **1987**.
   – A specialist review concerning macrocycles containing group 15 elements, but rich in template methodology.
14. D.H. Busch, *Chem. Rev.* **1993**, *93*, 847.
   – A personal view of the development of macrocyclic chemistry.
15. J.-M. Lehn, *Supramolecular Chemistry*, VCH, Weinheim, **1995**.
   – A state-of-the-art discussion of the journey from two-dimensional to three-dimensional chemistry
16. B. Dietrich, P. Viout, J.-M. Lehn, *Macrocyclic Chemistry*, VCH, Weinheim, **1993**.

7 The Three-Dimensional Template Effect, Supramolecular Chemistry and Molecular Topology

7.1 The Three-Dimensional Template Effect

In the previous chapter, we saw that it was possible to introduce three-dimensional structure into the macrocyclic ligands formed in template reactions, but the examples we considered had this structure incorporated into the ligand backbone. Typical examples of such structures were presented in **6.60** and **6.63**. However, the three-dimensional structure was not a direct consequence of the co-ordination geometry about the metal centre and we saw many examples of reactions involving six-co-ordinate nickel(II) or copper(II) starting materials that gave square-planar complexes of tetradentate macrocyclic ligands. Even if six-co-ordinate complexes were obtained, these consisted of tetradentate macrocyclic condensation products with two axial ligands. These results suggest that if we are to use the three-dimensional properties of metal templates, we should choose kinetically inert centres. We will take this as our starting point for the discussion of the three-dimensional template effect, although when we have developed an understanding of the basic principles, we will see that it is possible to utilise labile centres in appropriate cases.

7.2 Template Condensations at Kinetically Inert Octahedral Metal Centres

What should we do to observe a three-dimensional template effect? First, we should choose a reaction type that we know to be effective for the formation of macrocyclic ligands and extend the methodology to a kinetically inert d^3 or d^6 metal centre. Let us reconsider the reaction, that we first encountered in Fig. 6-11. In this reaction, a dioximato complex reacted with BF_3 to give the nickel(II) complex of a dianionic macrocycle (Fig. 7-1).

What would happen if we were to use kinetically inert cobalt(III) or iron(II) complexes instead of labile ones in these reactions? The reaction of tris(dimethylglyoximato)cobaltate(III) with $BF_3 \cdot Et_2O$ gives the cobalt(III) complex cation **7.1**, in which a three-dimensional ligand has been formed. The geometry about the metal centre is closest to trigonal prismatic. The three-dimensional ligand is formed as a consequence of the arrangement of the six reactive sites about the kinetically inert cobalt(III) centre (Fig. 7-2). The new ligand has an overall charge of -2. Exactly similar reactions occur with iron(II) dioximato complexes.

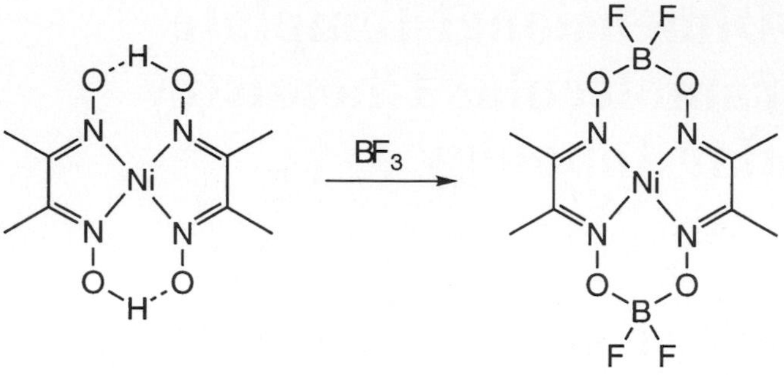

Figure 7-1. The reaction of a nickel(II) dioximato complex with BF₃ gives the nickel(II) complex of a new dianionic, macrocyclic ligand. Note that in the interests of clarity, throughout this chapter we use simple lines to represent co-ordinate bonds

In the formation of **7.1** we have utilised the three-dimensional properties of a kinetically inert metal centre to form a total of six new B–O bonds. The reaction involves essentially three components (the dioximato complex and two molecules of BF₃). Ligands such as **7.1** are known variously as encapsulating ligands, cages or cryptands. These encapsulating ligands bear the same relationship to macrocyclic ligands that the macrocycles bore to chelating polydentate ligands. The design of such encapsulating ligands has provided an interesting synthetic challenge to the preparative chemist. The important feature of a three-dimensional ligand is the possession of a cavity that can accommodate a metal centre. The ligand that we have seen initially is designed for a six-co-ordinate metal centre. Other ligand systems may be devised to incorporate a variety of other co-ordination geometries and numbers. At this point we must make a brief aside regarding topology in chemistry.

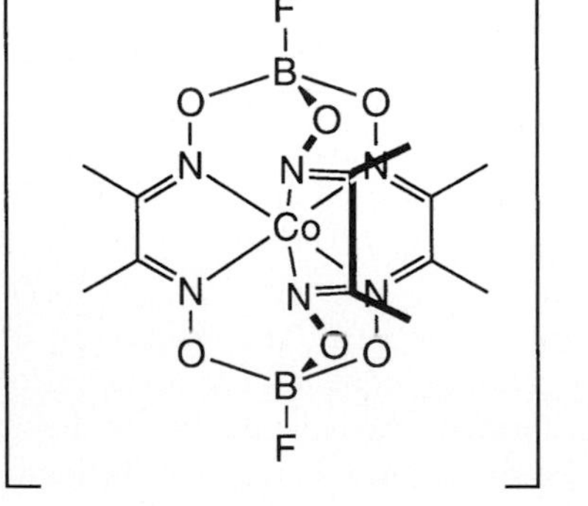

7.1

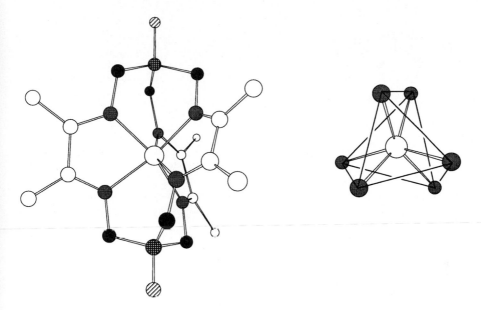

Figure 7-2. A representation of the cobalt(III) complex **7.1** formed from the reaction of tris(di-methylglyoximato)cobaltate(III) with $BF_3 \cdot Et_2O$. The three-dimensional structure observed in the crystal lattice of the tetrafluoroborate salt is also shown. The final view emphasises that the co-ordination geometry about the metal centre is distorted towards trigonal prismatic in these complexes.

7.2.1 Molecular Topology

Molecular topology is concerned with the invariant structural properties of a molecule. For the moment we will forget that molecules possess three-dimensional structure, and concern ourselves only with the bonding network within a molecule. We are allowed to make any changes to the shape of a molecule that does not involve the breaking of a bond. If as a result of any of this "pushing and pulling" of the molecule we can represent it unambiguously on a sheet of paper with *no bond crossings*, the molecule is topologically planar. Under this scheme, benzene, cyclohexane and bicyclo[2.2.2]octane are all topo-logically planar (Fig. 7-3) as they can be represented in two-dimensions with no bond crossings.

Figure 7-3. Some molecules that are topologically planar. Although cyclohexane and bicy-clo[2.2.2]octane both have defined three-dimensional structures, they may be represented in two dimensions with no bond crossings, and are hence topologically planar.

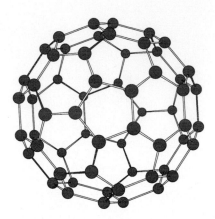 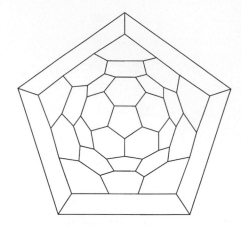

Figure 7-4. Two representations of the fullerene C_{60}. The upper representation shows the "normal" three-dimensional structure, whilst the lower one emphasises that this molecule is topologically planar. The fact that the various rings appear deformed is immaterial in this topological representation.

Even a molecule with as defined a three-dimensional structure as the fullerene C_{60} is topologically planar (Fig. 7-4). This type of representation is known as a *Schlegel diagram*.

Now let us consider the new ligand in complex **7.1**. Although this ligand possesses a defined three-dimensional structure, it is *topologically planar*, as shown in Fig. 7-5. But what about the complex **7.1** itself? However much we push or pull the bonds, there is no way in which we can draw this molecule without at least two bond crossings.

The importance of these topological concepts will become apparent as we progress through this chapter. However, for the moment, we will merely note that co-ordination to the metal centre results in the formation of a topologically non-planar complex **7.1** from a topologically planar ligand.

Figure 7-5. The ligand from complex **7.1** drawn to show that it is topologically planar. It is possible to draw the ligand in such a way that there are no bond-crossings.

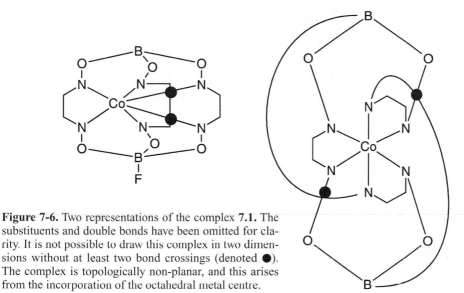

Figure 7-6. Two representations of the complex **7.1**. The substituents and double bonds have been omitted for clarity. It is not possible to draw this complex in two dimensions without at least two bond crossings (denoted ●). The complex is topologically non-planar, and this arises from the incorporation of the octahedral metal centre.

7.2.2 More Encapsulating Ligands from Two Reactants

Let us now consider another cyclisation that involves only two different types of reactant. The reaction of dimethylglyoximedihydrazone with formaldehyde in the presence of a nickel(II) template yields the complex **7.2**. What happens if we react the iron(II) complex of dimethylglyoximedihydrazone with formaldehyde. The reaction proceeds in an analogous manner to that giving **7.2** – the product is **7.3**. The ligand in **7.3** is derived from the condensation of three molecules of the dimethylglyoximedihydrazone with six molecules of formaldehyde. Notice that, once again, the free ligand is topologically planar but the iron complex must be drawn with at least two bond crossings, and is thus topologically non-planar (Fig. 7-7).

7.2

7.3

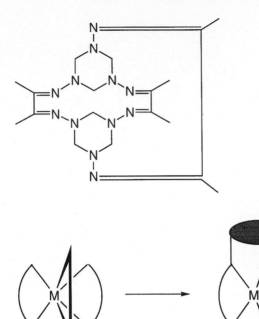

Figure 7-7. A representation of the ligand in complex **7.3** emphasising its topological planarity. It is not possible to draw the metal complex in two dimensions without at least two bond crossings, and the complex is topologically non-planar as a result of the incorporation of the six-co-ordinate metal centre.

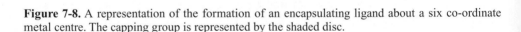

Figure 7-8. A representation of the formation of an encapsulating ligand about a six co-ordinate metal centre. The capping group is represented by the shaded disc.

Although we approached this section in terms of kinetically inert metal centres, it is also possible to build such encapsulating ligands about relatively labile metal ions. For example, although the square-planar nickel(II) complex **7.2** may be formed, this can react with an excess of dimethylglyoximedihydrazone and formaldehyde to give the nickel(II) analogue of **7.3**.

A cartoon representation of these capping reactions is presented in Fig. 7-8, and emphasises the formation of the capping groups above and below the metal centre. We can now extend the methodology to reactions involving more than two components, in the same way that we developed structural complexity in Chapter 6.

7.2.3 Encapsulating Ligands from More Than Two Components

The reactions discussed above are akin to those we considered in the early part of Chapter 6 and involve only two different types of reactant. In the final parts of the previous chapter, we illustrated the power of metal ions in building macrocyclic ligands from a variety of components in programmed, stepwise reactions. Can we apply the same principles to the assembly of encapsulating ligands?

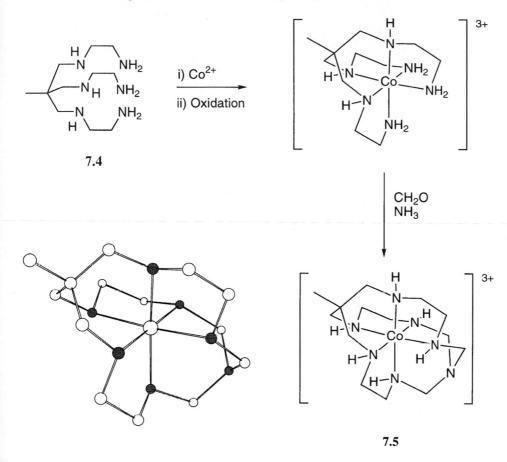

7.4

7.5

Figure 7-9. The condensation of the cobalt(III) complex of the hexadentate ligand **7.4**, which contains three primary amino groups, with formaldehyde and ammonia, gives the encapsulated complex **7.5**. A view of the cation **7.5** as found in the solid state structure of its perchlorate salt is also presented.

The answer is that we can. Consider the reaction that we introduced in Fig. 6-44, in which a nickel(II) complex with two terminal amino NH_2 groups reacted with formaldehyde and amine to give a macrocycle. The key steps involved the formation of imines (or the equivalent aminol) followed by nucleophilic attack by the amine. To make an encapsulating ligand by this methodology, we would need to have three terminal amino groups which could formally undergo reaction with formaldehyde to give the imine intermediates. This is the case in the cobalt(III) complex of the hexadentate ligand **7.4**, which has three terminal amino groups. The reaction of this cobalt(III) complex with a mixture of formaldehyde and ammonia yields the cobalt(III) complex **7.5**, which contains the encapsulating ligand **7.6** (Fig. 7-9). The reaction probably proceeds in a similar manner to those discussed in Chapter 6. Overall, the new ligand **7.6** is formed from one molecule of **7.4**, one ammonia and three formaldehyde molecules.

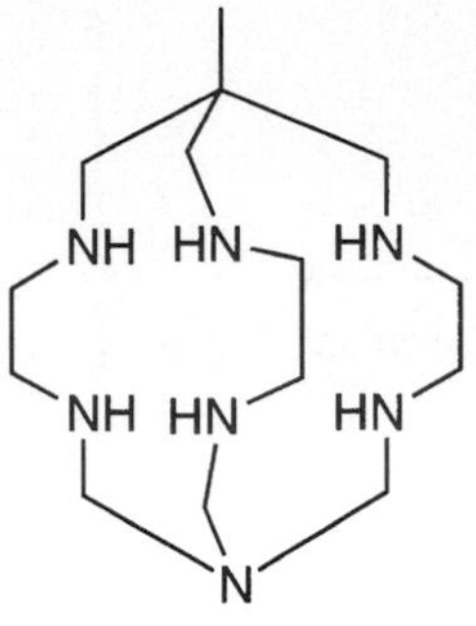

7.6

A simple structural change in the starting hexadentate ligand allows the preparation of complexes of encapsulating ligands, such as **7.7**. Complexes of the ligands **7.6** and **7.7** are prepared by the reaction of appropriate kinetically inert complexes with ammonia and formaldehyde. It is convenient to think of the ammonia and the three formaldehyde molecules as generating a 'cap' for the molecule. In this case, the capping atom is the nitrogen atom. When we considered the analogous macrocyclic reactions in Chapter 6, we found that it was possible to replace the ammonia by a range of RNH_2 species to give structural variety. This is not possible with the encapsulating ligands, since the three valence orbitals of the nitrogen are involved in bonding within the cage. However, it is possible to replace the ammonia by carbon nucleophiles, exactly as we did for the macrocyclic systems.

As an example, consider the reaction of the cobalt(III) complex of **7.8** with formaldehyde and nitromethane to give **7.9**, the cobalt(III) complex of ligand **7.10** (Fig. 7-10). Notice that **7.8** is the same precursor that we needed for the synthesis of **7.7**. In the new ligand, **7.10**, the cap is formed from three molecules of formaldehyde and one of nitromethane, and the capping atom is now the carbon of the CNO_2 group derived from the nitromethane.

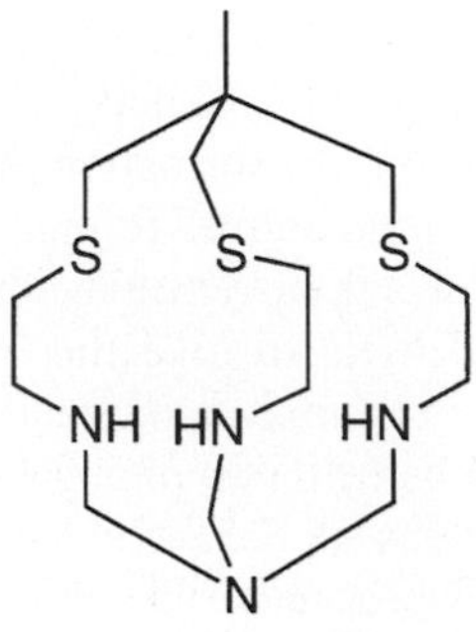

7.7

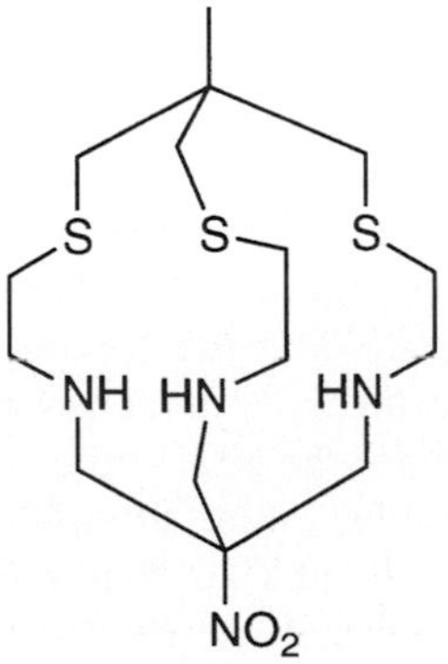

7.10

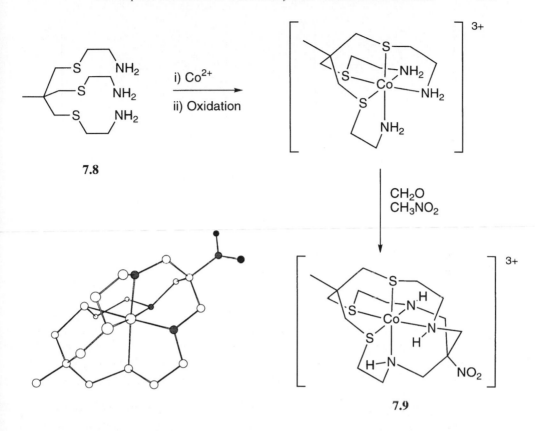

Figure 7-10. The condensation of the cobalt(III) complex of **7.8** with formaldehyde and nitromethane gives the encapsulated complex **7.9**. A view of the cation in the solid state is also presented.

It is also possible to make rather more dramatic changes in the structure of the pre-
formed capping group, and **7.11**, the cobalt(II) complex of a functionalised triaza
macrocyclic ligand, reacts with formaldehyde and nitromethane to give **7.12**.

A cartoon version of the methodology we have developed is shown in Fig. 7-11. However, it is completely reasonable to ask whether it is possible to use multi-component reactions to assemble top and bottom capping groups, as indicated in Fig. 7-8. In fact, the
genesis of this type of reactivity is found in Chapter 6; consider, for instance, the relationship between Figs. 6-46 and 6-47.

Perhaps the best examples of the assembly of an encapsulating ligand in which top and
bottom caps are built from multiple components come from the studies of Sargeson. As a
typical example, consider the reaction of $[Co(en)_3]^{3+}$ with formaldehyde and ammonia in
the presence of a base to give complex **7.13** (Fig. 7-12). This reaction proceeds in remarkably good yield to give the desired encapsulated complex. These ligands are regarded as
crypts (or tombs) in which the metal ion is placed, and Sargeson emphasises this in the
necrophilic nomenclature which he has adopted. Thus, complex **7.13** is known as
cobalt(III) sepulchrate, and the parent ligand **7.14** as sepulchrate. The top and the bottom

7.11   **7.12**

caps have been generated in this reaction, which involves one cobalt(III) centre, three 1,2-diaminoethane ligands, six formaldehyde and two ammonia molecules. It is remarkable that the assembly of all of these components leads to a single product in good to excellent yield. We will have more to say about sepulchrates and the mechanistic aspects of their formation in Section 7.2.4. At the moment we will concern ourselves with other developments of this methodology.

Unfortunately, not many other metal ions are effective as templates for the formation of sepulchrates, although other kinetically inert centres such as rhodium(III) have been used with varying degrees of success. The problem is associated with the effectiveness of the ligand in encapsulating the metal centre. A space-filling representation of the structure of cobalt(III) sepulchrate is presented in Fig. 7-13 and clearly reveals that the metal ion is "trapped" within the ligand cage.

As a consequence, there is a very significant energy barrier towards removal of the metal from the complex, although prolonged reaction with concentrated cyanide is sometimes successful. This kinetic stability of the complex leads to some remarkable properties

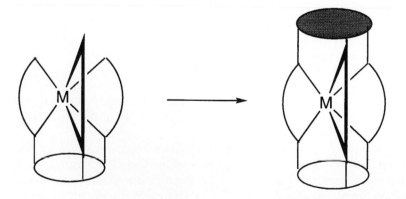

Figure 7-11. A cartoon representation of the multi-component assembly of an encapsulating ligand. The white disc represents a pre-formed cap within the starting ligand and the shaded disc the new cap that has been built from a number of components. For example, in the ligand **7.9**, the shaded cap is built from three formaldehyde molecules and a nitromethane molecule.

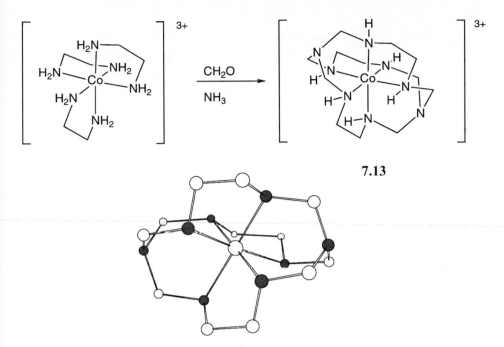

7.13

Figure 7-12. The reaction of [Co(en)$_3$]$^{3+}$ with formaldehyde and ammonia gives cobalt(III) sepulchrate **7.13**. A representation of the complex ion in the solid state is also presented.

of sepulchrates and related ligands, which the interested reader will find discussed in the references cited in the suggestions for further reading found at the end of this chapter. Just as there is a large energy barrier associated with the loss of the metal, so there is a barrier towards incorporation of a different metal, and transmetallation methods are not particularly rapid with these ligands.

Perhaps the obvious extension of the methodology is to replace the ammonia nucleophile with the carbon nucleophile nitromethane. The reaction of [Co(en)$_3$]$^{3+}$ with formaldehyde and nitromethane in the presence of a base gives complex **7.15** (Fig. 7-14). Con-

7.14

Figure 7-13. A space-filling representation of the cobalt(III) sepulchrate cation showing that the cobalt centre is "buried" deep within the ligand. The cobalt and nitrogen atoms are shaded. The ligand is oriented such that the two capping nitrogen atoms lie along the *x* axis.

tinuing with the funereal nomenclature, the basic cage system with the carbon caps is known as sarcophagine, and **7.15** is cobalt(III) dinitrosarcophagine.

The remarkable kinetic stability of the sepulchrate complexes has been fully exploited by Sargeson and his co-workers, and a chemistry reminiscent of that of benzenoid aromatics has been developed. The reduction of cobalt(III) dinitrosarcophagine (diNOsar) **7.16** by zinc and hydrochloric acid yields a cobalt complex of diaminosarcophagine, **7.17** (diamsar). The initial product is actually a cobalt(II) complex of **7.17**, but oxidation with hydrogen peroxide generates the cobalt(III) species. Small amounts of cobalt(III) sarcophagine (sarcophagine = **7.18**) are also obtained in this reaction. Finally, cobalt(III) di-

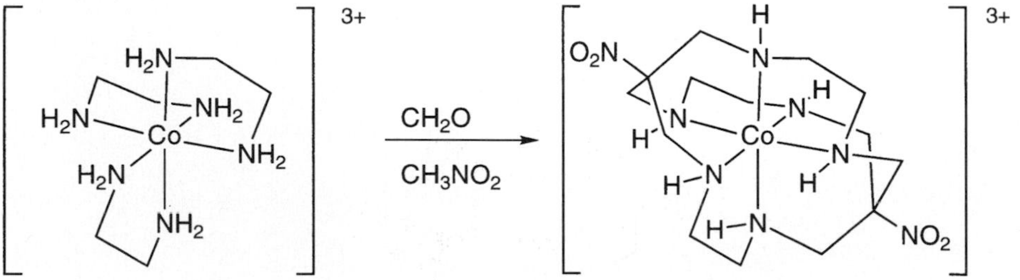

$$\left[\begin{array}{c} H_2N-\overset{\displaystyle H_2N}{\underset{\displaystyle H_2N}{}}\overset{\displaystyle |}{\underset{\displaystyle |}{Co}}\overset{\displaystyle -NH_2}{\underset{\displaystyle NH_2}{}} \end{array}\right]^{3+} \quad \xrightarrow[\text{CH}_3\text{NO}_2]{\text{CH}_2\text{O}} \quad \left[\begin{array}{c} O_2N \cdots \overset{\displaystyle H}{\underset{\displaystyle N}{}} \cdots Co \cdots NO_2 \end{array}\right]^{3+}$$

7.15

Figure 7-14. The reaction of [Co(en)₃]³⁺ with formaldehyde and nitromethane gives cobalt(III) dinitrosarcophagine **7.15**.

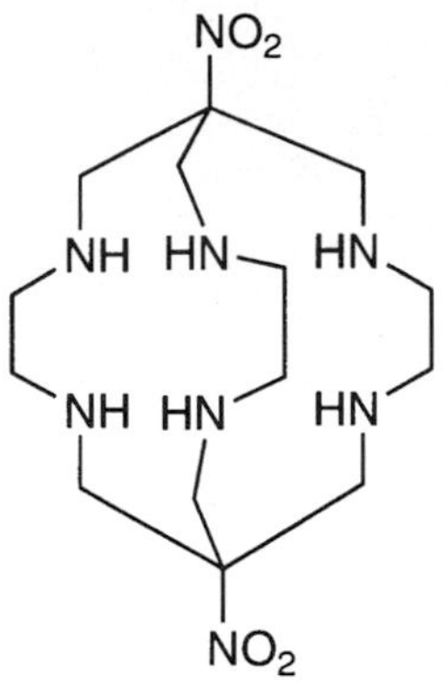

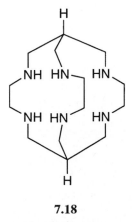

7.16

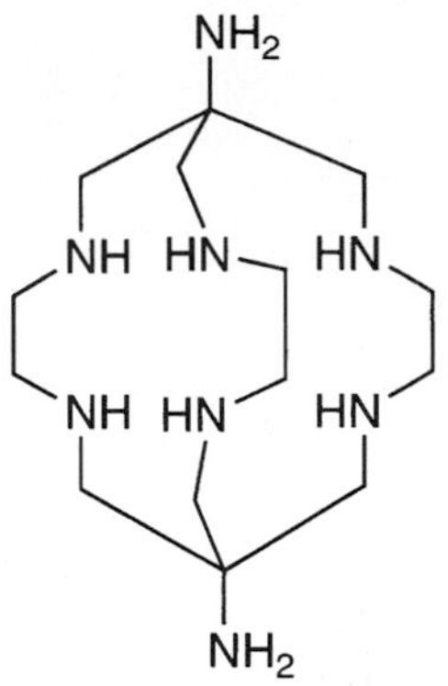

7.17

amsar undergoes a Sandmeyer-type reaction with sodium nitrite and hydrochloric acid to give dichlorsar, **7.19**. These reactions are summarised in Fig. 7-15

The reactions that we have considered have involved building the encapsulating ligand from a six-co-ordinate metal centre with one hexadentate ligand or three didentate ligands. Naturally, other combinations are possible. For example, it should be possible to build an encapsulating ligand about a six-co-ordinate metal centre with two tridentate ligands. However, in such cases the reactions are not always quite as simple as those that we have discussed above. As an example, consider the reaction of **7.20** with formaldehyde and nitromethane, which might be expected to yield **7.21**. In practice, a mixture of products, including **7.22**, **7.23**, **7.24** and **7.25**, is formed.

7.18

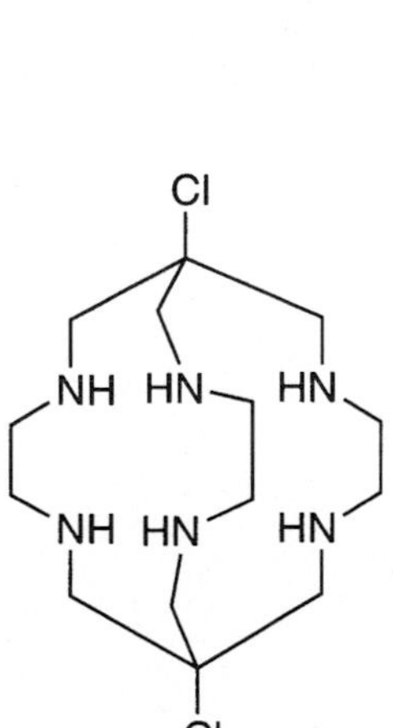

7.19

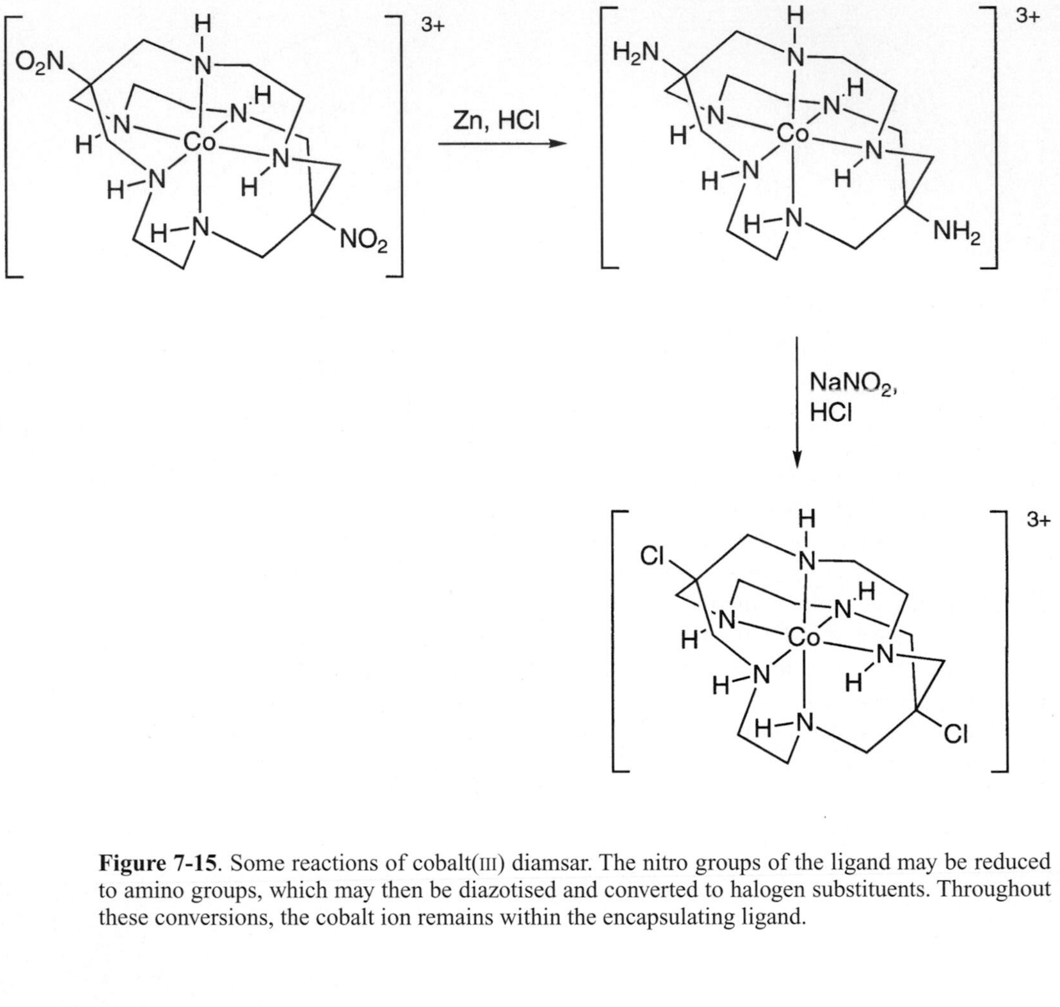

Figure 7-15. Some reactions of cobalt(III) diamsar. The nitro groups of the ligand may be reduced to amino groups, which may then be diazotised and converted to halogen substituents. Throughout these conversions, the cobalt ion remains within the encapsulating ligand.

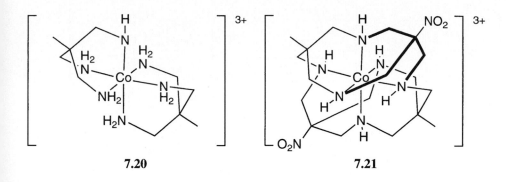

7.20

7.21

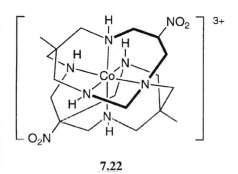

7.22

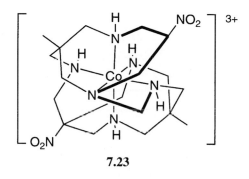

7.23

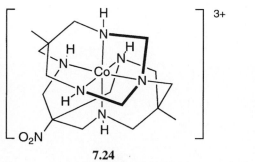

7.24

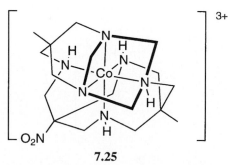

7.25

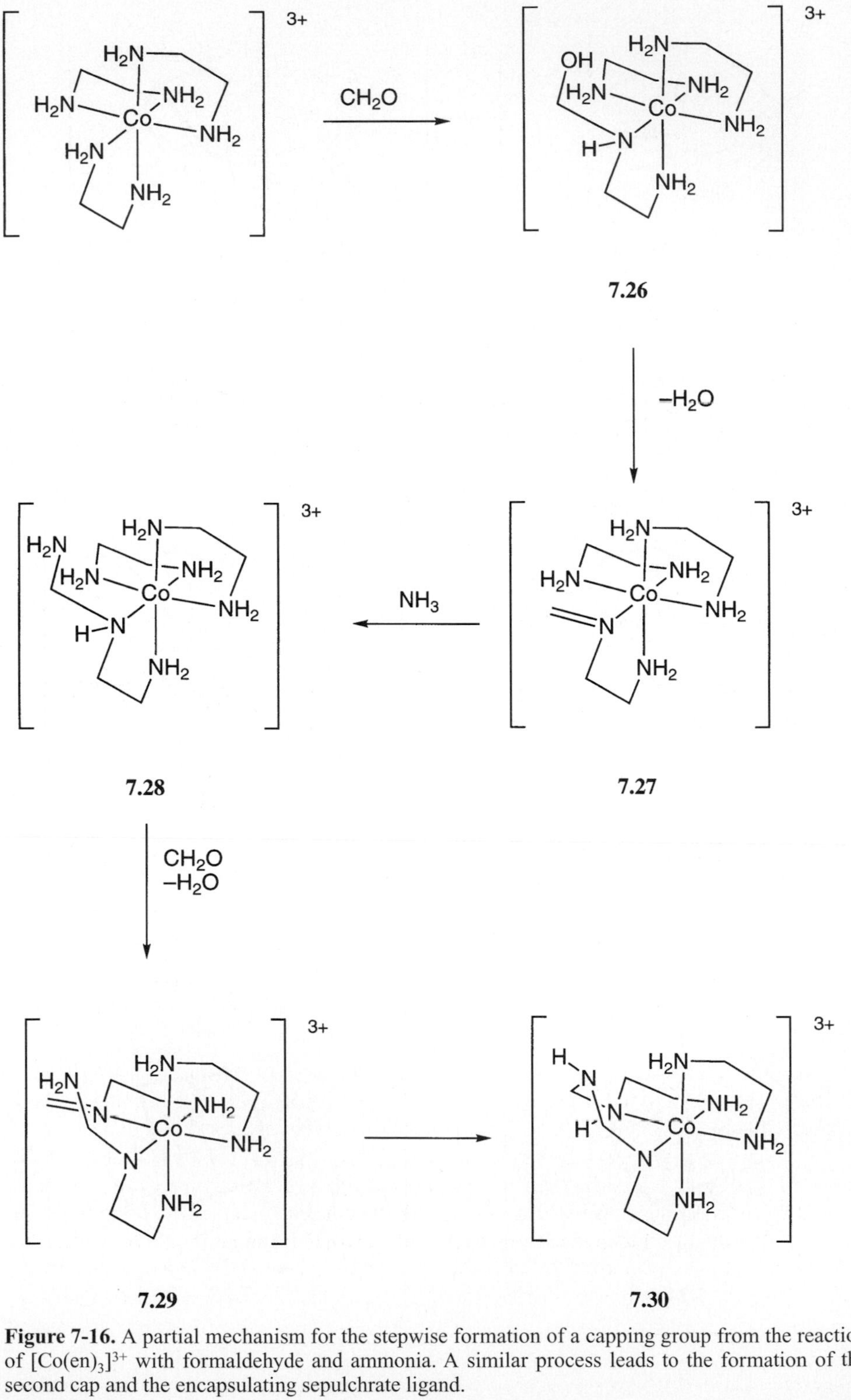

Figure 7-16. A partial mechanism for the stepwise formation of a capping group from the reaction of [Co(en)₃]³⁺ with formaldehyde and ammonia. A similar process leads to the formation of the second cap and the encapsulating sepulchrate ligand.

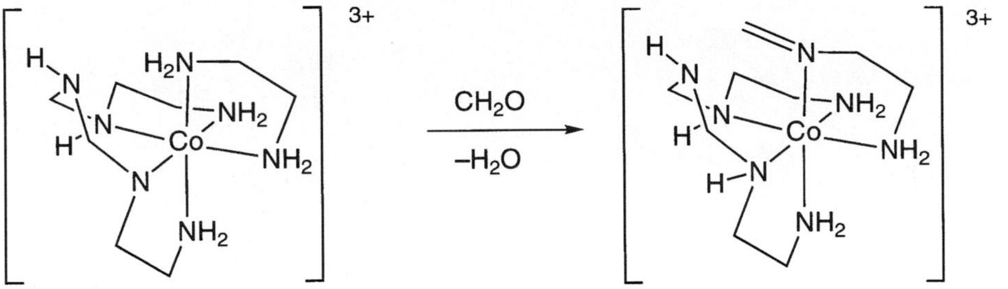

7.30

CH_2O

$-H_2O$

7.31

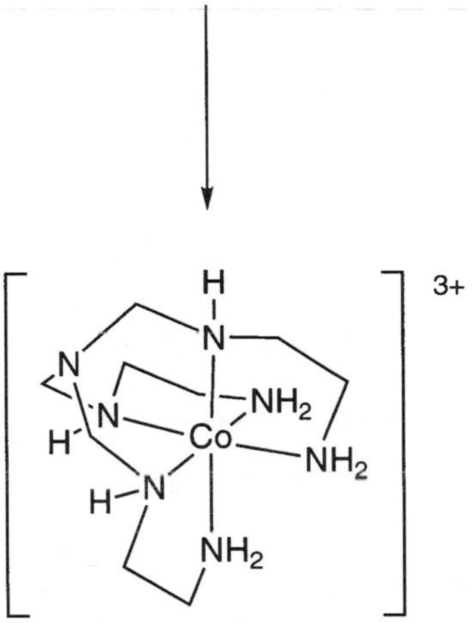

7.32

Figure 7-16. Continued.

7.2.4 Some Mechanistic and Stereochemical Aspects of Sepulchrates and Related Encapsulating Ligands

The reaction of $[Co(en)_3]^{3+}$ with formaldehyde and ammonia is remarkable in that it gives the desired sepulchrate complex in good yield. The building of the new ligand is programmed in a step-wise manner. Sargeson has made detailed mechanistic studies of these reactions, and although the scheme presented here may not be precise in all of its details, the general sequence is correct. Of the starting materials, ammonia or $[Co(en)_3]^{3+}$ could react with formaldehyde. We will first consider the consequences of the $[Co(en)_3]^{3+}$ ion reacting with the formaldehyde. The initial product would be an aminol **7.26**, which then dehydrates to give an imine **7.27** (Fig. 7.16). This co-ordinated imine can now be attacked by an ammonia molecule to give the aminal **7.28**. This is the reason that we can ignore the

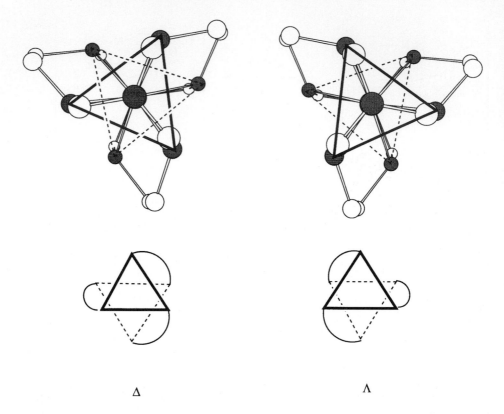

Δ   Λ

Figure 7-17. Views of the two enantiomeric Δ and Λ forms of the cobalt(III) sepulchrate cation. The ligand has been oriented such that the capping nitrogen atoms lie along the axis perpendicular to the page. The faces described by the two sets of nitrogen donor atoms in each case have been emphasised. The bold triangle is the face closest to the viewer and the dotted triangle is that furthest away.

alternative possibility of initial attack of ammonia upon the formaldehyde. The reaction of $[Co(en)_3]^{3+}$ with the imine, $HN=CH_2$ formed from the dehydration of the initially formed H_2NCH_2OH would give the same intermediate **7.28**. The next step probably involves the formation of a second imine from one of the 1,2-diaminoethane ligands. The intermediate **7.29** has the imine correctly oriented for intramolecular attack to give the secondary amine **7.30**. Finally, the first complete cap is formed by the condensation of a third molecule of formaldehyde to give the imine **7.31**, which undergoes an intramolecular attack by the capping nitrogen to give the "half-way" product, **7.32**. A similar sequence of events occurs at the three other primary amino groups to generate the second cap.

A second consequence of this sequential mechanism is even more surprising. The geometry about the metal centre in a sepulchrate is close to octahedral, and this is emphasised in Fig. 7-17. We discussed some of the stereochemical properties of complexes with three didentate chelating ligands in Chapter 2, and in the same way that the $[Co(en)_3]^{3+}$ cation may exist in the two enantiomeric Δ and Λ forms, so may the sepulchrate complex. These two enantiomers are shown in Fig. 7-17. The reaction of $[Co(en)_3]^{3+}$ with formaldehyde

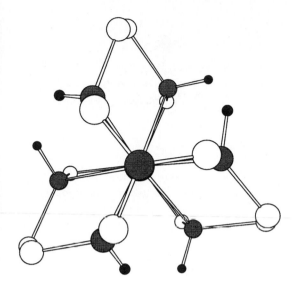

Figure 7-18. A view of the actual diastereomer of Δ-[Co(sep)]³⁺ which is formed from the reaction of Δ-[Co(en)₃]³⁺ with formaldehyde and ammonia. The methylene hydrogen atoms have been omitted for clarity, and the amine nitrogen atoms are shown in black. The cation is oriented in the same way as the Δ-enantiomer in Fig. 7-17

and ammonia is stereo-retentive; Δ-[Co(en)₃]³⁺ gives Δ-[Co(sep)]³⁺ and Λ-[Co(en)₃]³⁺ gives Λ-[Co(sep)]³⁺.

However, in Chapter 2 we saw that individual chelate rings could have δ or λ conformations, depending upon the relative stereochemistry at each of the nitrogen donor atoms. In the case of the sepulchrate, each of the six nitrogen donor atoms co-ordinated to the cobalt(III) could possess an *S* or an *R* absolute configuration. The cobalt(III) centre is non-labile, so these nitrogen atoms are unlikely to invert (racemise) rapidly. As a consequence, a large number of diastereomers are possible for each of the Δ and Λ configurations. In practice, the reaction of Δ-[Co(en)₃]³⁺ gives a single diastereomer of Δ-[Co(sep)]³⁺ and that of Λ-[Co(en)₃]³⁺ gives a single diastereomer of Λ-[Co(sep)]³⁺. This impressive stereocontrol is a result of the specific orientations of the various imine and other intermediates involved in the formation of the cap. Although we will not discuss the detailed origins in this book, Fig. 7-18 shows the configuration of the diastereomer of Δ-[Co(sep)]³⁺ that is actually formed.

7.3 Supramolecular Chemistry

In the past ten years, a new discipline known as supramolecular chemistry has come to the fore. Metal-directed chemistry has proved to be very important in supramolecular chemistry and for the remainder of this chapter we will enter this interesting and novel world.

7.3.1 What is Supramolecular Chemistry?

In many respects, the above question is like asking "How long is a piece of string?" – the answer will depend upon the precise circumstances and who is asked. It is probably true that most workers actively involved in this field have their own 'private' definition. The rapid development of this area over the past decade or so is due to the endeavours of Jean-Marie Lehn more than any other, and it seems appropriate to commence with his own definition.

"Supramolecular chemistry is the chemistry of the intermolecular bond, covering the structures and functions of the entities formed by the association of two or more chemical species ...."

This association of different chemical species allows the construction of highly organised molecular architectures from simple molecular components. The chemical species involved may be drawn from any of the traditional entities of inorganic, organic or biological chemistry, and the resultant molecular assemblies are studied by scientists from all disciplines ranging from chemistry, through biochemistry to engineering and physics. In the same way that an infinite variety of molecular components for supramolecular systems may be envisaged, so may a range of intermolecular bonding processes be considered. Many different types of interaction have been investigated. These include electrostatic and ion-pairing interactions between oppositely charged ionic species, hydrophobic or hydrophilic association of appropriate functional groups, hydrogen-bonding between complementary substituents, host–guest interactions (often combined with other intermolecular interactions beyond the simple "fit" of the guest in the cavity of the host), π-stacking between aromatic rings and other donor acceptor interactions between Lewis bases and Lewis acids. Examples of some of these types of interactions are indicated in Fig. 7-19.

It is important that we are able to recognise what is and what is not a supramolecular system. The critical distinction is between a large molecule and a supramolecule. A large molecule may not be broken down into simpler molecular components without the cleavage of covalent bonds. In contrast, a supramolecule may be broken down, in principle at least, into a number of simpler molecular components, each of which is capable of an independent existence (Fig. 7-20). We see, for instance, that the fullerenes C_{60} (**7.36**) and C_{70} (**7.37**) are merely large molecules. Similarly, polymers such as polyethylene glycol (**7.38**) or polystyrene are large molecules. In contrast, the adduct formed between a calixarene and acetonitrile (**7.40**) *is* a supramolecule, as is a catenane in which two cyclic molecules are inter-linked, but with no covalent bonds between them (**7.39**). Note that the two inter-linked rings in a catenane are each capable of indepen-

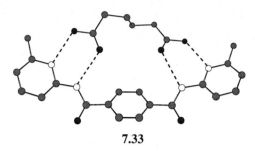

7.33

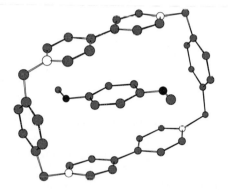

7.34

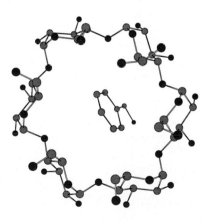

7.35

Figure 7-19. Three examples of supramolecules formed from the assembly of two molecular species. In **7.33**, the 1,4-butanedioic acid is specifically hydrogen bonded to the bisamide. In **7.34**, π-stacking interactions between the cationic cyclophane and the host 1,4-dimethoxybenzene stabilise the structure, whilst in **7.35** the cavity of the α-cyclodextrin host is the correct size for the benzaldehyde guest. In each case, hydrogen atoms have been omitted for clarity; carbon atoms are grey, nitrogen atoms white and oxygen atoms black.

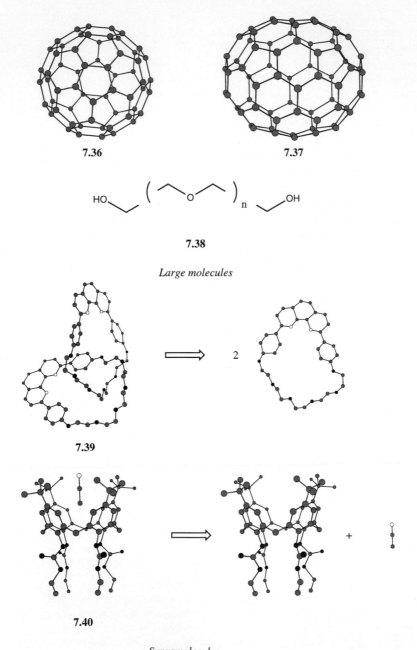

7.36   7.37

Large molecules

7.39

7.40

Supramolecules

Figure 7-20. Examples of large molecules and supramolecules showing how a supramolecule may, in principle, be split into two or more molecular components without the breaking of covalent bonds. In each case, hydrogen atoms have been omitted for clarity; carbon atoms are grey, nitrogen atoms white and oxygen atoms black.

dent existence, although in practice they may only be separated by the breaking of a covalent bond in one of the rings. The situation is perhaps more ambiguous when we consider large bio-molecules such as proteins or polypeptides. The primary sequence of a protein defines a large molecule – to obtain the component amino acids it is necessary to cleave covalent C-N bonds. However, the various interactions which define the secondary, tertiary and quaternary structures of proteins may be covalent or non-covalent in origin.

As no new covalent bonds are formed in the assembly of the supramolecule, the individual molecular components are expected to retain, essentially unchanged, their own molecular character and properties. However, the proximity and the spatial arrangement of the molecular components of the supramolecule may be such that additional interactions between them are optimised, promoted or even initiated. Herein lies one of the real promises of supramolecular chemistry – it allows the precise control of intermolecular processes and reactions by removing the usual requirements for the molecular reactants to form contact pairs with the correct mutual spatial orientation of functionality. In effect, the supramolecule encapsulates that contact pair.

From the above discussions it is apparent that there are two themes running through supramolecular chemistry – those of *structure* and of *function*. It is also apparent that different workers place different emphases upon these two factors, which in turn influence the ways of thinking about supramolecular chemistry (and ultimately its definition). In this book we will be concerned with supramolecular systems which are assembled by the interactions of ligands with metal centres. The metal–ligand bonding represents the non-covalent interaction in the definition of Lehn. Whilst one can argue about the description of metal–ligand bonding as non-covalent, it is undeniable that such interactions provide a powerful and novel tool in supramolecular chemistry. Before we consider this chemistry in action, we should discuss a fundamental concept of supramolecular chemistry — *molecular recognition*.

7.3.2 Molecular Recognition, Complementarity and Self-Assembly

The fundamental process in the formation of a supramolecule is the coming together of the molecular components as the intermolecular bonding develops. If this process is to be of real application, then the assembly must be *selective*. In anthropomorphic terms, the molecules must 'recognise' one another. *Molecular recognition* is an intrinsic feature of supramolecular and biological chemistry. Consider the process represented in cartoon form in Figure 7-21, in which two different types of molecule contain some feature which allows a specific interaction to form a new supramolecule. These interactions may be any of the types discussed in the previous section. The two molecular components are said to be *complementary*.

The importance of complementarity in the assembly of supramolecules is seen in cartoon form in Figure 7-22, where a third type of molecule has been added to the system. This third species does not contain features complementary to either of the other two components, and so only the desired supramolecule is formed. In principle, the presence of the third molecule has no effect upon the assembly of the supramolecule from the other two types of molecule.

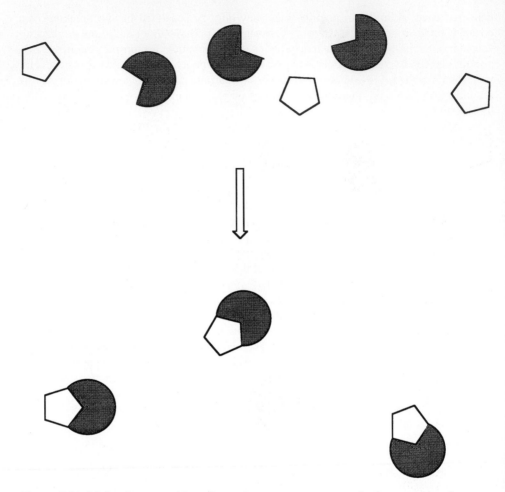

Figure 7-21. Molecular recognition of complementary components in the assembly of a supra-molecule.

In simple systems, electrostatic interactions are not usually sufficiently selective to allow highly specific complementary molecules to be designed. In general, any cationic species will tend to form an ion pair with an anionic species. If only singly-charged sites are involved, the recognition process is likely to be highly inefficient and unselective in the absence of any other intermolecular interactions. When two or more charged sites are introduced, a degree of selectivity may be observed (Fig. 7-23) and this is, in part, responsible for the binding of ionic species to protein surfaces.

Some of the simplest selective systems are based on hydrogen bonding interactions between molecules. Carboxylic acids hydrogen bond to form dimers (Fig. 7-24), although this process is relatively unselective in the absence of other interactions. The interaction is described as *self-complementary* – the carboxylic acid functionality contains both the carbonyl oxygen atom as a donor and the hydrogen atom as the acceptor.

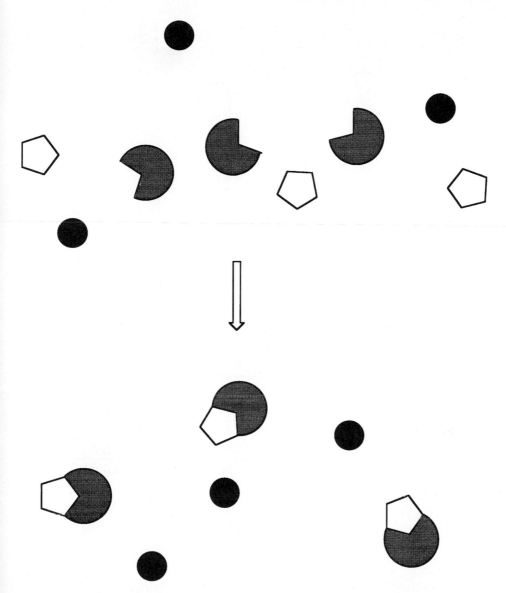

Figure 7-22. The assembly of a supramolecule from the same two complementary components shown in Figure 7-21, but in the presence of a third, non-complementary, species.

If two or more such carboxylic acid groups are present in a molecule, intricate solid-state structures may be assembled, as seen in the ribbon-like solid state structure of terephthalic acid (**7.41**).

With carboxylic acids, the simple dimeric species are reasonably persistent in solution (and the gas phase), but the more extended structures are generally limited to the solid state. Hydrogen bonding between carbonyl oxygen atoms and amino N–H groups has

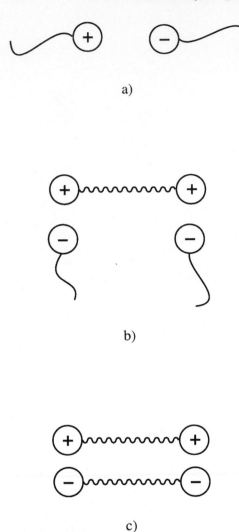

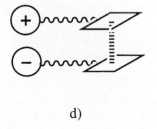

Figure 7-23. Supramolecules formed from charged molecular components: a) Unselective ion pairing of singly charged species, b) relatively unselective interactions between a polycation and monoanions, c) selective interactions between a complementary polycation and a polyanion, and d) highly selective interaction between two ions with another complementary intermolecular interaction.

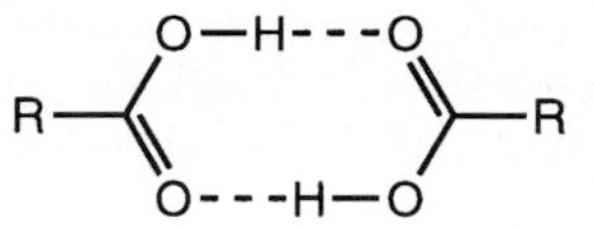

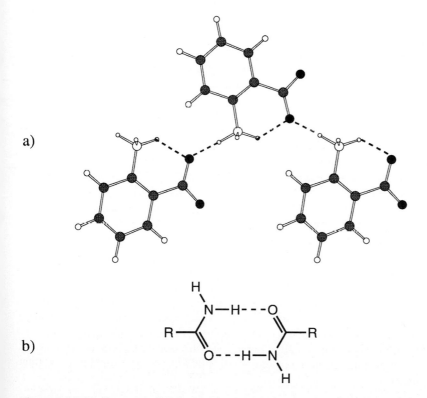

7.41

Figure 7-24. Self-complementary hydrogen bonding within a carboxylic acid allows the construction of intricate solid state structures, as seen in the crystal structure of terephthalic acid, **7.41**; carbon atoms are grey, hydrogen atoms white and oxygen atoms black

a)

b)

Figure 7-25. Hydrogen bonding between carbonyl groups and amino N–H groups is a useful structural feature for the assembly of supramolecules. Part of the hydrogen bonding network present in the solid state lattice of the zwitterionic compound anthranilic acid is shown in a). The self-complementary hydrogen bonding possible with an amide is shown in b).

proved to be altogether more profitable for the selective assembly of supramolecular structures. Part of the hydrogen bonding network present in the solid-state structure of anthranilic acid is shown in Figure 7-25. An amide represents a simple, self-complementary hydrogen bonding system with the carbonyl oxygen acting as donor and the amino hydrogen as the acceptor (Fig. 7-25)

To put the above discussion into other words, the molecular components of a supramolecular system contain *information* in the form of specific molecular recognition features. The object of the exercise is to ensure that these molecular recognition features are mutually complementary.

The last of the important concepts that we will consider is s*elf-assembly*. Most chemists have, at some time in their careers, wondered why molecules cannot just make themselves. The process by which molecules build themselves is termed *self-assembly* and is a feature of many supramolecular systems. If the molecular components possess the correct complementary molecular recognition features *and their interaction is thermodynamically favourable* then simply mixing them could result in the specific and spontaneous self-assembly of the desired aggregate. This assumes that there is no significant *kinetic* barrier to the assembly process. The recognition features within the components may be any of the intermolecular bonding processes mentioned above, but we will be concerned with interactions between transition metal ions and polydentate ligands.

7.3.3 Metal-Ion Complementarity and Metallosupramolecular Chemistry

Transition metals have characteristic and preferred co-ordination numbers (the *number* of ligands or donor atoms bonded to the metal) and co-ordination geometries (the *spatial arrangements* of the ligands and donor atoms in space). For example, nickel(II) centres are usually found to be four co-ordinate, with square-planar or tetrahedral co-ordination, or six-co-ordinate octahedral, whilst palladium(II) is found to be almost invariably four-co-ordinate square-planar. A large number of copper(I) complexes are found to be based upon four-co-ordinate tetrahedral structures. In the context of the previous section, we may think of these preferences in terms of information stored within the metal centre – they are specific molecular recognition features. In *metallosupramolecular chemistry* we are concerned with the matching of these recognition features of the metal with those of the ligand. In the same way that the preferred co-ordination properties of the metal centre are molecular recognition features, so are the inherent bonding properties of polydentate ligands (the *number*, *type* and *spatial distribution* of donor atoms). For example, a ligand such as 2,2':6',2"-terpyridine (**7.42**) presents a near-planar N_3 donor set to a metal that can only occupy three *meridional* sites in an octahedral complex whilst the *tris*(pyrazolyl)borate anion (**7.43**) possesses an N_3 donor set that occupies three *facial* sites (Fig. 7-26). In the remainder of this chapter, we will use this molecular coding in the metal and the ligand to control the assembly of supramolecular complexes with highly specific shapes and properties.

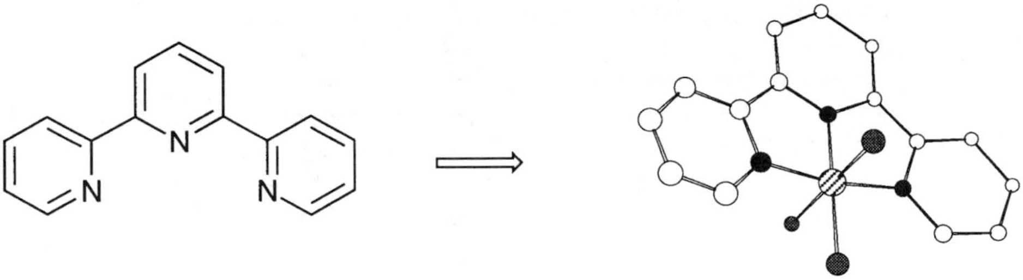

7.42

meridional

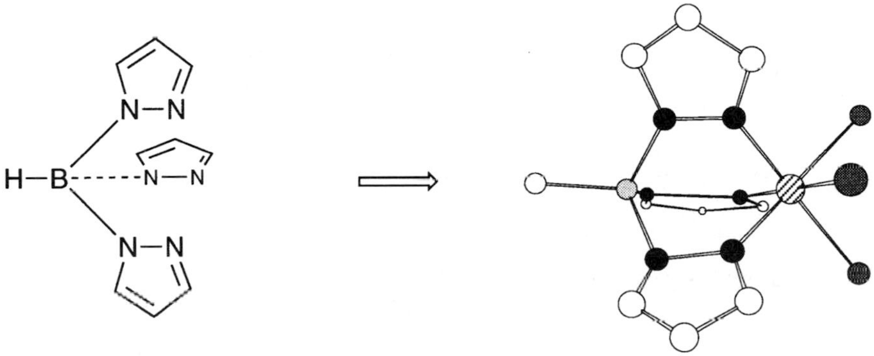

7.43

facial

Figure 7-26. The structure of complexes may be dictated by steric constraints within the ligands. 2,2':6',2"-Terpyridine (**7.42**) can only occupy the three *meridional* sites in an octahedral complex, whereas the *tris*(pyrazolyl)borate anion (**7.43**) must occupy three *facial* sites.

7.4 Building Helical Complexes

7.4.1 Double-Helical Structures

Let us now put the above principles into practice by considering the assembly of multiple-helical compounds. A simple chemical model for the formation of helicates involves the twisting of 'molecular threads', as shown in Figure 7-27. The incorporation of metal-binding domains into these threads allows the use of metal ions to control the twisting.

The trick lies in recognising that the crossing points of the molecular threads correspond to a point at which the two threads are co-ordinated to a single metal ion. This would mean that the two helical structures in Figure 7-27 would be achieved by the incorporation of one- and two metal–binding domains, respectively. The first structure (**7.44**) arises from interaction with a single metal ion, the second (**7.45**) from interaction with two metal ions (Fig. 7-28).

All that remains is to convert the cartoon structures into real molecules! Groups such as 2,2'-bipyridine and 1,10-phenanthroline have been popular choices for the metal-binding domains. The principles are actually very simple. If we incorporate a didentate metal-binding domain into the threads, then structure **7.46** simply corresponds to the bin-

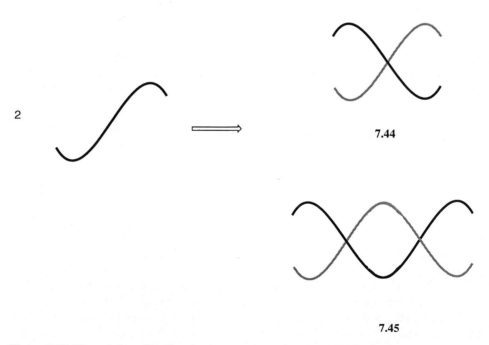

Figure 7-27. The twisting of 'molecular threads' allows the assembly of helical structures. In compound **7.44** there is a single crossing of the molecular threads, whereas in **7.45** there are two such crossings.

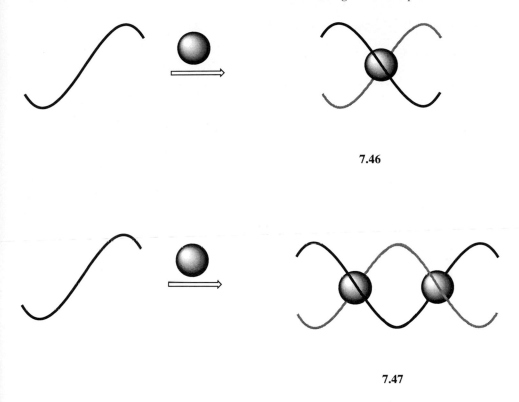

7.46

7.47

Figure 7-28. The use of metal ions to control the assembly of double-helical complexes. The twisting of the molecular threads is initiated by the co-ordination of metal-binding domains within the ligand to the metal ions. The assembly of the mononuclear compound **7.46** requires the incorporation of a single metal-binding domain in each molecular thread, whereas compound **7.47** requires two metal-binding domains per thread.

ding of a four-co-ordinate metal ion in a tetrahedral complex, whilst the more elegant dinuclear structure **7.47** will require the incorporation of two didentate domains per thread and the binding of two tetrahedral metal centres. Is it possible to convert these ideas into practice? In Figure 7-29, the generic features of a ligand containing two didentate metal-binding domains are presented, together with an actual molecule, **7.48**, which possesses this structure.

When **7.48** reacts with copper(I), which usually forms four-co-ordinate tetrahedral complexes, a double-helical species, **7.49**, is indeed formed (Fig. 7-30). This is a genuine self-assembly process – simply treating the ligand with the appropriate metal ion leads to the desired structure. A wide variety of other spacer groups have been incorporated between didentate domains. In practice, some consideration needs to be given to the nature of the spacer group that is selected. If it is too long, or too flexible, other co-ordination possibilities can occur.

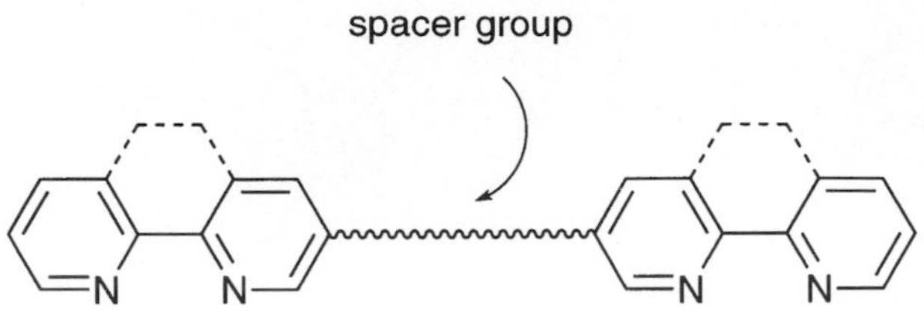

spacer group

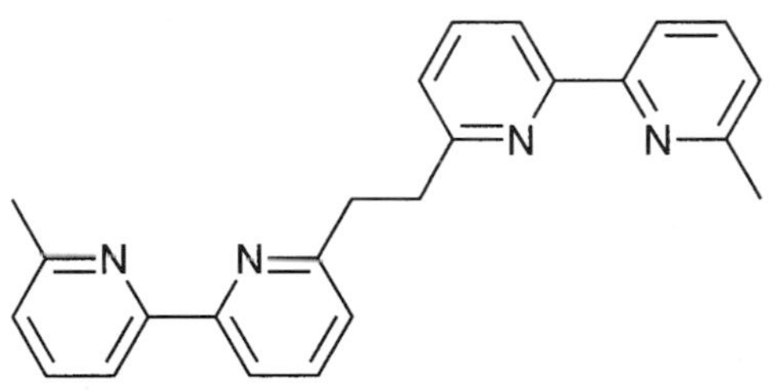

7.48

Figure 7-29. The generic features needed in a molecular thread designed to give structures such as **7.47**, and an actual example of such a ligand, **7.48**. A variety of spacer groups may be incorporated between the metal-binding domains.

Even a very simple ligand such as 2,2':6',2":6",2"'-quaterpyridine (**7.50**), gives double-helical complexes on reaction with copper(I) or silver(I) ions. This process is shown schematically in Figure 7-31. A tetrahedral metal ion cannot bond to all four nitrogen donors of **7.50**. Consider the co-ordination of the first two donor atoms to a tetrahedral centre; it is not possible for the two remaining pyridine donors to co-ordinate to the same metal centre if a tetrahedral geometry is to be maintained. In effect, we have partitioned **7.50** into two separate didentate metal-binding domains. However, the non-co-ordinated domain *can* bond to another metal centre. If this is in a second {M(**7.50**)} unit, the metal ion can achieve a near-tetrahedral geometry by binding two didentate bpy domains from two separate ligands . The inevitable consequence of such a bonding mode is the formation of a dinuclear double-helical complex.

Of course, it is quite possible to further extend these assembly processes to give double-helical complexes with even more bond crossings. For example, a double-helical complex with three bond-crossings should result from the reaction of a molecular thread containing three metal-binding domains with three tetrahedral metal ions (Fig. 7-32). An example of the assembly of such a trinuclear double-helical complex is seen in the formation of **7.52** from the reaction of **7.51** with silver(I) salts (Fig. 7-33).

Naturally, it is not necessary to limit the procedure to the use of tetrahedral metal centres. For example, it is possible to build double-helical structures from the interaction of molecular threads containing tridentate domains with metal ions possessing a preference for a six-co-ordinate octahedral geometry. An example of such a process is shown in Figu-

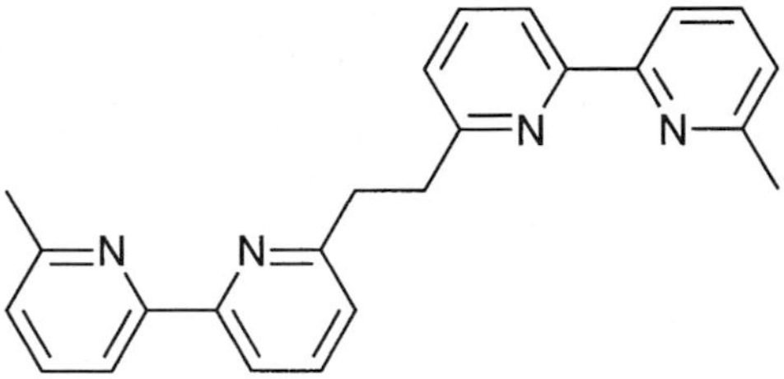

7.48

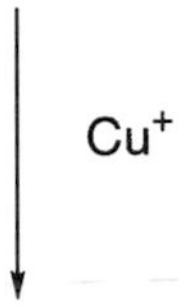

Cu⁺

7.49

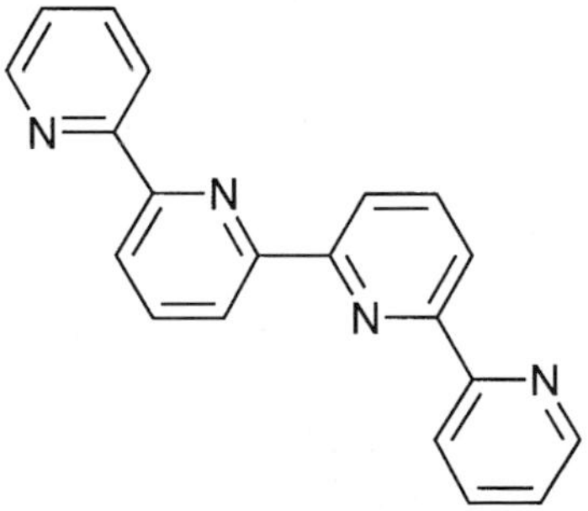

7.50

Figure 7-30. The reaction of **7.48** with copper(I) gives the desired double-helical dinuclear complex **7.49**. The two molecular threads have been shaded differently in **7.49** to emphasise the helical structure.

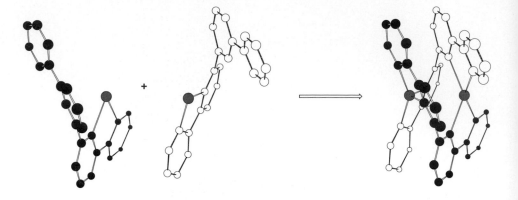

Figure 7-31. The reaction of **7.50** with tetrahedral metal ions gives dinuclear double-helical complexes. The interplay of the metal and ligand requirements are emphasised in this process. Once again, the molecular threads have been shaded differently to clarify the structure of the product.

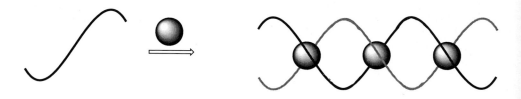

Figure 7-32. The interaction of a molecular thread containing three didentate metal-binding domains with tetrahedral metal ions should give a trinuclear double-helical complex.

re 7-34; the ligand **7.53** possesses a total of six nitrogen donor atoms. These may be partitioned into two tridentate metal-binding domains upon interaction with an octahedral metal ion. The consequence is the formation of double-helical complexes.

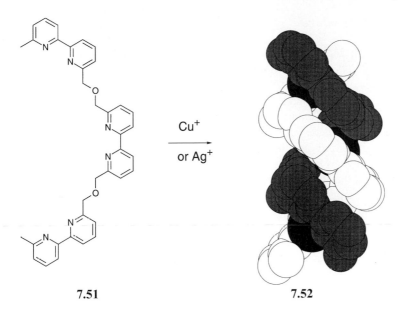

7.51   **7.52**

Figure 7-33. The interaction of the ligand **7.51**, which contains three didentate metal-binding domains with copper(I) or silver(I) ions, results in the assembly of a trinuclear double-helical structure, **7.52**.

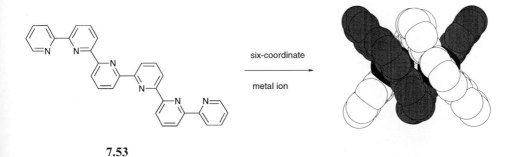

7.53

Figure 7-34. The interaction of the potentially hexadentate ligand **7.53** with octahedral metal ions results in a partitioning into two tridentate domains and the formation of dinuclear double-helical complexes.

7.4.2 Triple-Helical Structures

In the preceding section we showed how the interaction of tetrahedral metal ions with ligands containing didentate metal-binding domains could be used to control the assembly of double-helical structures. The key feature was the recognition that a crossing of molecular threads could be achieved by the co-ordination of these metal-binding domains to the metal centres. Consider what might happen when a molecular thread containing two didentate metal-binding domains reacts with an octahedral metal ion. Two possible structures are immediately apparent, as shown in Figure 7-35. In the first, **7.54**, a double-helical array is built, with four of the six co-ordination sites at each metal centre occupied by didentate domains from the two different threads, the co-ordination being completed by an additional two monodentate ligands (or an additional didentate ligand). In the second structure, **7.55**, a *triple-helical* complex is formed from the co-ordination of three didentate domains from each of three ligand threads to each metal centre.

The precise structure of the ligand thread allows some control over which of these structures (or indeed any other structure) is formed. An example of such a process is shown in Figure 7-36, in which a ligand containing two didentate domains, **7.56**, forms a triple-helical dinuclear complex upon co-ordination to two six-co-ordinate cobalt(II) centres.

Once again, it is possible to extend these ideas to the formation of complexes containing progressively more metal centres. As an example, consider the ligand **7.57**. This contains a total of three didentate 2,2'-bipyridine-like domains. Upon reaction with nickel(II) salts, a trinuclear triple-helical complex, $[Ni_3(7.57)_3]^{6+}$ **7.58**, is formed, in which each of the six-co-ordinate nickel(II) centres is co-ordinated to a didentate metal-binding domain from each of three ligand threads.

It should be stressed that the coding for the formation of these topologically complex molecules needs to be carefully controlled in order to obtain the desired structures. To illustrate this, consider ligand **7.59**, which contains two didentate metal-binding domains. This might be expected to react with octahedral metal ions to give a triple-helical dinuclear complex. Reaction with iron(II) does indeed give a species of stoichiometry $[Fe_2(7.59)_3]^{4+}$; however, the crystal structure reveals that an 'untwisted' complex, **7.60**, has been formed.

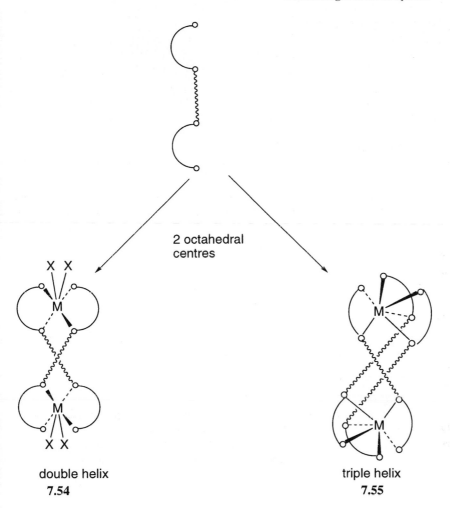

Figure 7-35. A ligand thread containing two didentate metal-binding domains can give a double-helical [M₂L₂] complex on reaction with tetrahedral metal centres. A double-helical [M₂L₂X₄] species, **7.54**, could also result from the interaction of this thread with an octahedral centre. In **7.54**, each metal centre is co-ordinated to a didentate metal-binding domain from each of two ligand threads and an additional two donors from other ligands. However, it is also possible for a triple-helical [M₂L₃] complex to be formed from the binding of a didentate domain from each of three ligand threads to each of two octahedral metal centres, **7.55**.

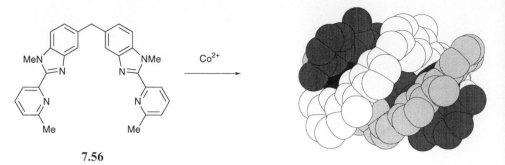

7.56

Figure 7-36. The ligand **7.56** contains two didentate metal-binding domains. On reaction with cobalt(II), a triple-helical dinuclear complex [Co₂(**7.56**)₃]⁴⁺ is formed, in which each six-co-ordinate cobalt(II) centre is co-ordinated to a didentate metal-binding domain from each of three ligand threads.

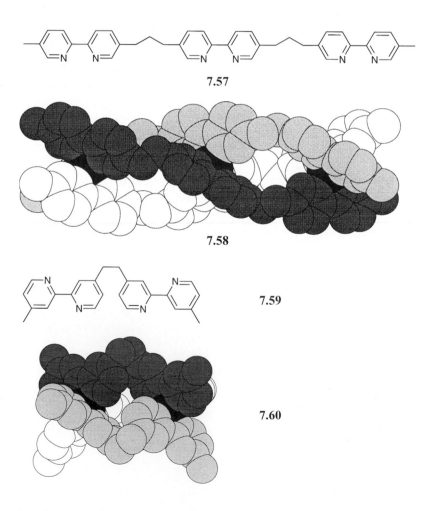

7.57

7.58

7.59

7.60

7.5 Catenanes – Molecular Daisy Chains

Although the formation of helical complexes described above demonstrates the subtle and elegant control that is possible over the conformation of co-ordinated ligands, we have not yet extended these effects to control of the reactivity.

Consider a difunctional molecular thread, with reactive sites at each end. These two reactive sites can react with some other difunctionalised molecule to give a cyclic structure (Fig. 7-37). This is exactly the type of process that we discussed at the beginning of Chapter 6.

Now let us consider what happens if two such molecular threads containing didentate metal-binding domains are twisted into a helical arrangement after co-ordination to a tetrahedral metal centre. Reaction with the difunctional reagent could proceed in several ways. For example, the result could be the formation of a [2+2] macrocyclic complex as a result of the difunctional reagent linking together the two molecular threads (Fig. 7-38).

However, let us choose molecular threads of such a structure and length, and difunctional reagents of such a length that it is only possible for reaction to give [1+1] cyclic products resulting from the linking together of the two ends of *one* molecular thread. This is most clearly seen in Figure 7-39, in which we have drawn the complex in a slightly different way. The consequence of this twisting of the molecular threads is the formation of a *catenane*. The two new macrocycles which have been formed are linked one through another. It is impossible to separate the two rings without breaking bonds. Removal of the metal centre will just leave the metal-free catenane.

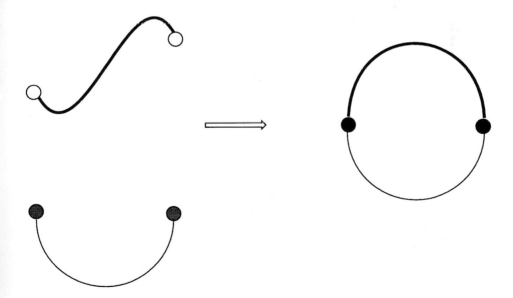

Figure 7-37. The reaction of a difunctionalised molecular thread with another difunctionalised compound to give a macrocyclic product.

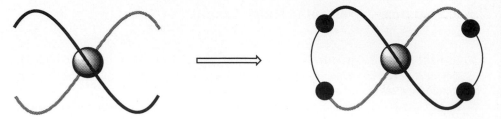

Figure 7-38. The formation of a co-ordinated [2+2] macrocyclic ligand in the reaction of a helical complex with a difunctional reagent. The difunctional reagent links together the two molecular threads. The black circles represent the sites at which the molecular thread has reacted with the difunctional reagent.

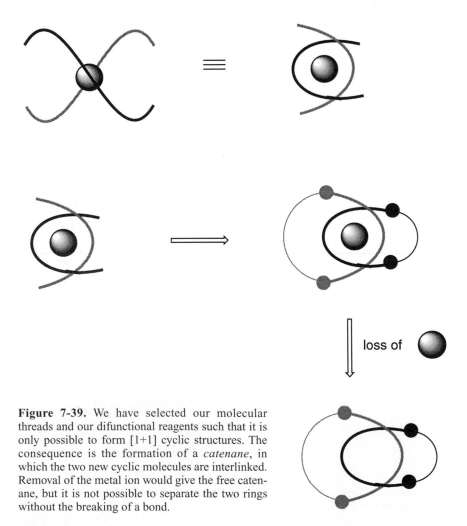

loss of

Figure 7-39. We have selected our molecular threads and our difunctional reagents such that it is only possible to form [1+1] cyclic structures. The consequence is the formation of a *catenane*, in which the two new cyclic molecules are interlinked. Removal of the metal ion would give the free catenane, but it is not possible to separate the two rings without the breaking of a bond.

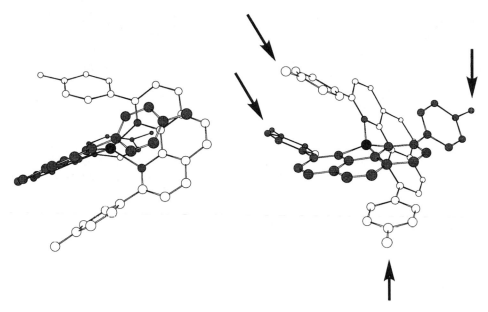

Figure 7-40. Two views of the copper(I) complex of ligand **7.61** showing the arrangement of the reactive sites (indicated by arrows) such that cyclisation to give [1+1] macrocyclic products is favoured. One of the molecular threads has been shaded in each case.

The only question is whether it is possible in reality to exercise such profound control over a cyclisation process. Although catenanes have been sought after for many years, until very recently they were only accessible in extremely low yields. The problems were associated with the need to inter-twist the molecular threads before the cyclisation process. A number of interesting approaches involving 'pure' organic chemistry have been recently developed, but none approach the elegance of metal-directed synthesis. We begin with a 1,10-phenanthroline derivative **7.61**, which contains two groups that can be used in subsequent cyclisations. The reaction of **7.61** with copper(I) gives a copper(I) complex, in which the desired twisting of the molecular threads has been achieved (Fig. 7-40). Notice that the ligands are arranged such that the two functional groups of each ligand are facing in the same direction. This will favour cyclisation processes leading to the [1+1] macrocycles if relatively short difunctional reagents are used. Only very long reagents would be able to span the distance between functional groups of different ligands.

When the copper complex of **7.62** reacts with $ICH_2(CH_2OCH_2)_4CH_2I$ in the presence of base, an intramolecular cyclisation occurs to form the macrocyclic ether **7.63**. However, because of the arrangement of the starting ligands about the copper(I) centre, the two macrocycles are interlinked, and the consequence is the formation of the copper(I) complex of the catenand (*catenand* = *caten*ane lig*and*) (Fig. 7-41).

These complexes of catenands are remarkably stable. However, treatment of the complex with cyanide results in demetallation and the formation of the free catenane, in which the two macrocyclic ligands are still interlocked. There is a conformational change upon demetallation and in the solid state the rings have 'slipped'. It is not trivial to estab-

7.61 R = Me

7.62 R = H

7.65 R = CH₂C≡CH

7.63

lish that the catenane has been formed, but one interesting proof of the interlinked structure for the complex and the free ligand comes from their mass spectra. The difficulty is to distinguish between the catenane, containing two interlinked macrocycles of mass M and a [2+2] macrocycle of mass $2M$. The fragmentation pattern in the mass spectrum of the [2+2] macrocycle would show a whole series of peaks descending in mass from m/z $2M$. In contrast, the first breaking of one of the macrocycles in the catenane results in a nicking of the chain, and the disruption of the interlinked structure. Accordingly, *no peaks will be observed with masses between* $2M$ *and* M (Fig. 7-42).

Naturally, it is possible to extend the 'daisy-chain'. The compounds formed from the interlinking of two rings are termed [2]-catenanes and we will conclude this section by showing one approach to a [3]-catenane. This illustrates another type of ring closure reaction, but is still dependent upon the conformational control of the metal ion in orienting the reactive ends of the precursors.

One synthetic strategy for the preparation of a [3]-catenane is shown in Figure 7-43. The first step involves the co-ordination of two different ligands to a metal centre. One of the ligands is a macrocycle and the other has two terminal acetylene end groups. Terminal acetylenes undergo a facile coupling reaction to buta-1,3-diynes upon reaction with copper salts and dioxygen. The co-ordinated metal controls the conformation of the open-chain ligand, such that the terminal acetylene groups are arranged so that intramolecular cyclisation or polymerisation is disfavoured and [3]-catenane is formed as a cyclic dimer.

In an example of a [3]-catenane formed by this methodology, the compound **7.64** is formed from the coupling of the copper(I) complex formed from the reaction of [Cu(MeCN)₄]⁺ with one equivalent of **7.63** and one equivalent of **7.65** in 58 % yield (Fig. 7-44)! The conformation of the [3]-catenane is partially controlled by intramolecular π-stacking interactions (Fig. 7-45). A number of other strategies have also been adopted

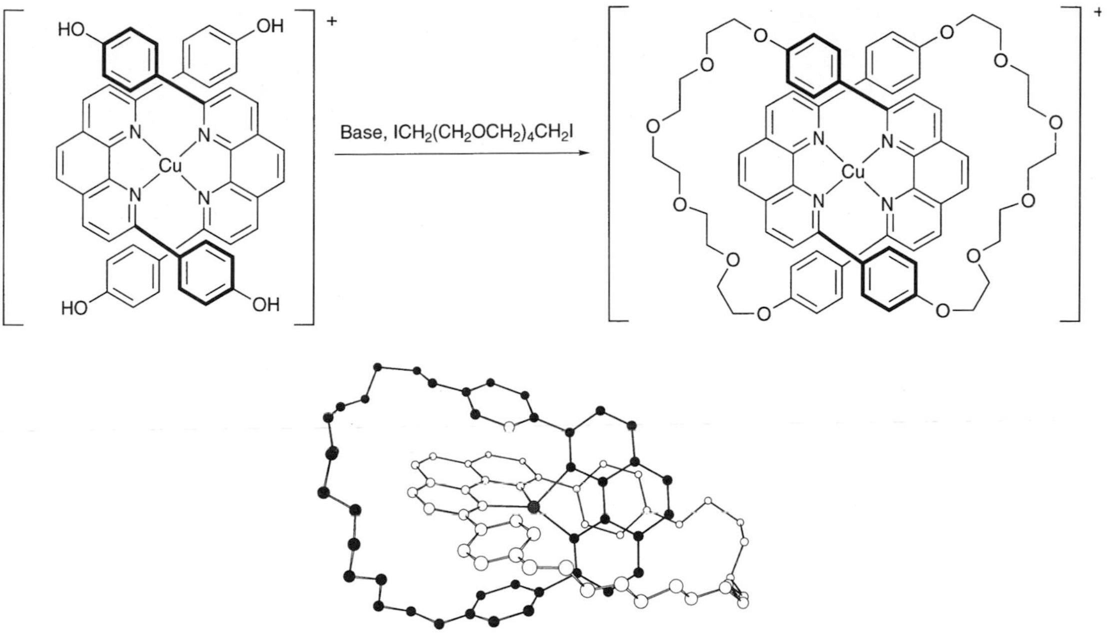

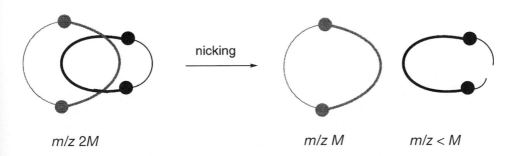

Figure 7-41. The reaction of the copper complex of **7.62** with $ICH_2(CH_2OCH_2)_4CH_2I$ in the presence of base results in the formation of the copper(I) complex of a catenand containing two interlocked macrocycles. The lower view shows the cation present in the solid state; the two interlocked macrocycles have been shaded to emphasise the structure.

m/z 2M   nicking   *m/z M*   *m/z < M*

Figure 7-42. Mass spectrometry allows us to distinguish between isomeric [2+2] macrocycles and catenanes of mass $2M$. The mass spectrum of a catenane should show no peaks between *m/z* $2M$ and M.

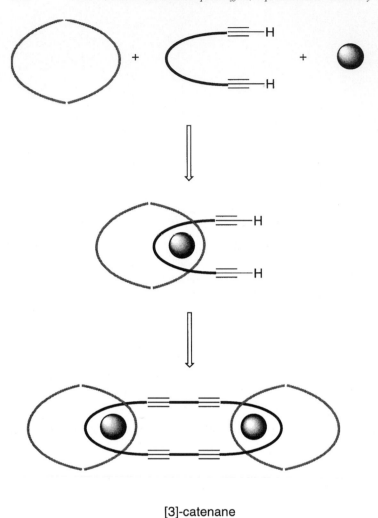

[3]-catenane

Figure 7-43. The formation of a [3]-catenane. The coupling of the terminal acetylene groups is achieved by reaction with copper salts and dioxygen.

and the reader will find more information about these in the suggestions for further reading at the end of this chapter. Needless to say, in all of these reactions the choice of the correct length molecular threads is critical for the successful assembly of the desired topology.

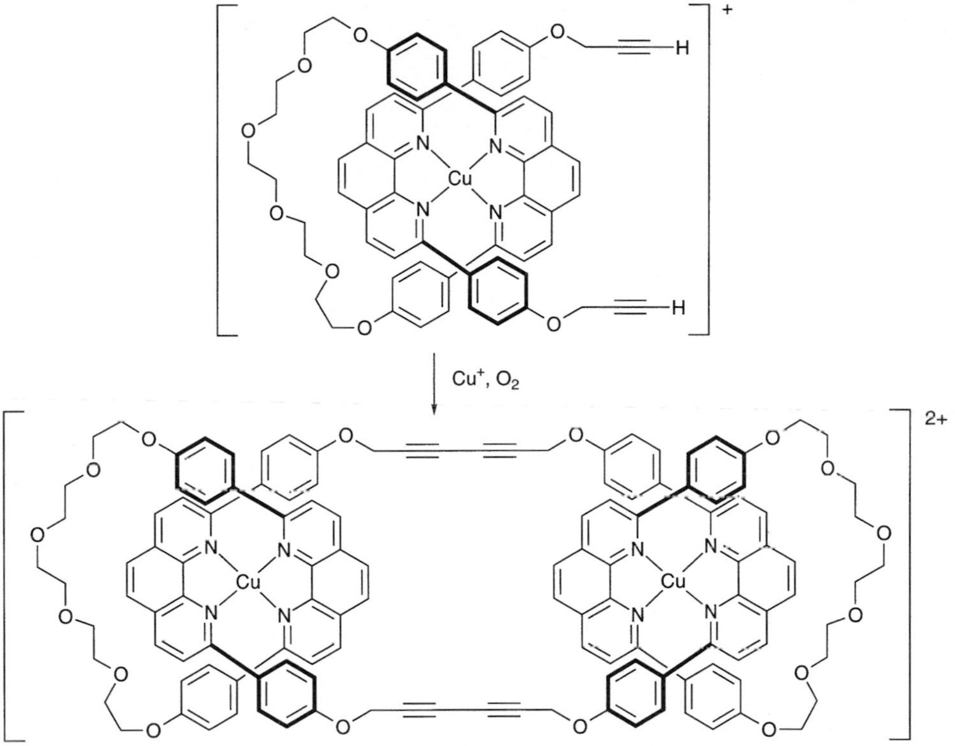

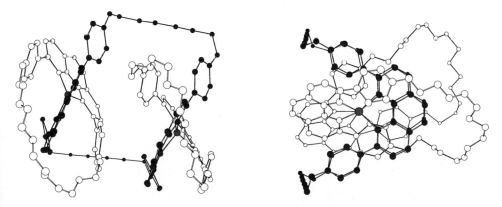

7.64

Figure 7-44. The synthesis of the dicopper(I) complex of a [3]-catenane using the general route shown in Figure 7-43.

Figure 7-45. Two views of the solid state structure of the [3]-catenane **7.64**. The large ring formed from the coupling of the **7.65** ligands has been coloured black and the two macrocyclic ligands **7.63** white. The second view, along the Cu⋯Cu axis, emphasises the folded structure of the large, central macrocyclic ring and shows some of the π-stacking interactions that are responsible for the adoption of this conformation.

7.6 Tying Molecules in Knots

To conclude this chapter, we will extend our investigation of metal-ion control over topology to the tying of molecular threads into knots. We commence by returning to some of our ideas about topology. The representation of a catenane in Figure 7-46 emphasises that a two-dimensional graph contains two points at which lines cross.

In the same way, the *precursor* for the catenane must also have two such crossing points at which the molecular threads are inter-twisted. This allows us to think about the topology in a slightly different way. The binding of *two* molecular threads, each containing a *single* didentate metal-binding domain, to a tetrahedral metal centre gives a mononuclear precursor with *two* crossing points (Fig. 7-47).

This leads us to the next question. What do we need to have in a molecular thread to give *three* crossing points and what sort of a structure results from the ring closure reactions? By analogy with Figure 7-47, we can see that molecular threads containing *two* didentate metal-binding domains should interact with *two* tetrahedral metal centres to give a *dinuclear* complex with *three* crossing points of two threads. Reaction of this with an appropriate difunctional reagent gives a new species with *three* crossing points (Fig. 7-48). Naturally, when it comes to molecular design, the choice of the right threads and the right difunctional reagents becomes absolutely critical - the structure shown in Figure 7-48 will only arise if the ends of the two different threads are linked in a particular way to maintain the three crossing points.

We are now left with another question - what is this new structure that we have drawn with the three crossing points? However we twist and turn the molecule (hint: try this with pipe-cleaners) the crossing points remain. If we start at any point on the structure and trace around we eventually end up at the same point. In other words, there is only a single cyclic molecule, but the twisting means that it is not the same as the [2+2] macrocyclic product. This view serves to emphasise the relationship with the double-helical structures that

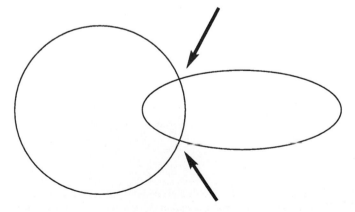

Figure 7-46. A cartoon view of a catenane. The view emphasises that it is impossible to draw a catenane in two dimensions without a minimum of two points at which lines cross. The crossing points are indicated by arrows.

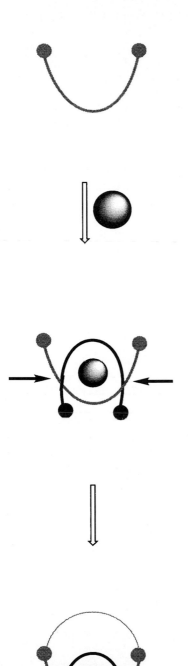

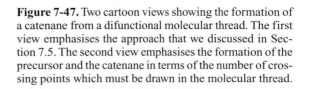

Figure 7-47. Two cartoon views showing the formation of a catenane from a difunctional molecular thread. The first view emphasises the approach that we discussed in Section 7.5. The second view emphasises the formation of the precursor and the catenane in terms of the number of crossing points which must be drawn in the molecular thread.

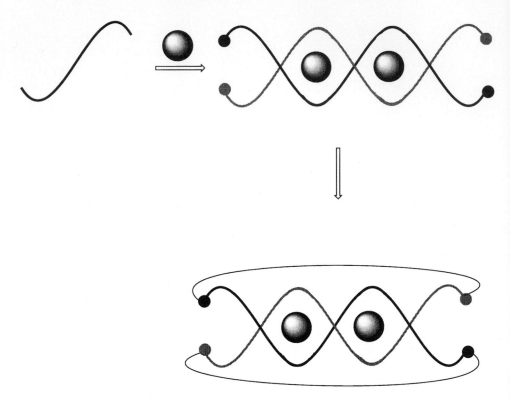

Figure 7-48. The interaction of a molecular thread containing two didentate metal-binding domains with tetrahedral metal centres to give a structure with three crossing points.

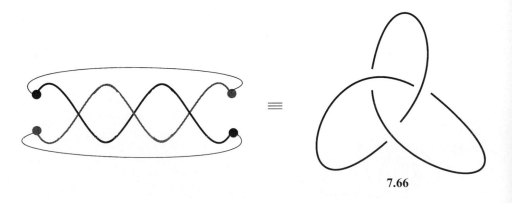

7.66

Figure 7-49. The cyclic structure with three crossing points corresponds to the trefoil knot, **7.66**. The identity is best seen with models made from pipe-cleaners.

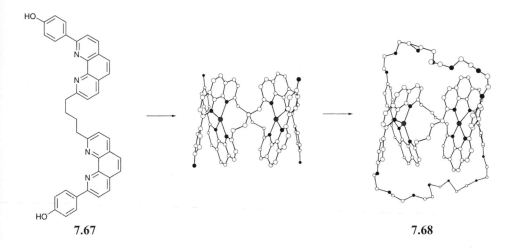

7.67   **7.68**

Figure 7-50. The reaction of the molecular thread **7.67** with copper(I) salts gives a double-helical precursor with three crossing points. Reaction with $ICH_2(CH_2OCH_2)_5CH_2I$ gives the dinuclear trefoil knot **7.68**.

we discussed in Section 7.4. The structure that we have produced is actually the simplest knot, **7.6**, known as the *trefoil knot* (Fig. 7-49). Clearly, the wrong design features in the threads will result in the formation of the [2+2] product rather than the knot.

We can now use similar methodology to that adopted for the synthesis of catenanes. The metal binding domains of choice are 1,10-phenanthrolines and the cyclisation will involve the formation of ether linkages from pendant phenol groups. A suitable ligand thread is seen in **7.67,** which contains two 1,10-phenanthroline metal-binding domains and two pendant phenol groups. Treatment of **7.67** with copper(I) salts results in the formation of the double-helical complex $[Cu_2(7.67)_2]^{2+}$, in which the pendant phenolic groups are arranged in the correct sites for connection by reaction with a suitable α,ω-dihalo compound. The choice of the spacer group between the 1,10-phenanthroline domains and the choice of the dihalide are critical, but small yields of the dinuclear trefoil knot **7.68** are obtained after reaction with $ICH_2(CH_2OCH_2)_5CH_2I$ (Fig. 7-50). Variations in the structure of the ligand and the alkylating agent have allowed yields of the trefoil knotted molecules to be increased.

7.7 Conclusions

We have come to the end of our journey through supramolecular chemistry. The bulk of the discussion in this chapter has centred upon the role of the metal ion in controlling the conformation of a co-ordinated ligand. We have seen that a variety of novel molecular architectures may be achieved from the simple principles that we have developed.

Suggestions for Further Reading

Many of the references quoted in Chapter 6 contain relevant material.

1. J.-C. Chambron, C. Dietrich-Buchecker, J.-P. Sauvage, *Topics. Curr. Chem.* **1993**, *165*, 131.
   – A short review which introduces many of the concepts of molecular topology.
2. V.I. Sokolov, *Russ. Chem. Revs.* **1973**, *42*, 452
   – A classical article on chemical topology.
3. C.O. Dietrich-Buchecker, J.-F. Nierengarten, J.-P. Sauvage, *Tetrahedron Letts.* **1992**, *25*, 3625; C.O. Dietrich-Buchecker, J.-P. Sauvage, *New. J. Chem.* **1992**, *16*, 277.
   – The art of making knots!
4. J.-P. Sauvage, *New. J. Chem.* **1985**, *9*, 299; C.O. Dietrich-Buchecker, J.-P. Sauvage, *Chem. Rev.* **1987**, *87*, 795
   – How to make catenanes.
5. A.M. Sargeson, *Pure Appl. Chem.* **1984**, *56*, 1603; **1984**, *58*, 1511; *Chem. Brit.* **1979**, *15*, 23.
   – Good reviews on the synthesis and properties of encapsulating ligands.
6. E.C. Constable, *Tetrahedron.* **1992**, *48*, 10013; *Progr. Inorg. Chem.* **1994**, *42*, 67.
   – Reviews of helication.
7. F. Vögtle, *Supramolecular Chemistry*, Wiley, Chichester, **1993.**
   – A good general introduction to the subject of supramolecular chemistry.
8. J.M. Lehn, *Supramolecular Chemistry*, VCH, Weinheim, **1995**.
   – An excellent overview.

8 Reactions of Aromatic and Heterocyclic Ligands

8.1 Introduction

The organic chemistry of aromatic and heteroaromatic compounds is particularly well-understood. The effects of substituents and ring heteroatoms on the rates and sites of attack by nucleophiles and electrophiles are both predictable and capable of rationalisation. What can the co-ordination chemist offer to an area as well-characterised as this? Perhaps because the organic chemistry of these systems is so well-behaved, the potential for modification of reactivity by co-ordination is great.

The electrophilic and nucleophilic substitution patterns of benzene derivatives are dictated by the σ- and π-interactions between the substituents and the ring π-cloud; co-ordination of a substituent to a metal ion may change the balance of these effects or even reverse them. Similarly, co-ordination of a heteroaromatic to a metal ion may open up pathways for electrophilic or nucleophilic attack that are relatively inaccessible in the free ligand. Paradoxically, although the potential for metal-directed reactivity in these systems is so great, there have been few systematic investigations. The majority of metal-directed reactions of benzene derivatives involve organometallic derivatives. These are extremely important synthetic methods, but the activation usually involves direct M–C interactions and will not be discussed at length in this book. Particularly useful systems involve iron or chromium carbonyl groups directly bonded to the aromatic rings. The interested reader is referred to the suggestions for further reading at the close of this chapter for further information on this important topic.

8.2 Electrophilic Aromatic Substitution

One of the commonest reactions shown by benzene derivatives is the electrophilic substitution of hydrogen by some other group. The mechanisms of these reactions are relatively well-understood and a vast range of electrophiles have been shown to be effective. The reaction involves the initial formation of a cationic intermediate which then rearomatises with the loss of a proton to give the substitution product (Fig. 8-1).

Many ligands contain phenyl groups, and it might be expected that co-ordination to a metal might in some respect modify the electrophilic reactions of these rings. The metal might be able to stabilise the cationic intermediate by back-donation of electron density to the π-cloud of the aromatic ring. In general, investigations into this area of ligand reac-

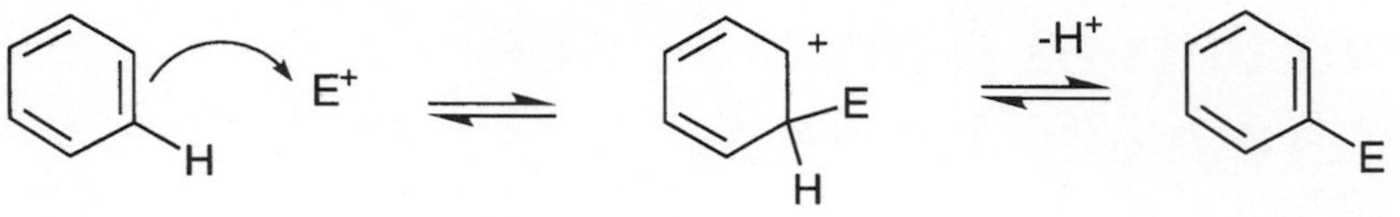

Figure 8-1. The usual mechanism for electrophilic substitution in a benzene derivative. The addition of the electrophile to the ring generates a short-lived cationic intermediate.

tivity have tended to give disappointing results. The pattern of substitution observed is not usually altered and the rate of electrophilic attack is either unchanged or reduced upon co-ordination. This is, of course, an effect that may be rationalised. The attack of the electrophile involves interaction with the π-electron density of the aromatic ring. Co-ordination to an electropositive or positively charged metal will result in a polarisation and reduction of this electron density. A typical example is seen in the bromination of the aniline ligand in the chromium(III) complex [Cr(PhNH$_2$)$_3$Cl$_3$], which proceeds smoothly to yield the corresponding tris(2,4,6-tribromoaniline) complex (Fig. 8-2). This is, of course, the same substitution pattern that is observed in the bromination of the free ligand. Note that it is necessary to use a kinetically inert metal centre with a relatively activated aromatic ligand. When the substitution pattern is the same as with the free ligand, it is essential to use a non-labile system, and to establish that the observed products do indeed arise from the reaction of the *co-ordinated* as opposed to the *free* ligand. The phenyl ring in aniline is activated towards nucleophilic attack by the conjugation of the lone-pair of electrons of the *sp^2* hybridised amino group.

Co-ordination of the nitrogen atom of the aniline to the metal ion reduces the lone-pair availability on the amine substituent and results in the formation of an *sp^3* hybridised nitrogen centre, so the phenyl ring in aniline complexes is expected to be slightly less prone to attack by electrophiles than the free ligand. This effect may be used to mediate the reactivity of highly reactive aromatics. It *is* possible to obtain 4-bromoaniline and 2,4-dibromoaniline from the reaction of [Pd(PhNH$_2$)$_2$Br$_2$] with bromine, whereas the 2,4,6-tribromo compound is the usual product obtained from the direct bromination of aniline.

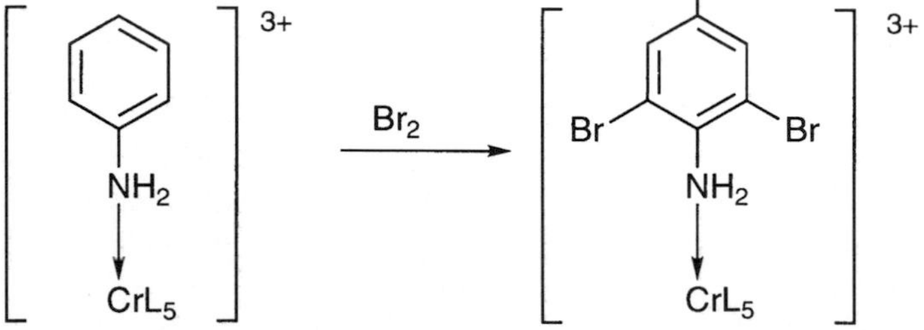

Figure 8-2. The bromination of aniline in the complex [Cr(PhNH$_2$)$_3$Cl$_3$] to give a complex containing 2,4,6-tribromoaniline. The use of a kinetically inert metal centre allows us to ensure that the reaction is one of the co-ordinated ligand, even though the organic product is the same as that obtained with the free ligand.

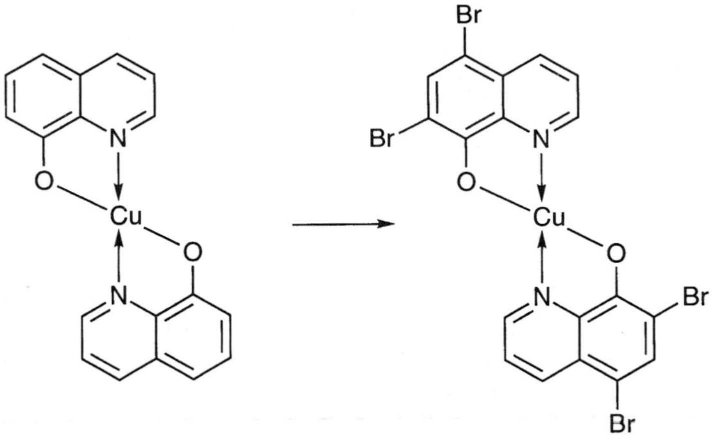

Figure 8-3. The bromination of the copper(II) complex of 8-hydroxyquinoline. The sites of bromination are the same as observed in the reactions of the free ligand.

This pattern of reactivity is reflected in most reactions of electrophiles with complexes containing aromatic ligands; the rates of reaction are modified but the position of substitution is unchanged with respect to the free ligand. The reactivity of a range of quinoline complexes with electrophiles has been studied in some detail and the products have been shown to be substituted in exactly the same sites as the free ligands. For example, di(8-oxyquinolinato)copper(II) reacts with molecular bromine to yield di(5,7-dibromo-8-oxyquinolinato)copper(II) (Fig. 8-3).

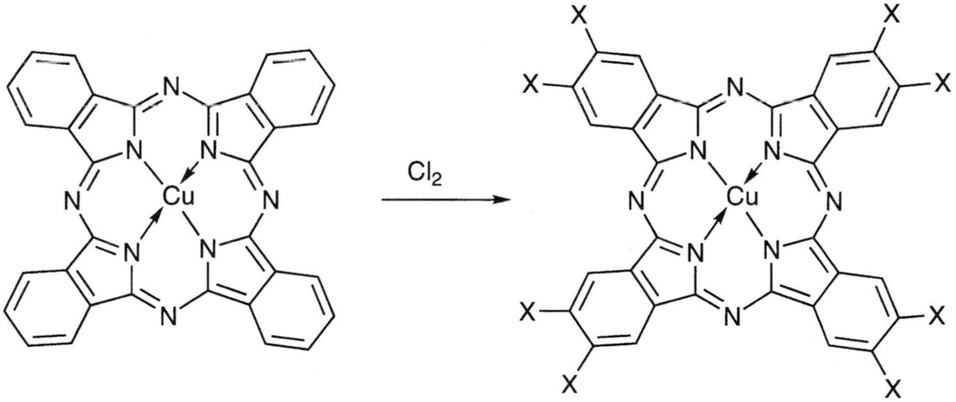

8.1   X = H or Cl

Figure 8-4. The sequential chlorination of phthalocyaninatocopper(II) results in a change in colour from blue to green. The metal centre remains within the macrocycle during the chlorination reaction.

An electrophilic substitution reaction of an aromatic ring provides one of the more spectacular examples of the extraordinary stability of phthalocyanine complexes. The macrocyclic complex phthalocyaninatocopper(II), **8.1**, undergoes sequential substitution of the benzene ring hydrogen atoms upon chlorination in relatively vigorous conditions (Fig. 8-4). A total of up to eight chlorine atoms may be introduced around the periphery of the macrocyclic ligand, with a progressive change in colour from blue to green with increasing chlorine content. These green compounds have found some application as coloured pigments and are potential intermediates for the synthesis of other substituted phthalocyanines.

The special case of metal-mediated stoichiometric electrophilic oxygen atom transfer to aromatic ligands will be discussed in more detail in the next chapter. Suffice it to say here, that numerous examples of transition metal-mediated (particularly copper and iron) oxygen atom transfer are known, although many of these processes are catalytic.

8.3 Nucleophilic Aromatic Substitution

Whereas co-ordination to a metal is seen to have relatively little effect upon the reactivity of aromatic ligands with electrophiles, this is not the case in reactions with nucleophiles. In the same way that we might expect co-ordination to a metal ion to reduce the tendency for attack of an aromatic ring by electrophiles, we might expect the attack of nucleophiles to be enhanced. The majority of studies in this area have been concerned with the dramatic effects that copper salts have on the reactivity of aromatic ligands. Copper salts are

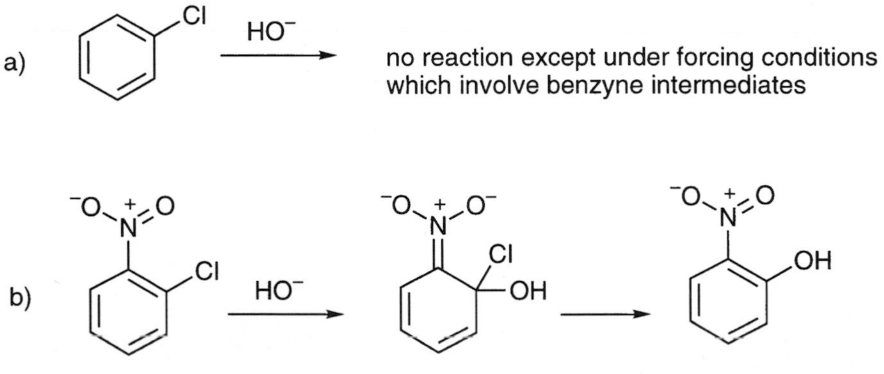

Figure 8-5. Reactions of halobenzene derivatives with nucleophiles. a) Unactivated compounds are extremely inert and only react by mechanisms that involve the formation of benzynes. b) The presence of the electron-withdrawing nitro group, which can stabilise an anionic intermediate by the delocalisation of the charge onto the electronegative oxygen atoms, allows facile nucleophilic substitutions.

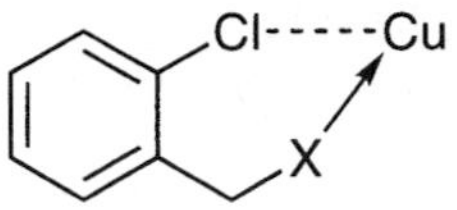

8.2 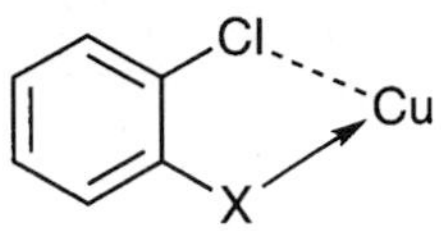   **8.3**

known to have profound effects upon the reactions of aromatic compounds, but in many cases the precise function of the metal in these reactions is not fully understood. We will concern ourselves primarily with stoichiometric reactions of metal complexes with nucleophiles.

Under normal circumstances, the nucleophilic displacement of halide from halobenzenes is not a facile process. Attack by the incoming nucleophile is unfavourable as a result of its repulsion by the π-cloud of the arene. Rapid displacement of halide *is* observed if strongly electron withdrawing substituents which can delocalise charge in an intermediate or transition state, are present. The most common such substituent is a nitro group, but we could envisage other groups, co-ordinated to a metal ion, fulfilling this role (Fig. 8-5).

Copper complexes of substituted haloarenes have been shown to be particularly prone to displacement of halide by nucleophiles. Most commonly, copper(II) species are involved, but in a few cases copper(I) has also been shown to be effective. This is not surprising in view of the facile inter-conversion of copper(I) and copper(II) in aerobic condi-

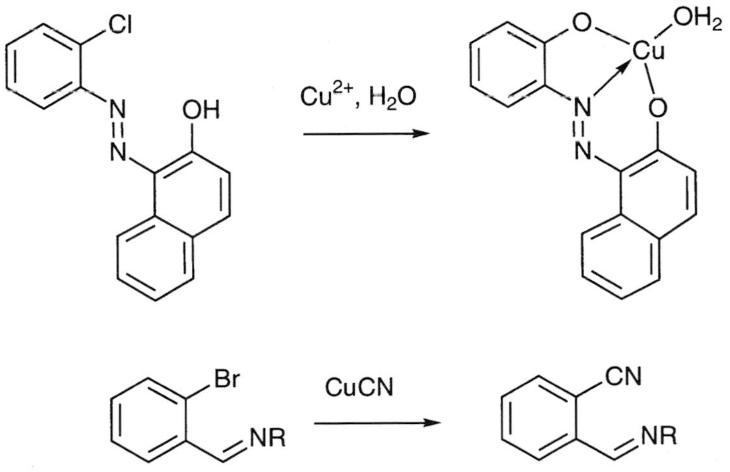

Figure 8-6. Examples of copper-mediated reactions of haloaromatics. In each case, the halogen is replaced by a nucleophile. The mechanisms probably involve intermediates related to **8.3**.

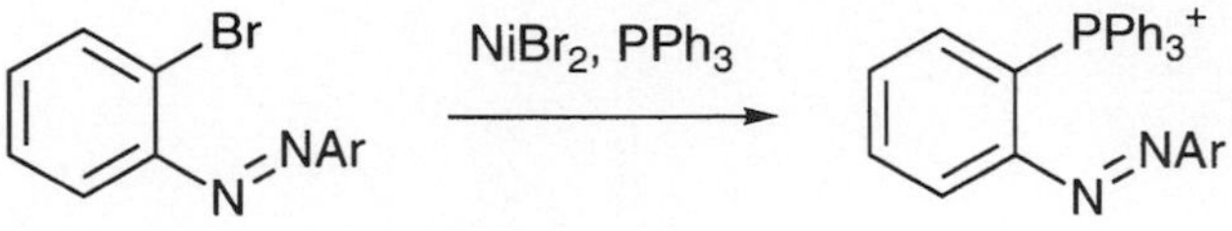

Figure 8-7. A nickel-mediated phosphanation of a bromoarene. The precise mechanistic details of this reaction are not known.

tions. The basic requirements for the reaction appear to be a potentially co-ordinating substituent *ortho* to the halo group. It is thought that the metal shows significant interactions with the halo group in these reactions. Evidence for this comes from the observation that the highest reactivities are observed when the donor atom in the substituent is arranged so that upon co-ordination a five- or six-membered chelate ring (**8.2** or **8.3**) may be formed with the halo group. Reactions may either be catalytic in copper or may involve stoichiometric complexes.

The commonest reactions involve the displacement of halide by hydroxide or cyanide ion to yield co-ordinated phenols or nitriles. Once again, the metal may play a variety of different functions. The polarisation of the C–Cl bond is the most obvious, but stabilisation of the product may be of equal importance, as could the involvement of a metal co-ordinated nucleophile. The availability of a one-electron redox inter-conversion between copper(II) and copper(I) also opens up the possibilities of radical mechanisms involving homolytic cleavage of the C–Cl bond. All of these different processes are known to be operative in various reaction conditions. In other cases, organocopper intermediates are thought to be involved.

Two examples of typical copper-mediated reactions are presented in Fig. 8-6, and illustrate the replacement of halide by hydroxide or cyanide. Yields in these reactions are frequently excellent, even if the precise mechanistic details are often somewhat unclear!

Reactions of this type have proved to be of considerable commercial significance in the industrial scale preparation of diazo dyes. Although the bulk of reactions involve copper

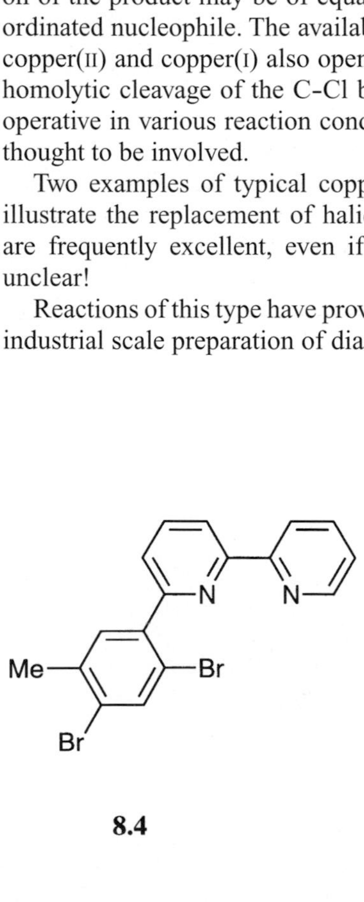

8.4

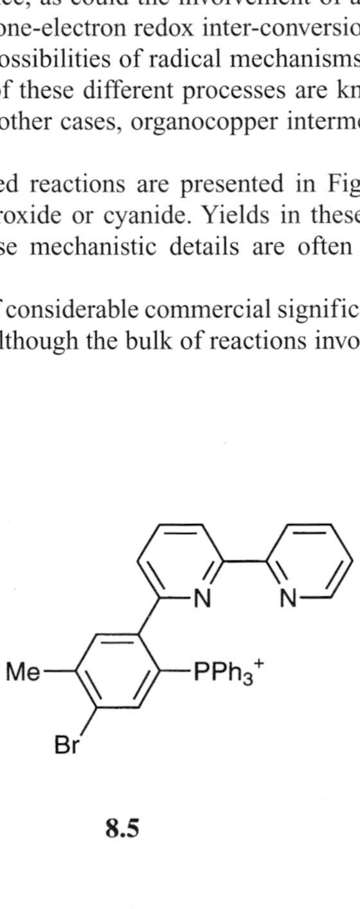

8.5

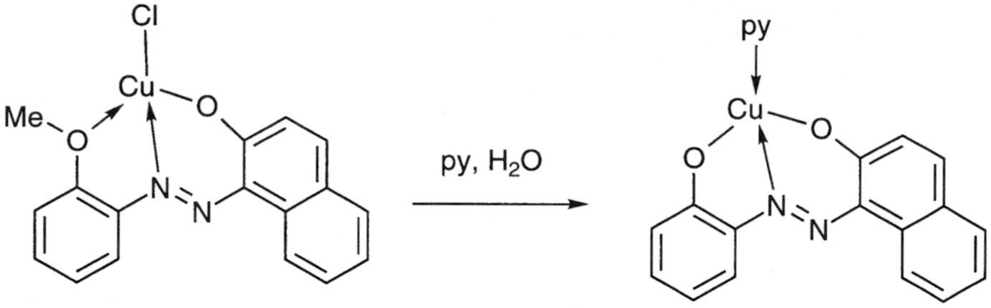

Figure 8-8. The copper-mediated conversion of an anisole derivative to a co-ordinated phenoxide. This provides a useful alternative to conventional methods of ether hydrolysis, which involve treatment with Lewis acids such as boron tribromide, hydriodic acid or pyridinium chloride.

complexes and the hydroxylation or cyanation of an aromatic ring, there are indications that these reactions may be rather more facile and widely distributed than presently thought. For example, co-ordination to nickel(II) has been shown to be effective in activating haloarenes towards displacement of halide by phosphines (Fig. 8-7). This reaction is thought to involve the intermediacy of an organonickel intermediate.

A related reaction of a bromoarene with a phosphine is seen in the conversion of **8.4** to **8.5**, which occurs in the presence of copper or nickel salts. Organometallic intermediates are once again implicated in this process.

In another related process, aryl ethers have been shown to undergo a facile cleavage reaction upon treatment with copper salts in the presence of an amine (Fig. 8-8). The driving force for the reaction is primarily the stabilisation of the phenoxide by co-ordination to the metal. Simple azo complexes have been shown to undergo these reactions under very mild conditions. The process is somewhat reminiscent of the Arbuzov reactions discussed in Chapter 4. The pyridine probably functions as both a ligand and as a base in this reaction. Reactions of this type are the basis of a useful conversion of a methoxy-substituted dye, **8.6**, to the corresponding phenol, **8.7**, in the presence of copper(II) salts and ammonia.

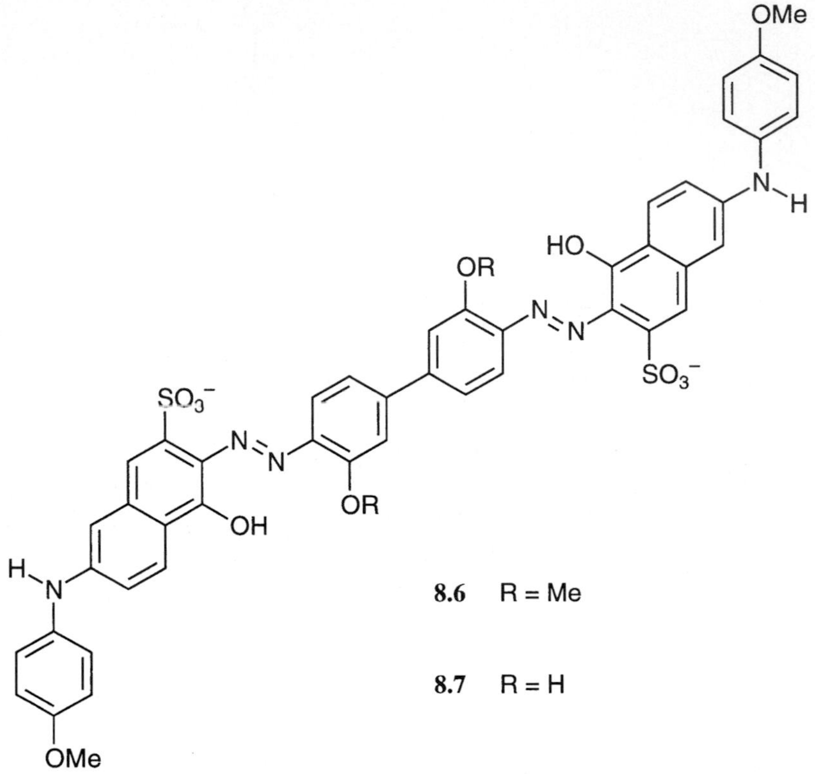

8.6   R = Me

8.7   R = H

8.4 Heteroaromatic Ligands

Heterocyclic ligands such as imidazole, pyridine, 2,2'-bipyridine and 1,10-phenanthroline (Fig. 8-9) have played a formative role in the development of co-ordination chemistry. Many hundreds of thousands of complexes and complex ions containing these ligands, such as $[Co(NH_3)_5(Himid)]^{3+}$, $[Ag(py)_4]^{2+}$ and $[Ni(bpy)_3]^{2+}$, have been prepared and characterised. The variety of spectroscopic properties and stoichiometries observed led to an improved understanding of the geometry and bonding in complexes and provided a touchstone for bonding theories.

It is convenient to consider heteroaromatic ligands in two classes – π-excessive, five membered rings typified by pyrrole, furan and thiophen, and π-deficient six-membered rings typified by pyridine. The π-excessive heterocycles are usually extremely reactive towards electrophilic attack and, with the exception of thiophen, do not exhibit the chemical inertness often associated with aromatic benzene derivatives. Conversely, the π-deficient heterocycles are extremely inert with respect to electrophilic attack. Paradoxically, it is the high reactivity of the five-membered rings and the inertness of the six-membered rings that give rise to common synthetic problems. The usual methods for the

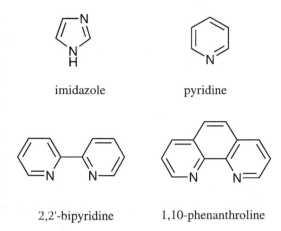

Figure 8-9. Some heterocyclic ligands of importance to co-ordination chemists

preparation of substituted benzene derivatives involve electrophilic (or more rarely nucleophilic) substitution of the ring. Such methods are not usually of use with heterocycles – they are either too vigorous or not sufficiently forcing! The outcome is that the preparation of derivatives of heterocycles is usually achieved by the synthesis of the ring from open-chain precursors bearing the appropriate functional groups in the desired positions.

8.5 Reactions of Five-membered Heterocycles

The reactivity of the five-membered heterocycles pyrrole, furan, thiophen and imidazole (Fig. 8-10) is characterised by interactions with electrophilic reagents. The precise nature of these reactions depends upon the particular ring system. Thiophens undergo facile electrophilic substitution, whereas the other compounds exhibit a range of polymerisation and other Lewis acid-initiated reactions upon treatment with electrophiles. We saw a number of examples of Lewis acid-promoted reactions of furans and pyrroles in Chapter 6. Although reactions of complexes of five-membered heterocyclic ligands have not been widely investigated, a few examples will illustrate the synthetic potential.

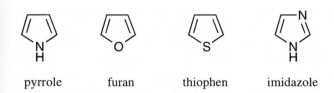

pyrrole   furan   thiophen   imidazole

Figure 8-10. Some π-excessive five-membered ring heterocyclic ligands of importance in co-ordination chemistry.

The imidazole complex [Co(NH₃)₅(HImid)]³⁺ (**8.8**) undergoes facile reaction with electrophilic reagents to give substitution products of the imidazole ring (Fig. 8-11). Thus, the nitration of [Co(NH₃)₅(HImid)]³⁺ with a mixture of concentrated nitric and sulfuric acids gives almost quantitative yields of the 4-nitroimidazole complex (**8.9**), as depicted in Fig. 8-11. The use of the kinetically inert cobalt(III) centre precludes complication by side-reactions involving the free ligand, and also ensures that no competing metal-centred oxidation processes occur.

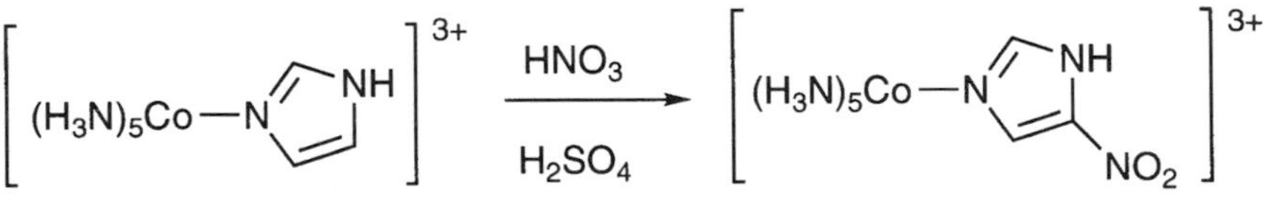

8.8   **8.9**

Figure 8-11. The nitration of an imidazole ligand co-ordinated to a kinetically inert cobalt(III) centre.

Similar electrophilic reactions occur when **8.8** is treated with bromine water. At low temperature the 4,5-dibromoimidazole complex **8.10** is obtained, whilst at higher temperature and after longer reaction times the 2,4,5-tribromoimidazole complex **8.11** is formed. Detailed mechanistic studies suggest that the important step is the reaction of a *deprotonated* co-ordinated imidazole ligand with molecular bromine.

When pyrroles are treated with strong acids, a mixture of tarry polymeric products is obtained, whilst the reaction of the complex **8.12** with a mixture of concentrated nitric and sulfuric acids yields the trinitrated product **8.13**. The deprotonation of the co-ordinated pyrrole is, once again, important. The complex **8.12** also undergoes a facile deuterium exchange to give **8.14**.

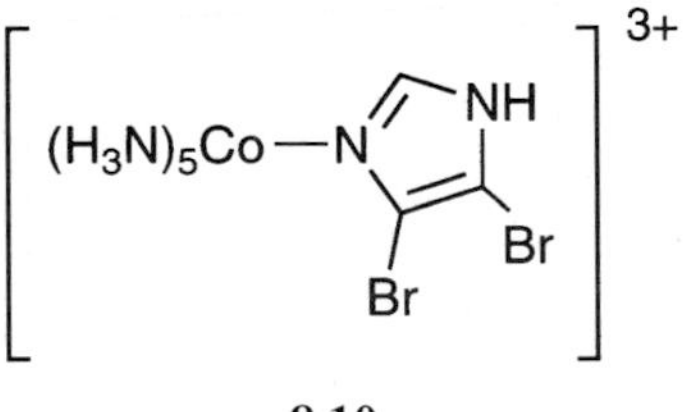

8.10 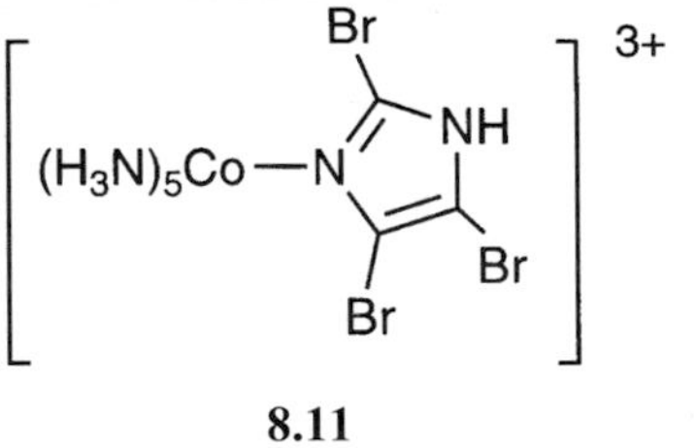   **8.11**

Although brominated derivatives of the five-membered heterocycles may be prepared by reactions of the co-ordinated ligands, these may then undergo further reactions with nucleophiles. As an example, the nucleophilic displacement of bromide from **8.15** by sulfide has been used to form new macrocyclic systems (Fig. 8-12). The palladium probably serves a dual function in this reaction. First, it organises the open-chain ligand such that the two reactive sites are held in proximity, so allowing the intramolecular formation of the sulfide and, second, it may activate the pyrrolic ring to nucleophilic displacement of bromide.

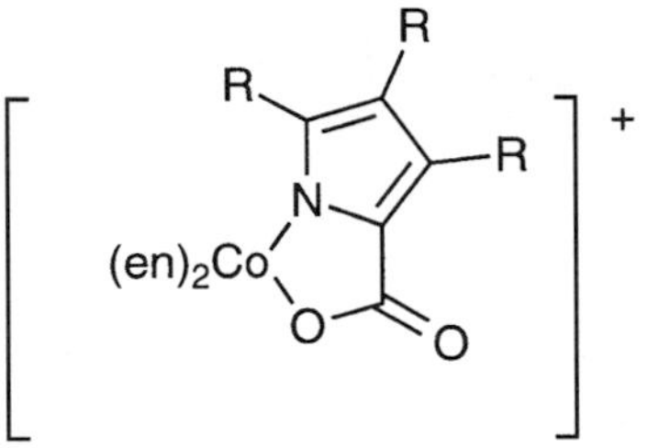

8.12	R = H
8.13	R = NO$_2$
8.14	R = D

The activation of the co-ordinated ligands towards attack by nucleophiles is also seen in the displacement of nitrite from **8.16** by the very weakly nucleophilic chloride ion (Fig. 8-13). This displacement occurs when the complex is treated with aqueous sodium chloride solution!

Actually, a wide range of cyclisation reactions of complexes involving five-membered rings and leading to porphyrins and related macrocyclic ligands have been reported. In general, these involve C-C bond formation of the type discussed in earlier chapters. The primary function of the metal centre is most likely to provide conformation control of the open-chain precursor or precursors. In general, the reactivity is basically that of the highly active pyrrollic ring systems. These metal-directed syntheses of porphyrins and porphinoids have achieved considerable successes in the synthesis of natural and unnatural macrocyclic ligands.

The formation of a porphyrinatocopper(II) complex from a biladiene derivative **8.17** is illustrated in Fig. 8-14. This mechanistically complex reaction involves a variety of metal-induced dehydrogenation and deprotonation steps. A similar strategy is used in the synthesis of the didehydrocorrin **8.18**, although in this case an enamine containing the correct number of carbon atoms is used as the precursor (Fig. 8-15). A wide range of related reactions involving metal-directed syntheses of porphyrins and similar ring systems from tetra-, di- and monopyrrolic precursors is also known.

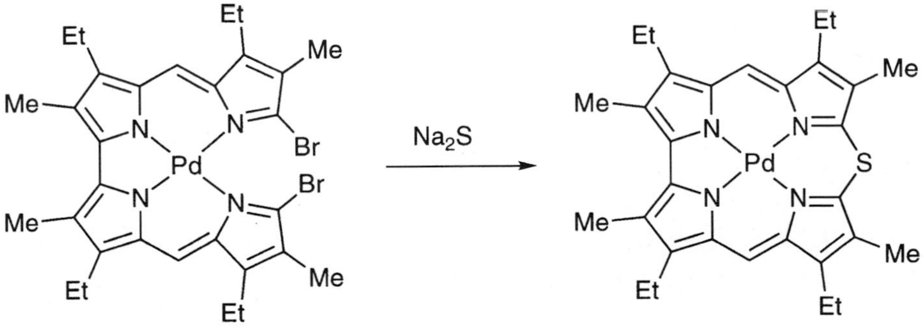

Figure 8-12. The use of a sulfide nucleophile to form a new tetrapyrrole macrocycle by the displacement of bromide from co-ordinated bromopyrroles.

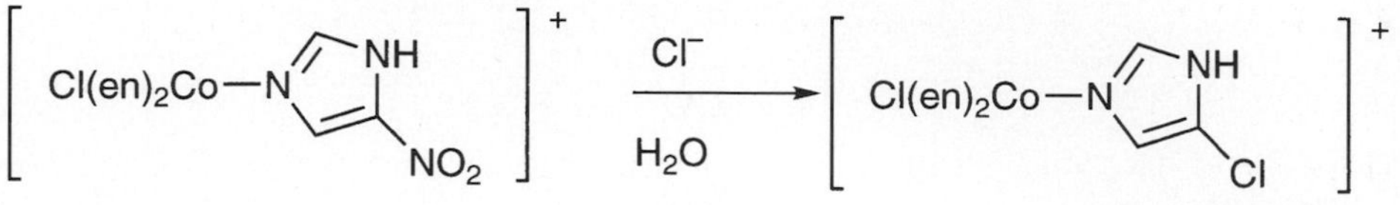

8.16

Figure 8-13. The nitro group in 4-nitroimidazole is activated towards attack by nucleophiles upon co-ordination to cobalt(III). In this case, the very weak nucleophile chloride ion displaces nitrite from **8.16** to give a co-ordinated 4-chloroimidazole ligand.

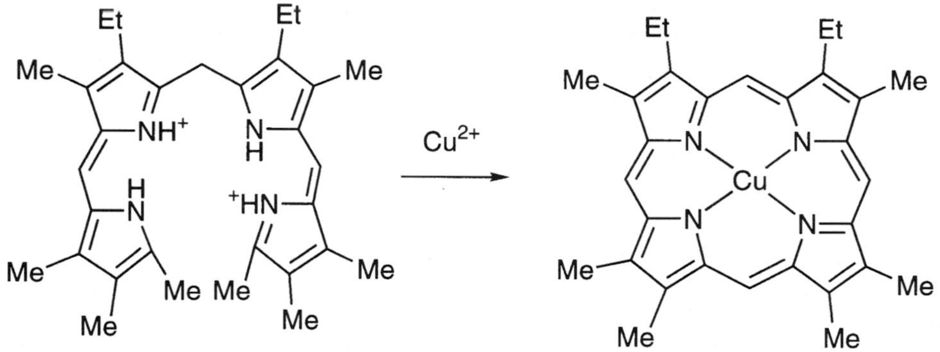

8.17

Figure 8-14. The copper-mediated formation of a porphyrin from a tetrapyrrole precursor.

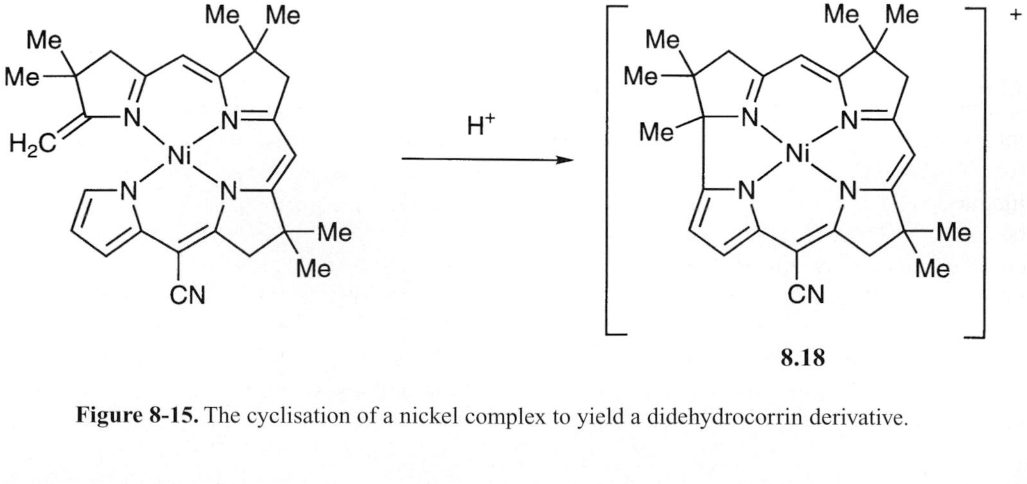

8.18

Figure 8-15. The cyclisation of a nickel complex to yield a didehydrocorrin derivative.

8.6 Reactions of Co-ordinated Pyridines with Nucleophiles

There has been a lively (and at times acrimonious) debate over the manner and degree to which the reactions of pyridines are modified upon co-ordination to a metal ion. Formally, co-ordination of a pyridine to a metal ion is analogous to quaternisation by an alkylating agent. This analogy between, say, the *N*-methylpyridinium cation **8.19** and a co-ordinated pyridine is emphasised in the representation of the co-ordination bond in Fig. 8-16. Notice that this valence bond representation of the M–N bond implies the development of positive charge at the 2- and the 4-positions of the co-ordinated pyridine ring. This build-up of positive charge on the ring should activate the pyridine towards nucleophilic attack.

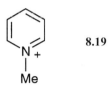

8.19

The ligand 2,2'-bipyridine (bpy) has played an important role in the development of co-ordination chemistry and, over the years, an increasing number of 'anomalies' in the reactivity of chelated complexes containing bpy or 1,10-phenanthroline (phen) have been noted. In part, these have been made obvious by the kinetic and thermodynamic stability of the chelated complexes that these strong field π-acceptor ligands form with transition metal complexes. In particular, the kinetic stability of complexes with d^3 and d^6 metal centres has allowed their reactions under relatively forcing conditions to be investigated. Detailed kinetic studies have shown that a remarkable variety of reactions, including hydrolysis, racemisation and electron transfer processes of these complexes show a strong and marked pH dependence. Specifically, rate equations such as Eqs (8.1) or (8.2) are frequently observed.

$$\text{rate} = k_1[M(bpy)_3] + k_2[M(bpy)_3][OH^-] \tag{8.1}$$
$$\text{rate} = k_1[M(bpy)_3] + k_2[M(bpy)_3][H^+]^{-1} \tag{8.2}$$

Rate equations of this type are normally associated with the formation of associative intermediates *or* the involvement of deprotonated ligand forms in reactions with two (or more) competitive pathways. However, in the case of octahedral metal complexes of ligands such as 2,2'-bipyridine or 1,10-phenanthroline, such mechanisms do not appear to be likely. Associative mechanisms would involve seven-co-ordinate intermediates, which are likely to be sterically strained and electronically disfavoured on ligand field grounds. Furthermore, this type of ligand does not appear to contain any strongly acidic protons which are likely to be involved in reactions with aqueous hydroxide ion (but see later). The suggestion has been made by Gillard that all of these observations may be conveniently explained in terms of nucleophilic attack by hydroxide ion on the co-ordinated ligand, which is rendered more electrophilic by co-ordination to the charged metal ion.

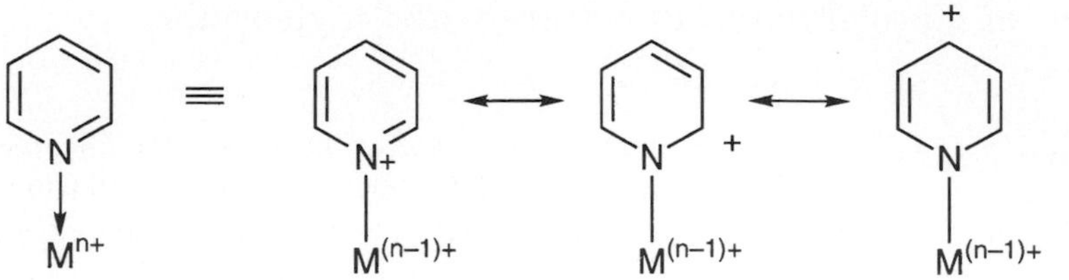

Figure 8-16. A valence bond representation of a co-ordinated pyridine. The positive charge is delo-calised and the 2- and the 4-positions of the ligand develop electrophilic character.

The specific proposal is that the hydroxide attacks the 2- or the 4-position of the co-ordi-nated pyridine to form a hydroxy-substituted 1,2- or 1,4-dihydropyridine (Fig. 8-17). These hydroxy species are known as pseudo-bases. This behaviour is fully in accord with the known behaviour of *N*-alkylpyridinium cations. Although the pseudo-bases of simple *N*-alkylpyridinium cations are not dominant solution species under aqueous conditions, those derived from a variety of other nitrogen heterocycles are readily formed and are well-known. The suggestions had the advantage of linking the apparently divergent fields of heterocyclic and co-ordination chemistry by explaining some well-documented anomalies in the reactivity of pyridine complexes.

Unfortunately, this proposal has generated a remarkably large amount of dissension, in part, at least, due to a strain of conservatism amongst co-ordination chemists! The problems arise from the way in which the ligand polarisation will express itself in the reac-tivity of the co-ordinated ligand. If we consider the free ligand pyridine, the electron den-sities alternate around the ring, with the 2- and 4-positions bearing slight positive charges, and the nitrogen atom and the 3-position having negative charge (Fig. 8-18).

It is convenient to separate the total electron density at each atom into σ- and π-com-ponents. It is likely to be the π-density that will be important in reactions with nucleophi-les, since in an orbitally controlled reaction (Chapter 1) the donor orbital of the incoming nucleophile will initially interact with the lowest vacant π^*-orbital. The overall pattern of charge alternation is repeated in both the π- and the σ-electron densities, and nucleophiles are expected to attack at the 2- or 4-positions. This is exactly the pattern that is seen in

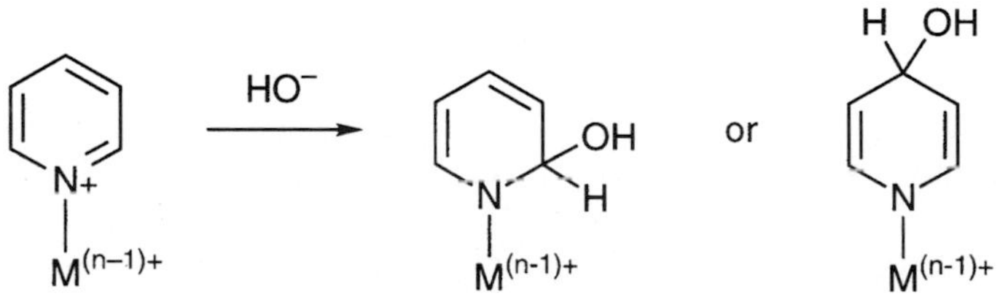

Figure 8-17. Gillard's proposal for the formation of pseudo-bases of co-ordinated pyridine ligands. It was suggested that the formation of these species allowed a simple explanation for the observed pH dependence of a variety of reactions of complexes of pyridine derivatives.

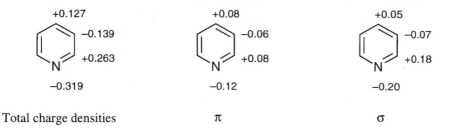

+0.127	+0.08	+0.05
−0.139	−0.06	−0.07
+0.263	+0.08	+0.18
−0.319	−0.12	−0.20

Total charge densities   π   σ

Figure 8-18. Charge densities in the pyridine molecule, calculated using the Fenske-Hall method. The total charge density indicates that pyridine might be expected to react with electrophiles at C(2) and C(4). Also shown are the residual charges associated with the gain or loss of electrons as a result of σ- and π-bonding at each carbon and nitrogen atom within the molecule.

the Chichibabin reaction of a pyridine with sodium amide to yield a 2-aminopyridine (Fig. 8-19).

The bulk of the discussion of Gillard's hypothesis has revolved around the relative contributions to the reactivity of the polarising effect of the metal ion, which will be predominantly expressed in the σ-orbitals, and any back-donation from the metal, which will be observed in the π-levels. Clearly, the overall charge distribution within the co-ordinated ligand will represent the balance of these two competing effects. However, we should also note that it may be the character of individual orbitals that is the critical factor. Electron-withdrawal from the σ-orbitals is expected to further activate the co-ordinated ligand towards attack by nucleophiles, whereas back-donation into the π-orbitals is expected to deactivate it. Calculations at various levels have purported to answer this question; contradictory results have demonstrated variously that the ligand will be activated, deactivated or unchanged upon co-ordination to a metal ion! Let us move to the reaction chemistry of these ligands to see if there is any support for the initial proposal.

Many of the studies purporting to show nucleophilic attack upon co-ordinated heterocyclic ligands have been performed with complexes of 5-nitro-1,10-phenanthroline, **8.20**. Unfortunately, this ligand proves to be a less than optimal choice, as nitroalkenes undergo facile reactions with nucleophiles and, in contrast to simple pyridines, the free ligand reacts with hydroxide ion. Hydroxide ion and other nucleophiles react with **8.20** at the 6-position to give the stabilised anion **8.21** (Fig. 8-20). It is almost certain that the majority of the reactions of 5-nitro-1,10-phenanthroline complexes with hydroxide

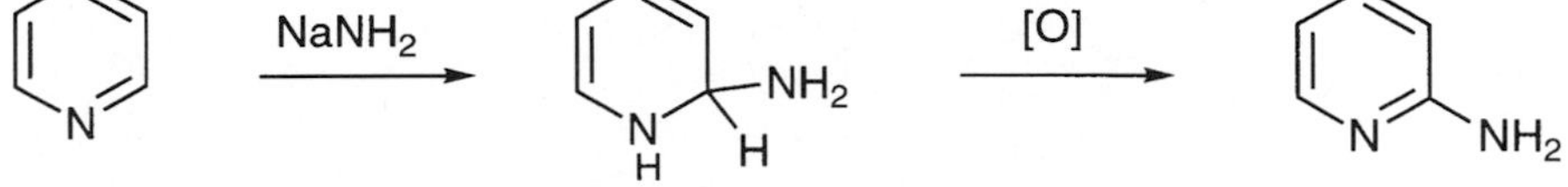

Figure 8-19. In the Chichibabin reaction a pyridine reacts with sodium amide in a solvent such as N,N-dimethylaminobenzene. The initial product is a dihydropyridine, which is oxidised to give a 2-aminopyridine. The attack of the amide nucleophile occurs predominantly at the 2-position of the pyridine.

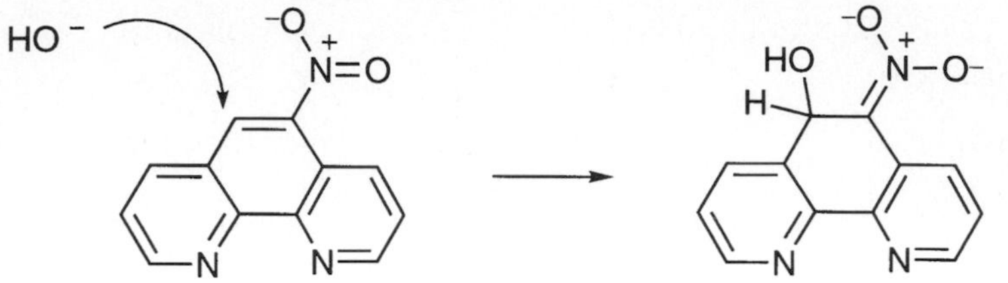

8.21

Figure 8-20. The reaction of hydroxide with the free ligand **8.20** gives a stabilised anion.

ion can be explained in terms of nucleophilic attack at the 6-position of the ligand rather than at the 2-position of the pyridine ring, as expected on the basis of the quaternisation analogy (Fig. 8-21). These reactions are extremely interesting and the influence of the metal ion upon the rate of attack at the 6-position provides valuable information about metal activated processes. No direct evidence for attack at the *pyridine* ring has been presented, although there have been suggestions that the kinetic product involves attack at the pyridine 2-position, followed by rearrangement of the nucleophile to the 6-position. Clearly, kinetic studies cannot unambiguously determine the site of attack. A wide variety of square-planar complexes of d^8 metal ions and kinetically inert octahedral complexes of d^3 and d^6 metal ions containing **8.20** as a ligand undergo this reaction with a range of nucleophiles, including hydroxide, alkoxide, amines, cyanide and nitromethane.

Accordingly, in order to establish activation of the *pyridine* ring towards nucleophilic attack, it is necessary to look at reactions of complexes containing the *unactivated* ligands, such as pyridine, bpy or phen, rather than **8.20**.

However, even in those cases involving unactivated ligands, it is still possible for numerous ambiguities to arise. Initial studies of the reactions of the complexes $[Pt(phen)_2]^{2+}$ and $[Pt(bpy)_2]^{2+}$ with nucleophiles such as hydroxide and cyanide indicated clearly that rapid and reversible spectroscopic changes occurred upon the addition of the nucleophile. These results were consistent with the formation of an associative intermediate and were interpreted in terms of attack of the nucleophile at the carbon atom adjacent to the nitrogen donor of the ligand (Fig. 8-22).

It has now emerged that these changes *are* due to the reversible addition of the nucleophile to the complex, *but* the site of addition in the products is at the metal. In other words, a five co-ordinate complex has been formed instead of the species in which the hydroxide is covalently attached to the pyridine ring (Fig. 8-23). It is, of course, impossible to distin-

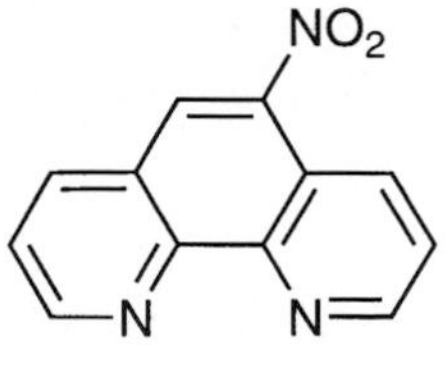

8.20

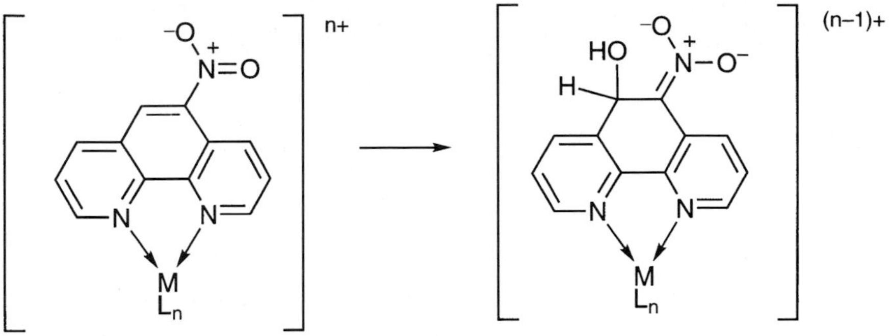

Figure 8-21. Nucleophiles also react with 5-nitro-1,10-phenanthroline when it is co-ordinated to a metal centre. The addition is favoured by the build-up of charge upon the ligand after co-ordination.

guish between these two processes on the basis of kinetic or simple spectroscopic data and there is no guarantee that intermediates involving attack at the ligand are not involved.

Accordingly, the search for unambiguous data leads to studies of kinetically inert octahedral metal complexes or pyridine, bpy and phen, which are not expected to form seven co-ordinate products. There have been numerous studies of $[M(bpy)_3]^{n+}$ and $[M(phen)_3]^{n+}$ complexes, but it is probably fair to state that no definitive evidence for activation towards attack of hydroxide at the *ligand* has been presented. There is no doubt, however, that sufficient kinetic anomalies exist for further study to be worthwhile, and the importance of $[M(bpy)_3]^{n+}$ and $[M(phen)_3]^{n+}$ complexes in photocatalytic systems should justify this. A *possible* chemical consequence of the reaction of a nucleophile with a co-ordinated heterocycle is seen in the reaction of *tris*(2,2'-bipyrimidine)ruthenium(II) and hydroxide; the products of the reaction are derived from hydroxide attack upon the co-ordinated pyrimidine (Fig. 8-24).

Probably the best characterised example of a product containing a ligand formally resulting from the addition of hydroxide ion to a co-ordinated heterocyclic ligand is seen in the reaction of the quinoxaline **8.22** with $[ReOCl_4]^-$ (Fig. 8-25). The product has the

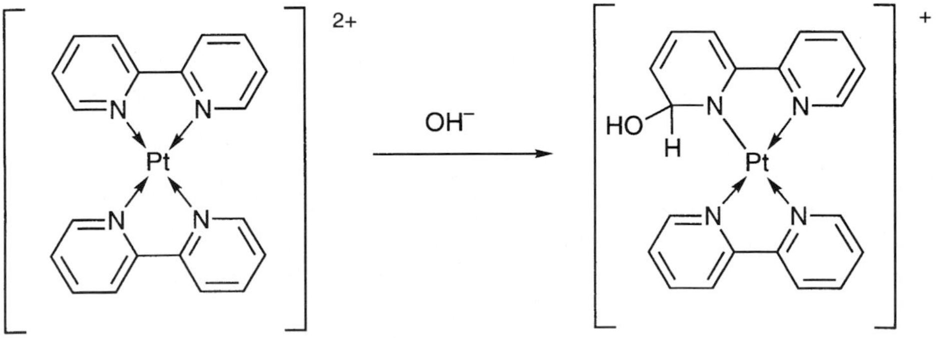

Figure 8-22. The Gillard mechanism for attack of hydroxide at a bpy ligand in a square-planar platinum(II) complex.

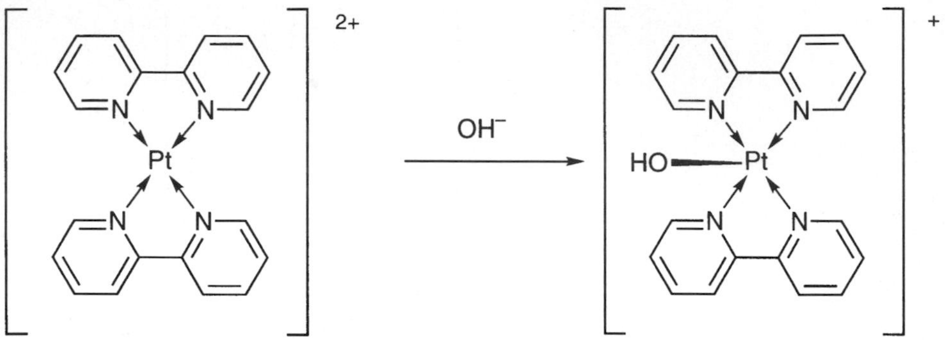

Figure 8-23. The formation of a *five-co-ordinate* platinum(II) complex after the addition of hydroxide ion to [Pt(bpy)₂]²⁺. This is thought to be the final product of the reaction.

structure **8.23** and formally arises from the addition of hydroxide to the quinoxaline. Whilst there is no doubt about the structure of the complex, the interpretation is not unambiguous. Unco-ordinated quinoxalines are known to react with water or hydroxide ion to form addition products, and the isolation of this product from a reaction of the free ligand with the rhenium complex could be interpreted in terms of addition of the water *before* or *after* co-ordination of the quinoxaline. Furthermore, the site of attack *is not* at a carbon adjacent to a nitrogen co-ordinated to a metal ion, as predicted by the Gillard mechanism. Finally, the new C–O bond could originate from the Re=O oxo group rather than from water.

The jury is probably still out of court regarding the reactions of simple nucleophiles such as hydroxide with co-ordinated pyridine ligands. Paradoxically, although it has proved to be very difficult to design experiments to unambiguously establish nucleophilic attack by hydroxide ion at the ligand, some of these self-same experiments have established a new reaction mode for co-ordinated pyridines. Treatment of [Ru(bpy)₃]²⁺ or [Os(bpy)₃]²⁺ salts with strong base in deuterated solvents results in a specific deuterium exchange reaction of the 3 and 3' positions of the ligand, followed by a slower exchange of the 5 and 5' protons (Fig. 8-26). A related reaction has been observed for [Rh(bpy)₃]³⁺

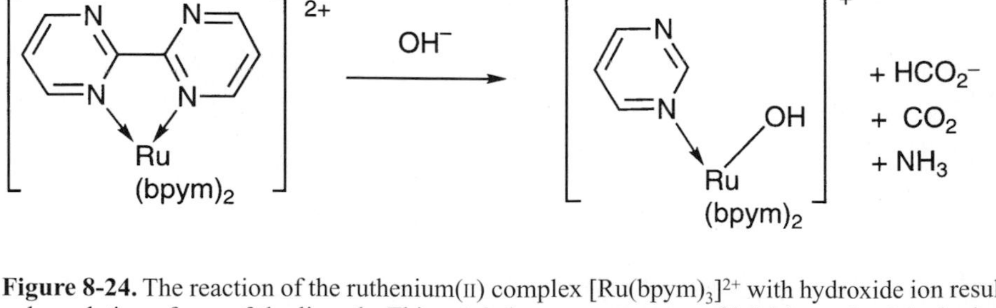

Figure 8-24. The reaction of the ruthenium(II) complex [Ru(bpym)₃]²⁺ with hydroxide ion results in a degradation of one of the ligands. This *might* be a consequence of initial attack of the hydroxide ion upon the 2,2'-bipyrimidine, although other possibilities cannot be ruled out.

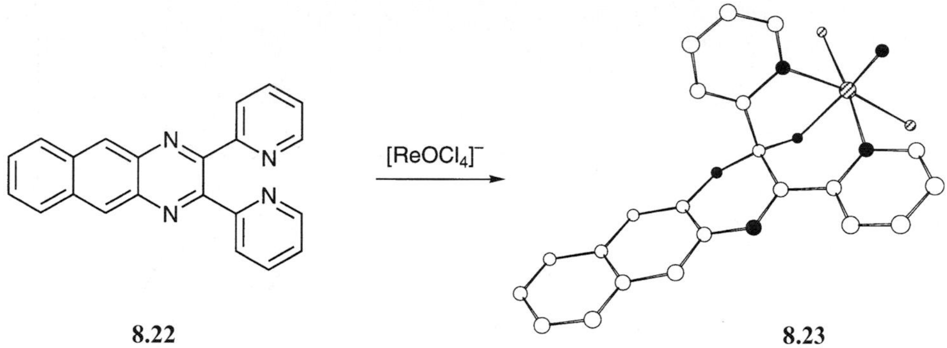

8.22

[ReOCl₄]⁻ →

8.23

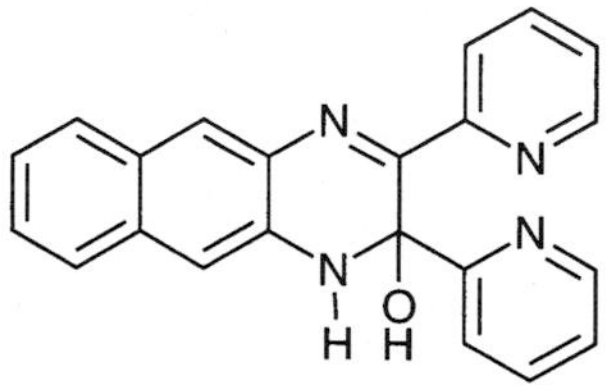

8.24

Figure 8-25. The reaction of the quinoxaline **8.22** with [ReOCl₄]⁻ gives the complex **8.23** containing the ligand **8.24**. Although **8.23** is formally the product from the addition of water to the co-ordinated ligand, little is known about the mechanism of formation of the products.

salts, but in this case the first site to exchange is the 6-position, followed by the 3-position. Clearly, there is a charge effect upon this process. The ruthenium(II) 2,2':6',2"-terpyridine complex [Ru(tpy)$_2$]$^{2+}$ undergoes similar reactions, with exchange occurring specifically at the 3,3',5' and 3"-positions to give **8.25**.

Why do these exchange reactions occur? The reaction almost certainly proceeds by deprotonation of the co-ordinated bpy ligand to generate a co-ordinated pyridyl anion. The simplest mechanism involves a simple deprotonation of the ligand, with the hydroxide ion acting as a base (Fig. 8-27). The difficulty with this mechanism is that it is not

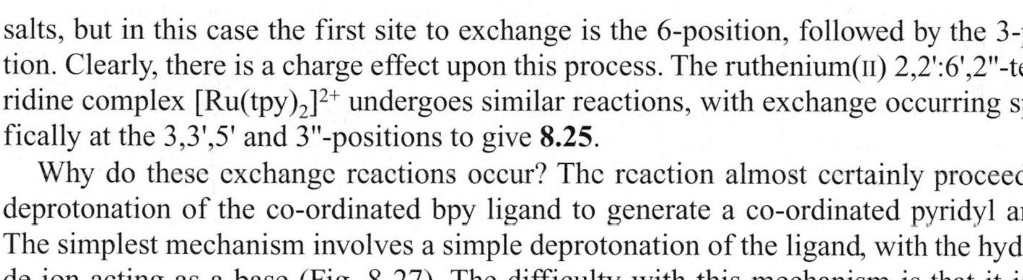

Figure 8-26. The [Ru(bpy)$_3$]$^{2+}$ and [Os(bpy)$_3$]$^{2+}$ cations undergo a specific deuteration at the 3-position of the pyridine ring. Further, slower, reactions occur to give sequential deuteration at the 5-, 6- and 4-positions.

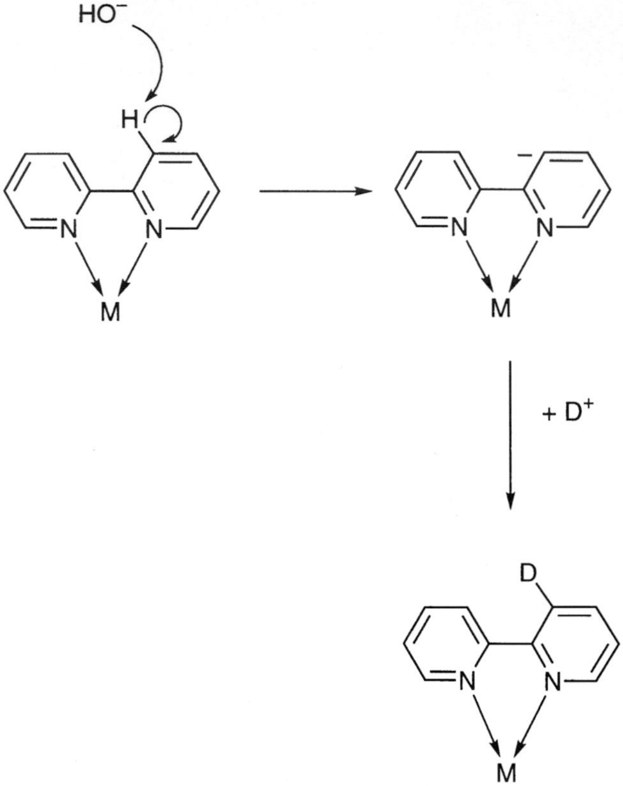

Figure 8-27. A possible mechanism for the deuterium exchange reaction in [Ru(bpy)$_3$]$^{2+}$ and [Os(bpy)$_3$]$^{2+}$ complexes. It is not obvious why deprotonation of the 3-position should occur selectively.

immediately obvious why the 3-position should be selectively deprotonated. Indeed, if anything, the charge distributions in the pyridine might suggest that deprotonation should occur at the 4- and the 6- positions.

In principle, exchange at the 3- or the 5- positions could be a result of nucleophilic attack by the hydroxide at the 6- or 4- position. Nucleophilic attack would occur at the electrophilic 4- or 6-positions of the pyridine rings, with subsequent H–D exchange at the 3- and the 5- positions (Fig. 8-28). The selectivity for exchange at the 3-position may be rationalised in terms of the relief of steric strain in the co-ordinated ligand. The 3- and 3'-protons are close together in the normal [M(bpy)$_3$]$^{2+}$ complexes – this strain should be relieved upon the addition of hydroxide and the formation of an sp^3 hybridised centre at C(3).

This mechanism does not explain what is happening in the rhodium(III) complex, [Rh(bpy)$_3$]$^{3+}$, where exchange occurs first at the 6-position. Increasing the charge on the metal ion will increase the tendency towards attack by a nucleophile and also the acidity of the pyridine ligand. Nucleophilic attack will always occur at the 4- and 6-positions of

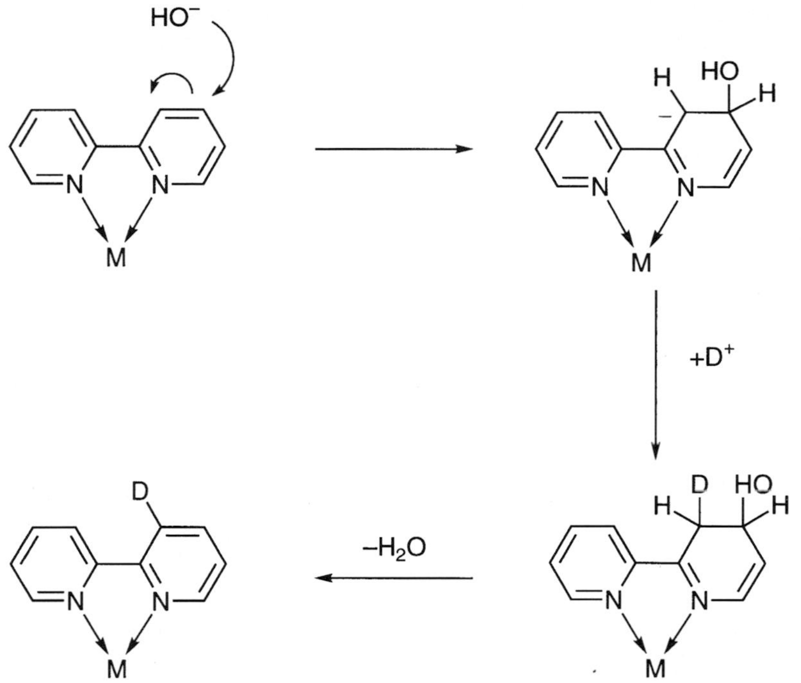

Figure 8-28. A mechanism for the deuteration of bpy complexes involving nucleophilic attack by the hydroxide ion. The hydroxide attacks the most electrophilic 4- and 6-positions of the co-ordinated pyridine ring. Specificity for the 3-position may be associated with steric relief as a result of the formation of an new *sp³* hybridised carbon centre.

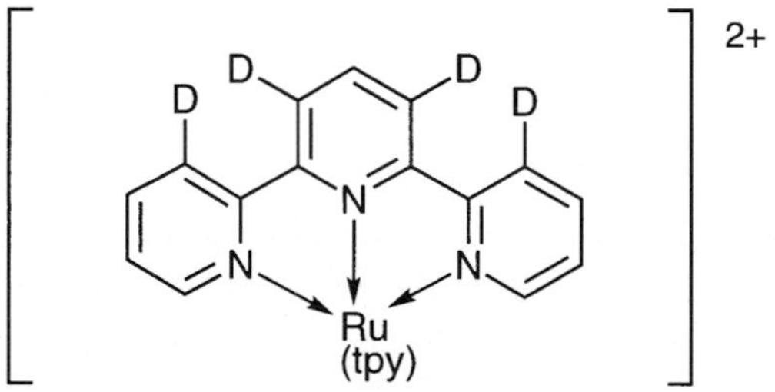

8.25

the ligand and this mechanism can *only* account for exchange at the 3- and 5-positions. In contrast to this, deprotonation is a function of the ability of the ligand to stabilise the anion through, essentially, the σ-bonding framework of the molecule. This is a consequence of our formally placing the charge into an *sp²* hybrid orbital lying in the plane of the ligand.

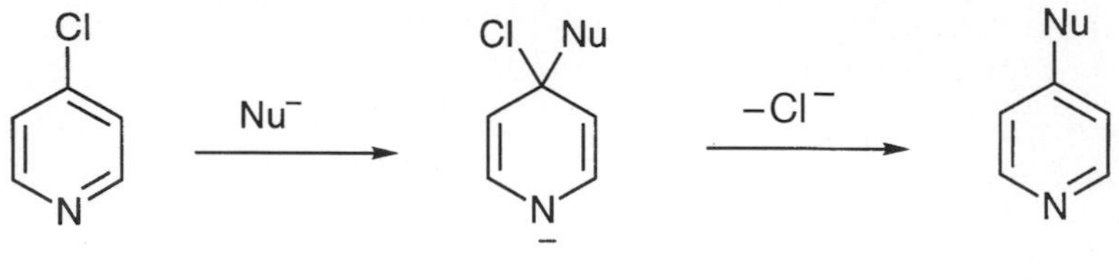

Figure 8-29. The addition-elimination process for the displacement of halide ion from a 4-halo-pyridine. The same mechanism is operative for 2-halopyridines. The key feature is stabilisation of the charge on the electronegative nitrogen atom in the intermediate.

Increasing the charge on the metal ion will affect the σ electron density at those sites closest to the metal more than at those remote from it. This is presumably the origin of the selective exchange at the 6-position in the case of the rhodium(III) complex. These tentative results suggest that two mechanisms may be operative for these exchange processes.

We noted above that the charge distribution on the ring atoms of pyridine is not uniform, and the pattern is such that the 2- and the 4-positions are slightly electrophilic. If good leaving groups are attached to these sites, they may be readily displaced by nucleophiles in an addition–elimination process (Fig. 8-29). The displacement of halide from free 2- or 4-halopyridines is facile, and is the basis for the preparation of derivatives functionalised at these sites. In contrast, the 3-position is not activated towards nucleophilic attack and it is extremely difficult to derivatise pyridines at this site (Fig. 8-30).

The co-ordination of a pyridine ligand to a metal ion might be expected to sufficiently polarise the ligand to enable nucleophilic displacement of halo- or other substituents at the 3-position. Unfortunately, this is not the case and there appear to have been no successful metal-directed reactions of 3-halopyridines (Fig. 8-31).

However, the reactions of 2- and 4-halopyridines *are* modified by co-ordination to a metal. The most unambiguous studies are concerned with the relative rates of reaction of free and co-ordinated ligands. Co-ordination of 4,4'-dichloro-2,2'-bipyridine to a ruthenium(II) centre dramatically activates the halo groups towards nucleophilic displacement. The rate of displacement of halide is many times greater than that of the free ligand (Fig. 8-32). Nucleophiles such as HSO_3^-, which do not react with the free ligand, are effective in these reactions and this provides a useful method for the preparation of pyridine-4-sulfonic acids.

The mechanism of these displacement processes presumably involves a metal-assisted addition–elimination reaction, in which the metal ion stabilises the charge in the intermediate (Fig. 8-33). A number of useful synthetic applications of this methodology are

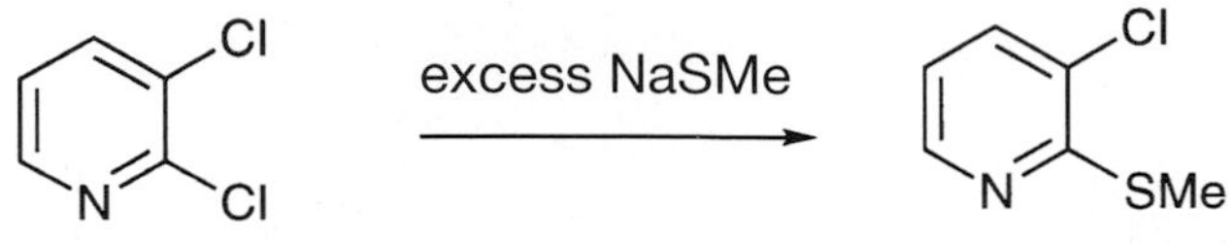

Figure 8-30. Only halogen groups attached to the 2- and 4-positions of pyridines are activated towards nucleophilic displacement. The 3-halopyridines are substitutionally inert.

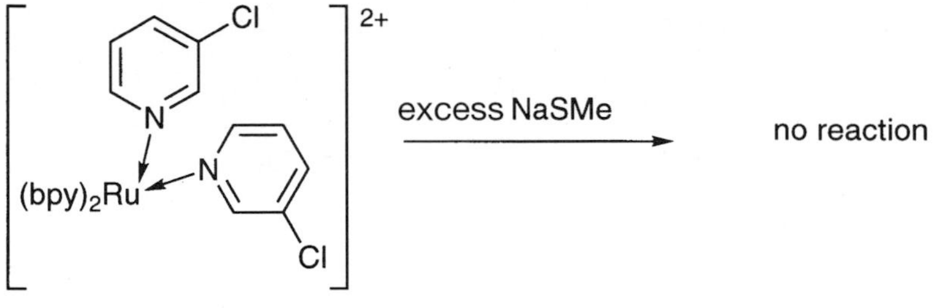

Figure 8-31. Even after co-ordination to a metal ion, 3-halopyridines are inert towards displacement of the halide by nucleophiles.

known. For example, **8.26** does not react with dimethylamine, whereas the iron complex $[Fe(\mathbf{8.26})_2][PF_6]_2$ reacts instantly at 0 °C to give the complex $[Fe(\mathbf{8.27})_2][PF_6]_2$; oxidation to the labile iron(III) complex $[Fe(\mathbf{8.27})_2][PF_6]_3$ followed by treatment with base allows the free ligand **8.27** to be isolated in near-quantitative yield. Similarly, **8.26** only reacts sluggishly with **8.28**, and **8.29** is only formed after four days in boiling dmf! In contrast, the complex $[Ru(\mathbf{8.26})_2][PF_6]_2$ reacts smoothly with **8.28** over a matter of hours to give the complex $[Ru(\mathbf{8.29})_2][PF_6]_2$.

Some particularly elegant applications of this methodology are used in the assembly of dendritic systems. For example, the ruthenium(II) complex of 4,4'-dichloro-2,2'-

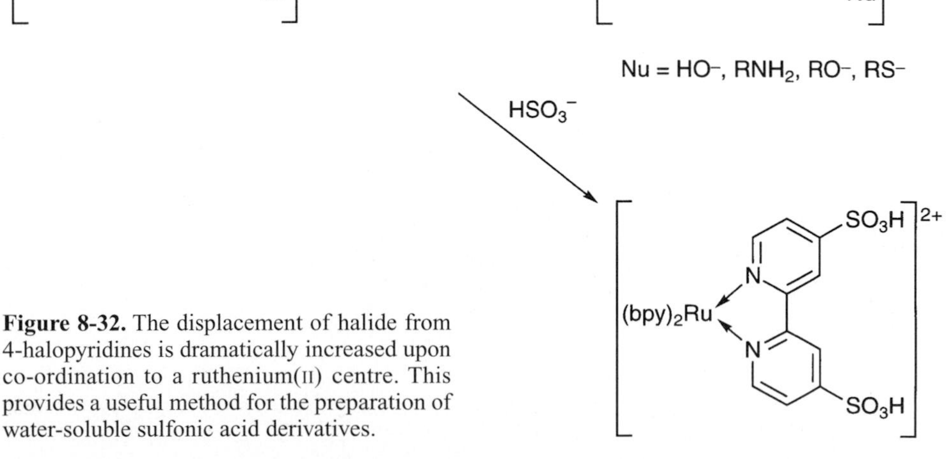

$Nu = HO-, RNH_2, RO-, RS-$

Figure 8-32. The displacement of halide from 4-halopyridines is dramatically increased upon co-ordination to a ruthenium(II) centre. This provides a useful method for the preparation of water-soluble sulfonic acid derivatives.

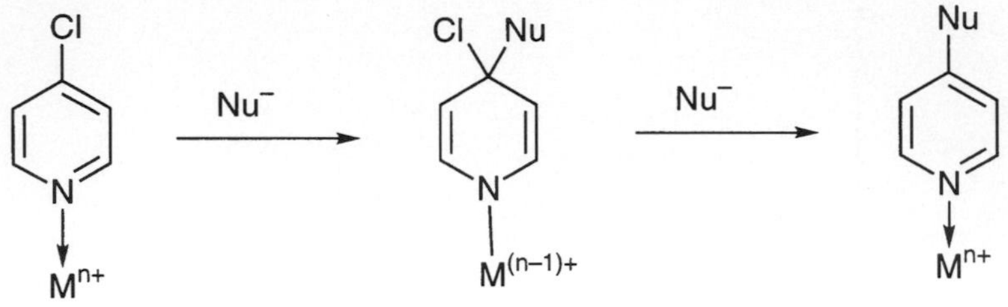

Figure 8-33. The metal-assisted addition–elimination reaction by which co-ordinated 2- and 4-halopyridines are activated.

bipyridine reacts smoothly with **8.28** to give a new complex with six 2,2':6',2"-ter pyridine groups arrayed around the central metal and that can be used to bind another six metal ions to give a heptanuclear complex (Fig. 8-34).

The displacement of halide ion by a wide variety of other nucleophiles has also been demonstrated. A particularly dramatic example is seen in the reaction of 6,6'-dichloro-2,2'-bipyridine (or 2,9-dichloro-1,10-phenanthroline) with ammonium tetrachlorozincate.

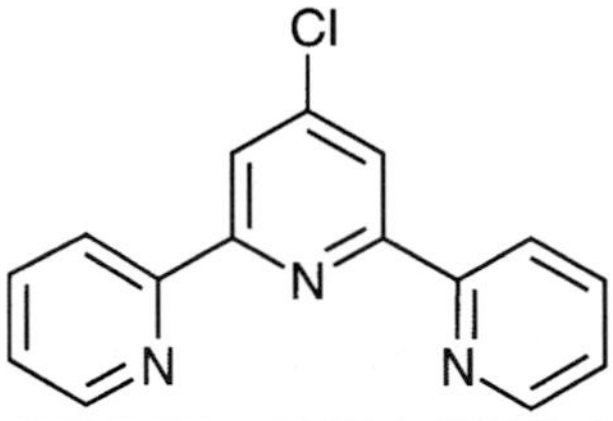

8.26

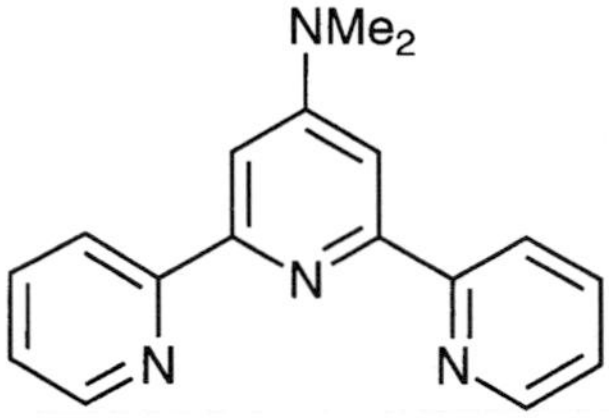

8.27

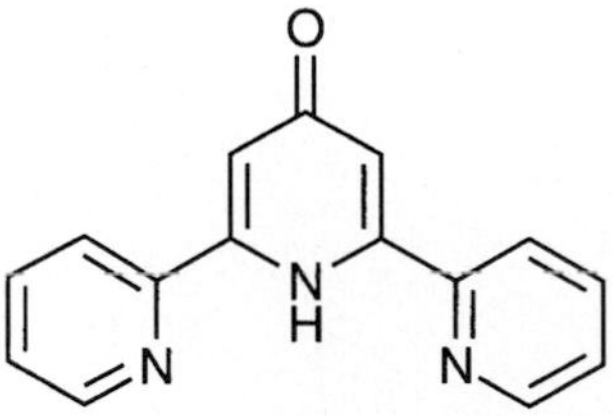

8.28

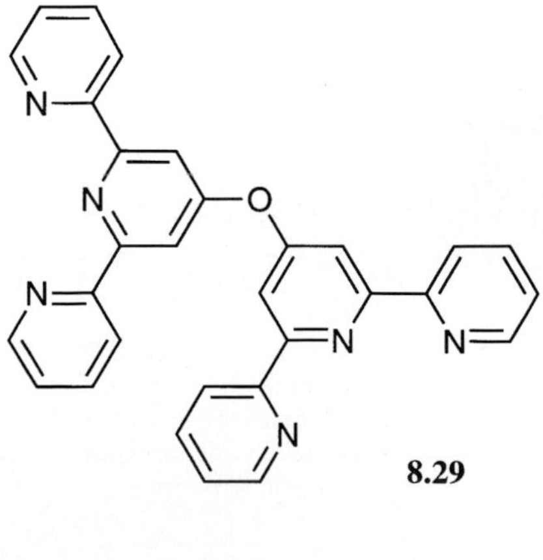

8.29

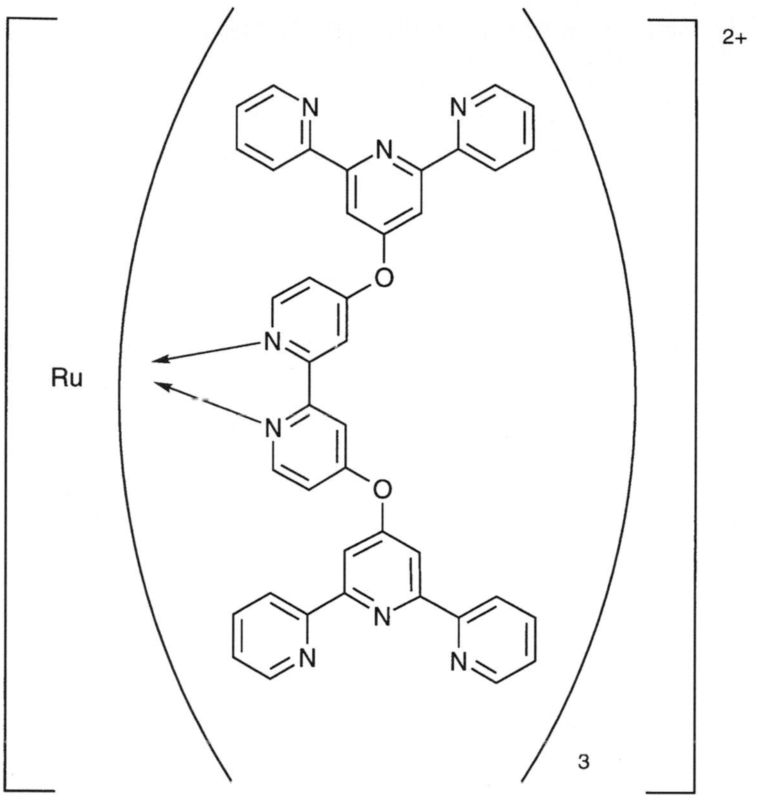

Figure 8-34. A new ligand with six unoccupied 2,2':6',2"-terpyridine metal-binding domains, prepared from the reaction of **8.28** with the complex [Ru(4,4'-Cl$_2$bpy)].

Co-ordination of the ligand to the zinc activates it towards nucleophilic attack by ammonia and the product is the zinc complex of a conjugated dianionic tetraaza macrocycle (Fig. 8-35). This ligand is of some interest in forming a very stable lithium complex, derivatives of which have found some application in photoaddressable devices.

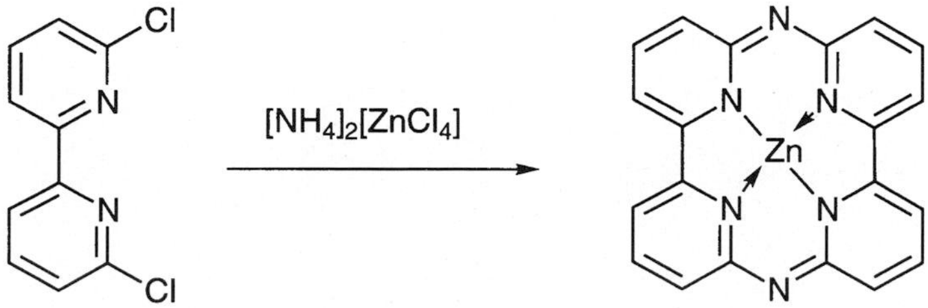

Figure 8-35. The formation of a macrocycle by the displacement of halogen from the 6-position of a pyridine.

8.7 Electrophilic Reactions of Co-ordinated Pyridines

In general, free ligand pyridines show a great reluctance to take part in electrophilic substitution reactions. Forcing conditions are frequently required, and low yields and specificity are normally observed. In principle, co-ordination to a metal ion capable of back-donation should increase the tendency for electrophilic attack, since back-donation results in an increase in π electron density.

The direct halogenation of pyridine ligands co-ordinated to palladium has been described. The chlorination of $[Pd(py)_2Cl_2]$ can give yields of 2-chloropyridine of the order of 90 % (Fig. 8-36). The mechanism of this reaction has not been studied in any depth. Co-ordinated pyridine ligands may even be nitrated in forcing conditions, although these reactions have received remarkably little recent attention. However, it is reported to be possible to nitrate pyridine ligands co-ordinated to ruthenium(II) to give 3-nitropyridine complexes (Fig. 8-37). In general, pyridine complexes are relatively inert towards electrophilic reactions and a series of kinetically inert chromium(III) and cobalt(III) complexes have been shown to resist nitration.

Very much easier nitration reactions are observed with cobalt(III), iron(III) and ruthenium(II) complexes of 1,10-phenanthroline, which undergo facile reactions with mixed acid to give complexes of 5-nitro-1,10-phenanthroline. This is probably the best way to prepare complexes of this ligand, which played a very important (although slightly misleading) part in the investigations of nucleophilic attack upon co-ordinated heterocycles discussed earlier. In this case, the nitration is associated with the benzene-like aromatic ring, and the generalisations in terms of rate and position of substitution mentioned in Section 8.2 apply. The position of substitution is the same as that in the free ligand, but the rate is significantly faster (Fig. 8-38). The increase in rate is thought to be a function of the metal ion protecting the 1,10-phenanthroline nitrogen atoms from pro-

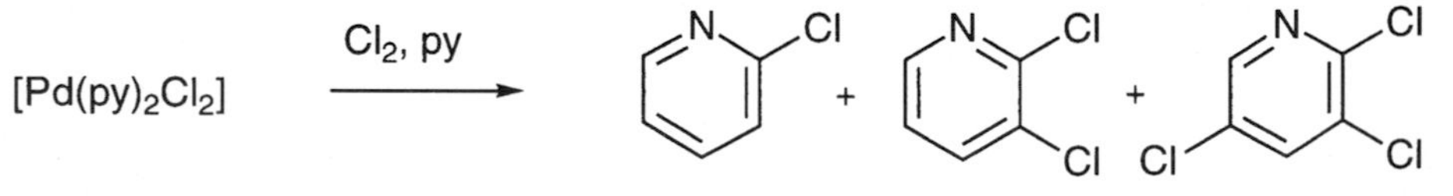

Figure 8-36. The chlorination of pyridine co-ordinated to palladium can give excellent yields of chloropyridines.

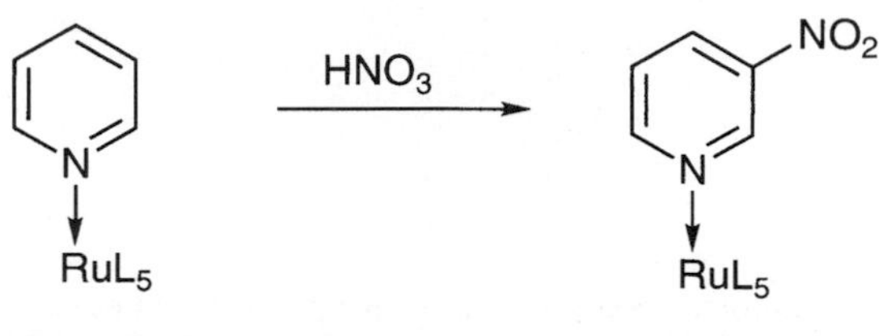

Figure 8-37. Co-ordinated pyridine ligands may be nitrated.

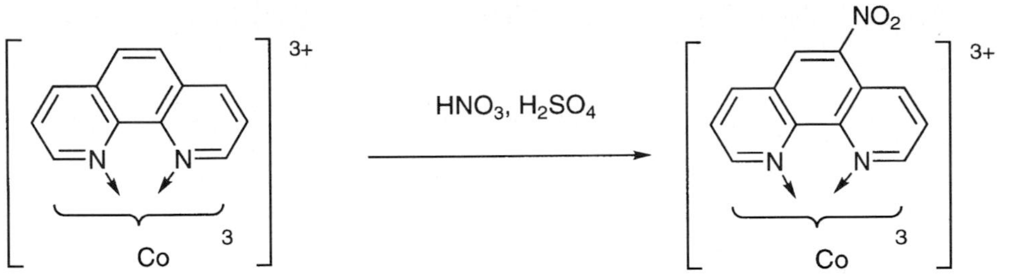

Figure 8-38. The nitration of [Co(phen)$_3$]$^{3+}$ provides a convenient method for the preparation of the corresponding 5-nitro-1,10-phenanthroline complex.

tonation. Protonation is a purely deactivating σ-bonding effect, whereas co-ordination to a metal ion involves the competing π-bonding component.

A similar pattern of reactivity is observed in quinoline complexes and very detailed studies have confirmed that 8-hydroxyquinolinesulphonic acid undergoes extremely rapid halogenation at the 7-position in the presence of metal ions (Fig. 8-39). Once again, the site of attack is the same as in the free ligand. Similar reactions of 8-hydroxyquinoline complexes occur with a variety of electrophiles.

The electrophilic reactions of co-ordinated 1,10-phenanthrolines are not always as simple as might be expected. Thus, the nitration of cobalt(III) 1,10-phenanthroline complexes yields 5-nitro-1,10-phenanthroline derivatives at low temperature, but prolonged reaction in hot solution leads to further reaction and oxidation of the ligand to give excellent yields of 1,10-phenanthroline-5,6-quinone complexes (Fig. 8-40). Even after the formation of the quinone, the complexes may exhibit further reaction. For example, reaction of the 1,10-phenanthroline-5,6-quinone complexes with base results in the formation of a complex of 2,2'-bipyridine-3,3'-dicarboxylic acid (Fig. 8-41)

The formation of this dicarboxylic acid derivative is of some interest. The free ligand quinone is known to undergo a benzilic acid rearrangement with concomitant ring contraction, followed by decarboxylation to yield a diazafluorenone on treatment with base (Fig. 8-42).

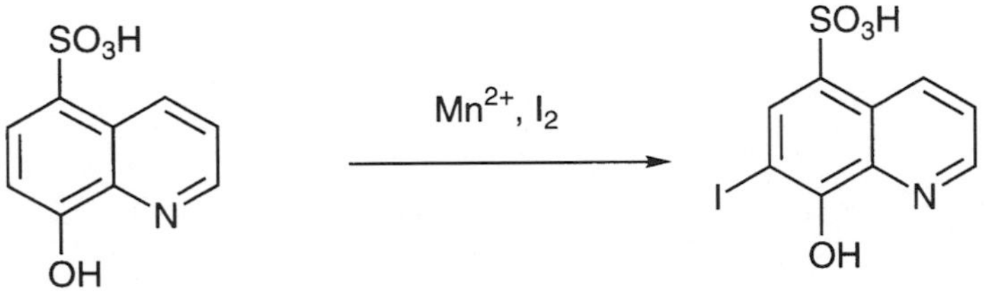

Figure 8-39. The halogenation of 8-hydroxyquinolines may be controlled by co-ordination to a metal ion.

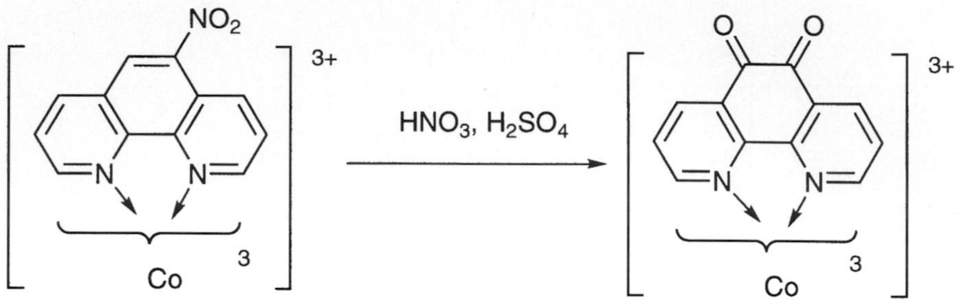

Figure 8-40. The prolonged reaction of [Co(phen)$_3$]$^{3+}$ or [Co(5-nitrophen)$_3$]$^{3+}$ with a mixture of nitric and sulfuric acids yields a quinone, derived from the oxidation of the benzene ring of the ligand.

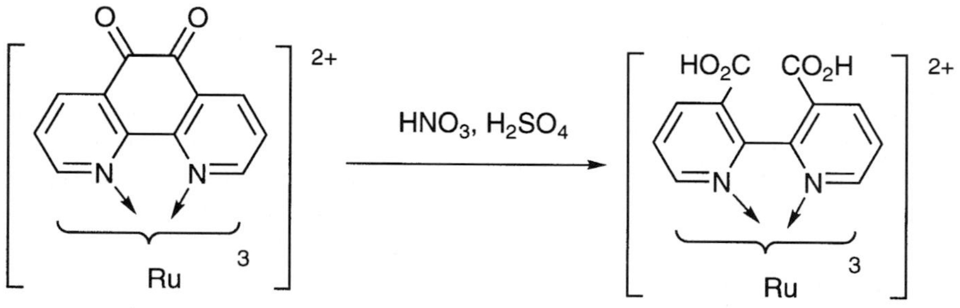

Figure 8-41. The metal complexes of 1,10-phenanthroline-5,6-quinone are not overly stable, treatment with base results in the formation of complexes of 2,2'-bipyridine-3,3'-dicarboxylic acid.

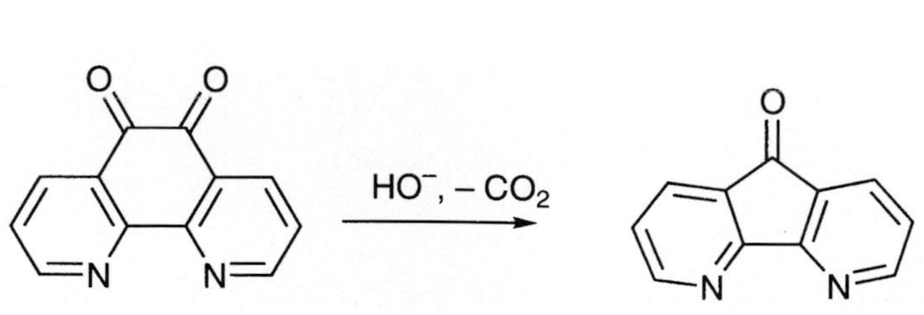

Figure 8-42. The benzilic acid rearrangement of 1,10-phenanthroline-5,6-quinone gives an intermediate which decarboxylates to a diazafluorenone.

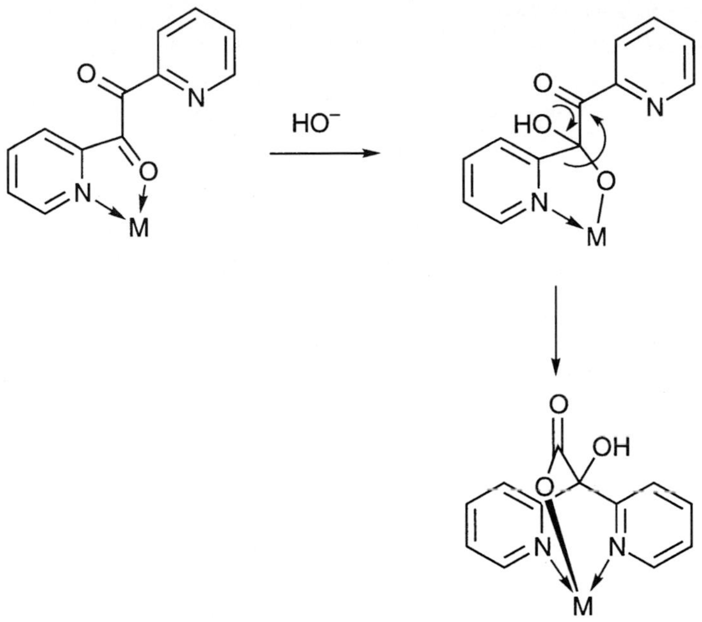

Figure 8-43. In the case of complexes of 2,2'-pyridil, the benzilic acid rearrangement may be followed and complexes of the intermediate may be isolated.

The oxidation of the metal complexes of 1,10-phenanthroline-5,6-quinone is thought
to proceed in a similar manner, with the first step being a benzilic acid rearrangement.
Rearrangements of this type may also be followed directly in nickel(II) and cobalt(III)
complexes of 2,2'-pyridil. The first step of the reaction involves nucleophilic attack on an
O-bonded carbonyl group to form a hydrate, followed by a benzilic acid rearrangement.
In this case, the benzilic acid rearrangement products may be isolated as metal complexes
(Fig. 8-43).

8.8 Summary

In this chapter we have made a very brief survey of the reactions of co-ordinated heterocyclic ligands. It is clear that the reactivity of these ligands is profoundly affected by co-
ordination to a metal ion, and dramatic reactions may sometimes be observed. It is para-
doxical that the precise mechanisms of some of the simplest reactions, such as the
interaction of hydroxide with pyridine complexes, remain a mystery.

Suggestions for further reading

1. A. Blackman, *Adv. Heterocycl. Chem.* **1993**, *58,* 123.
    – A short review describing many interesting reactions of complexes of heterocyclic ligands.
2. O. Mønsted, G. Nord, *Adv. Inorg. Chem. Radiochem.* **1991**, *37*, 381.
    – An article describing some reactions of metal diimine complexes.
3. N. Serpone, G. Ponterini, M.A. Jamieson, F. Bolletta, M. Maestri, *Coord. Chem. Rev.* **1983**, *50*, 209.
    – A critical review of the Gillard mecahnism.
4. R.D. Gillard, *Coord. Chem. Rev.* **1975**, *16*, 67.
    – The origin of the Gillard mechanism.
5. E.C. Constable, *Polyhedron* **1983**, *2*, 55.
    – A less critical discussion of the Gillard model.

9 Oxidation and Reduction of Co-ordinated Ligands

9.1 Introduction - The Meaning of Oxidation and Reduction

In this chapter we will briefly consider the ways in which co-ordinated ligands may undergo metal-promoted redox reactions. The concepts of oxidation and reduction in organic chemistry are rather more complex than those of simple electron-gain or loss that are used in transition metal chemistry. The principal problem is associated with the oxidation state of carbon. What, for example, are the oxidation states of the carbon atoms in methane and tetrachloromethane? Methane may be converted to tetrachloromethane by reaction with dichlorine (Fig. 9-1 a). The conversion of the dichlorine to hydrogen chloride suggests a reduction, with a resultant *oxidation* at the carbon. Similarly, the sequential conversion of an alcohol to an aldehyde and a carboxylic acid upon reaction with chromium(VII) compounds is usually thought of as an oxidation (Fig. 9-1 b). What are the oxidation states of the carbon in each of these compounds? Finally, how do we quantify the sequential conversion of ethyne, to ethene and then ethane upon reaction with dihydrogen (Fig. 9-1 c)? This is a process that must, surely, be regarded as a reduction.

In practice, it is quite easy, but not always intuitive, to assign oxidation states to carbon centres on the basis of the electronegativity values of the attached atoms. For example, in methane the electronegativity of hydrogen ($\chi = 2.2$) is lower than that of carbon ($\chi = 2.6$) and the bonds are polarised in the sense $C^{\delta-}$—$H^{\delta+}$. The formal oxidation state of the carbon is –4. Chlorine ($\chi = 3.2$) has a higher electronegativity than carbon and in tetrachloromethane the bond is polarised in the sense $C^{\delta+}$—$Cl^{\delta-}$; the formal oxidation state of the carbon is +4. Similarly, in CH_3Cl, CH_2Cl_2 and $CHCl_3$ the formal oxidation states of the carbon are –2, 0 and +2, respectively. Oxygen ($\chi = 3.4$) is also more electronegative than carbon and in the sequence methanol – methanal – methanoic acid the oxidation state of the carbon changes from –2, to 0 to +2. Finally, upon passing from ethyne to ethene to ethane, the oxidation state of the carbon changes from –1 to –2 to –3. To confuse matters more, a variety of different practical definitions are in use, and we tend to tailor our definition to the individual case being discussed.

In practical terms, a number of definitions are in common usage and these are presented in Table 9-1. Using these criteria, we see that the reactions presented in Figs. 9-1 b and 9-1 c are unambiguously oxidations and reductions respectively.

In this chapter, we will concentrate upon the oxidation reactions, which have proved to be very sensitive to metal-ion control.

a) $CH_4 + 4Cl_2 \longrightarrow CCl_4 + 4HCl$

   -4 $+4$

b) $CH_3OH \xrightarrow{\ \ K_2Cr_2O_7\ ,\ H^+\ \ } CH_2O$

   -2 0

   $CH_2O \xrightarrow{\ \ K_2Cr_2O_7\ ,\ H^+\ \ } HCO_2H$

   0 $+2$

c) $HC{\equiv}CH \xrightarrow{\ \ H_2\ \ } H_2C{=}CH_2$

   $-1\ -1$ $-2\ -2$

   $H_2C{=}CH_2 \xrightarrow{\ \ H_2\ \ } H_3C{-}CH_3$

   $-2\ -2$ $-3\ -3$

Figure 9-1. A number of "organic" redox reactions, showing the formal oxidation state of the carbon in each of the compounds.

Table 9-1. Some useful definitions of oxidation and reduction which are of use in different circumstances.

Oxidation	Reduction
Loss of electrons	Gain of electrons
Removal of hydrogen atoms	Addition of hydrogen atoms
Addition of oxygen atoms	Loss of oxygen atoms

9.2 Oxidation by Loss of Electrons

In some respects, the gain or loss of electrons is the commonest type of redox reaction observed in purely 'inorganic' systems and one of the least common associated with 'organic' systems. Whereas many metal ions may undergo oxidation state changes associated with the gain or loss of *one* electron, those observed for most organic compounds are found to involve *two* electrons. This is a manifestation of the fact that carbon-centred radicals are usually high-energy, electron-deficient species. In fact, this is one of the reasons why thermodynamically unstable "organic" compounds are kinetically stable in air. Redox processes involving the coupling of one-electron and two-electron reactants are generally slow – dioxygen is a triplet diradical and reaction processes generally involve the sequential transfer of single electrons. In some cases, however, a judicious combination of metal and organic compound may be used to effect a formal one-electron process.

One of the best examples of this type of process is seen in the copper(II)-catalysed oxidation of ascorbic acid (vitamin C). Co-ordination of ascorbic acid (in the enediolate form) to copper(II) is a very favourable process and results in the formation of complex **9.2** containing a five-membered chelate ring (Fig. 9-2).

The complex **9.2** may be written as a copper(II) complex of the enediolate (**9.3**) or as a *copper*(I) complex of a radical ligand (**9.4**). The two representations **9.3** and **9.4** are different valence bond structures (Fig. 9-3).

The radical form **9.4** has an unpaired electron and may undergo fast reactions with redox partners that also undergo one-electron processes. Such a redox partner is the triplet radical, dioxygen. The copper complex of ascorbic acid undergoes rapid aerial oxidation to give the dione, dehydroascorbic acid, which may be viewed as being derived by electron loss from the radical (Fig. 9-4).

This redox system is genuinely catalytic in copper, since the reduction product of the dioxygen is superoxide, O_2^-. The superoxide ion undergoes a redox reaction with copper(I) to regenerate copper(II) and peroxide ion, O_2^{2-}. The reaction is best regarded as an oxidation purely by electron transfer, since no C–H bonds are broken, and the O–H bond breaking is essentially a prerequisite for interaction with the copper(II). It is worthy of note that the same transformation of enediolates to diones is achieved in biological systems by the trinuclear copper enzyme, ascorbic acid oxidase. In general, metal ions

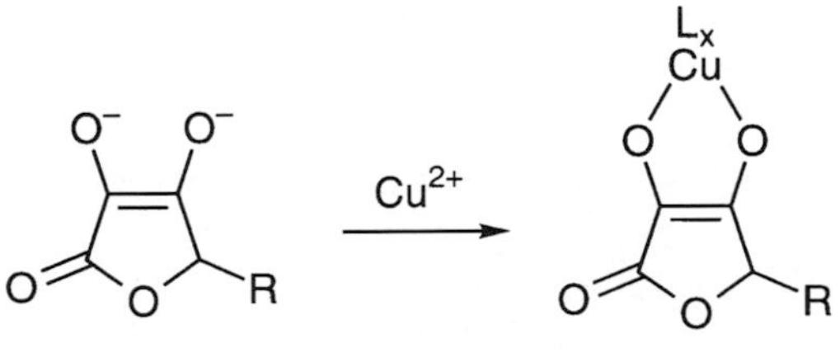

9.1   **9.2**

Figure 9-2. The co-ordination of copper(II) to ascorbic acid (**9.1**) generates a complex of the ligand in its enediolate form.

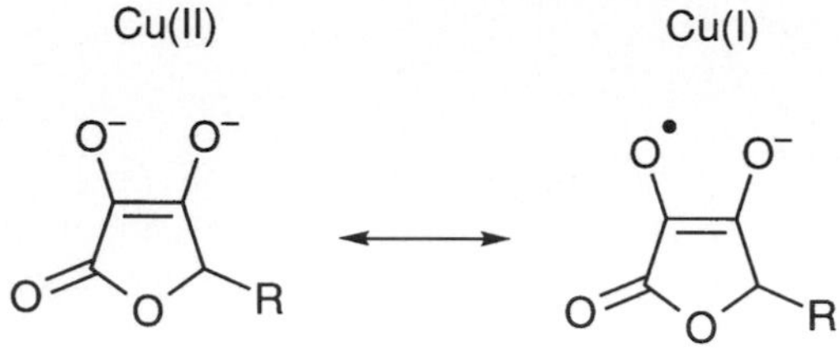

Cu(II)   Cu(I)

Figure 9-3. The copper complex of ascorbic acid may be written in two different resonance forms. In **9.3** it is written as a copper(II) complex of a dianionic ligand, whereas in **9.4** it is viewed as a copper(I) complex of a radical anion.

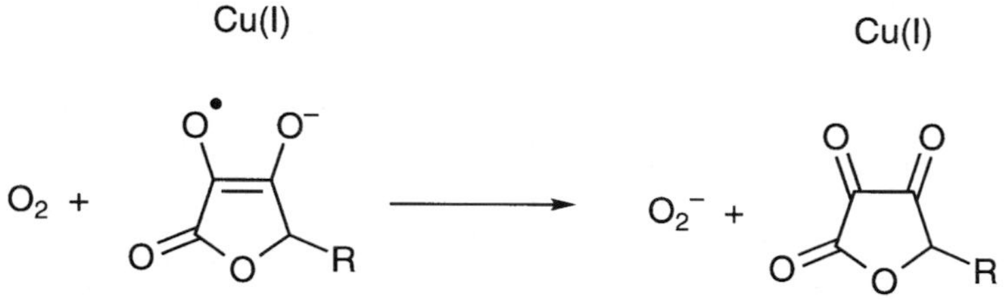

9.3   9.4

Cu(I)   Cu(I)

Figure 9-4. The reaction of dioxygen with the copper complex of ascorbic acid generates a copper complex of dehydroascorbic acid.

that have two readily accessible adjacent oxidation states are expected to be effective – the copper(I)/copper(II) and iron(II)/iron(III) couples are of particular importance.

The oxidation of benzene-1,2-diols to benzo-1,2-quinones is also a process of considerable biological importance, and solutions of copper compounds in organic solvents frequently act as catalysts for the aerial oxidation of such compounds (Fig. 9-5). These reactions almost certainly involve sequential one-electron processes, as indicated in Fig. 9-6. In some cases, the semiquinone forms may be isolated.

Similar processes are observed with transition metal dithiolene complexes, in which a range of formal oxidation states of metal ions may be accommodated. In reality, this is

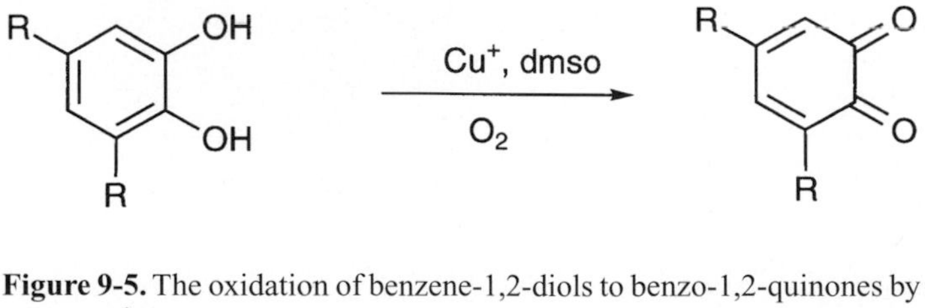

Figure 9-5. The oxidation of benzene-1,2-diols to benzo-1,2-quinones by dioxygen is catalysed by copper salts.

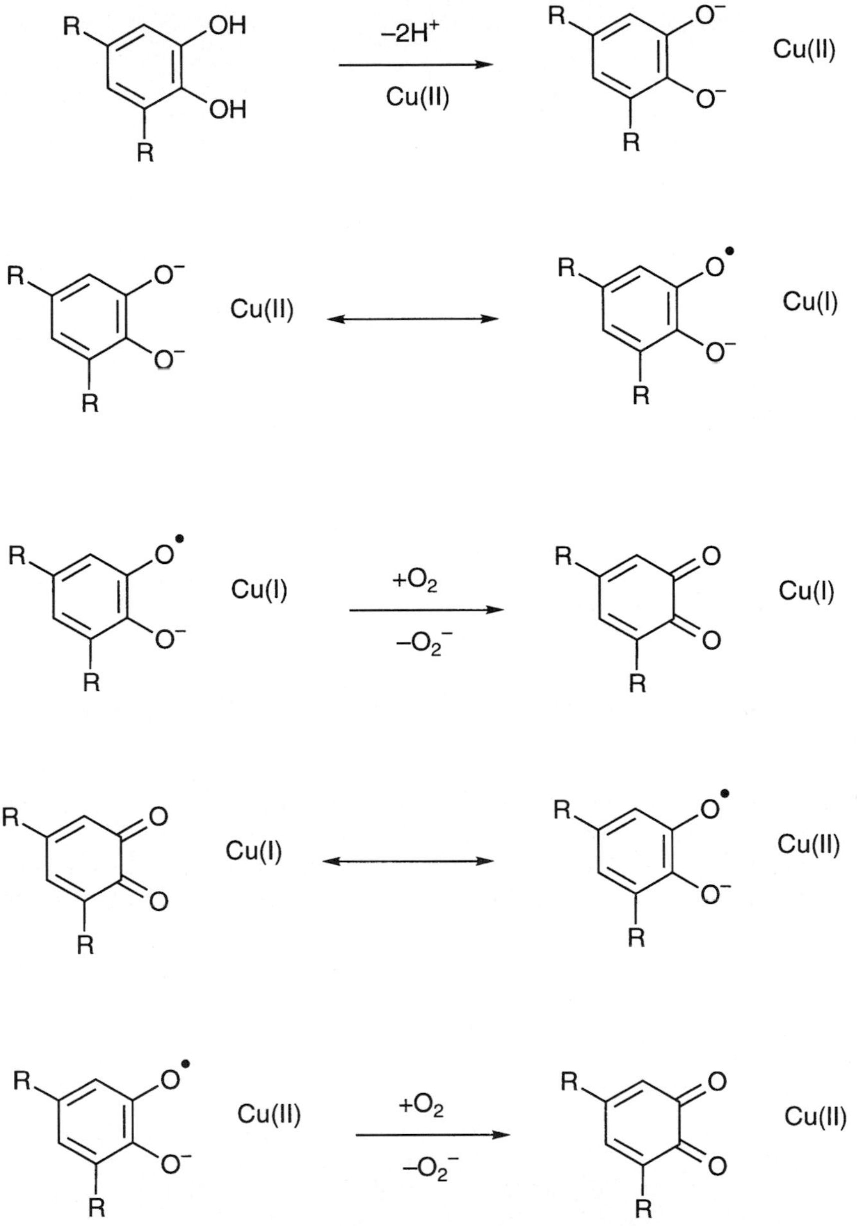

Figure 9-6. A sequential mechanism for the oxidation of benzene-1,2-diols to benzo-1,2-quinones by dioxygen and copper salts. Note the involvement of the radical semiquinone forms.

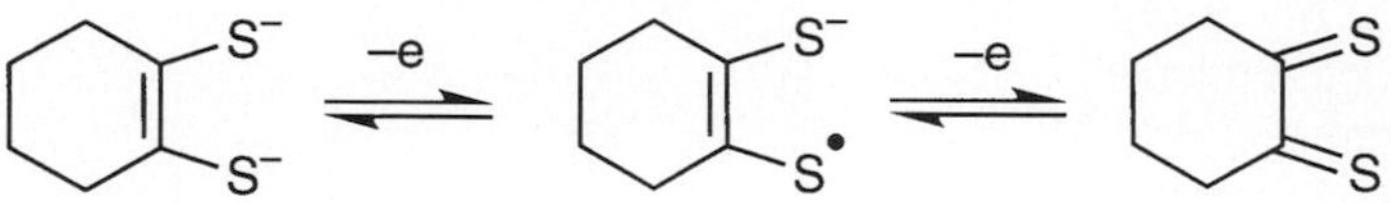

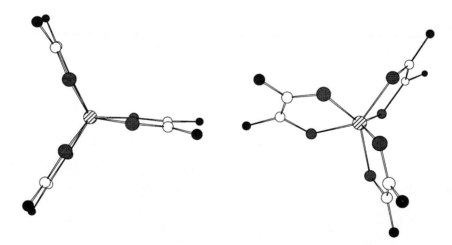

Figure 9-7. The various oxidation levels which are possible in a 1,2-dithiolene ligand.

achieved by transformation between neutral dithione, monoanionic radical anion, and dianionic dithiolate tautomers of the ligand (Fig. 9-7).

Ligands of this type have proved to be of interest to co-ordination chemists for two reasons. First, complexes of the type $[ML_3]^{n+}$ tend to adopt trigonal prismatic rather than octahedral co-ordination geometries (Fig. 9-8) and also because of the ambiguity in describing the oxidation states of the complexes formed. An array of three ligands could all be in the 1,2-dithione forms (neutral), in the enedithiolate form (a total charge of –6) or at any state in-between. However, when the oxidation state of the metal may also vary, it is frequently ambiguous exactly which form of the ligands are associated with which oxidation state (Fig. 9-9).

Another common type of reaction involving sulfur compounds is the oxidation of thiols or thiolates to disulfides. This process is found to be very sensitive to the presence of metal ions. The metal can act as the primary oxidant, or dioxygen may be involved in a reaction with a co-ordinated thiol (Fig. 9-10). Very often, oxidation reactions involving metal ions and thiols are catalytic in the metal.

Reactions of this type are very often associated with copper(II) or iron(III), and it is likely that copper redox chemistry plays a role in the biological transformations of thiols

Figure 9-8. Two views of the complex [Re(PhCSCSPh)$_3$ emphasising the trigonal prismatic structure. The phenyl groups have been represented by the black spheres to simplify the structure. It is thought that interactions between the sulfur atoms are responsible for the adoption of the trigonal prismatic geometry.

L = (NC)CSCS(CN)

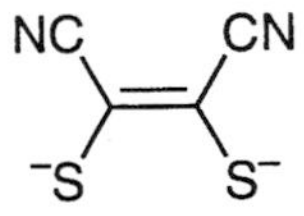

9.5 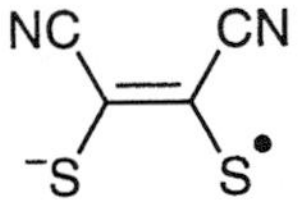   9.6 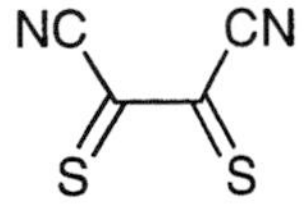   9.7

$$[CrL_3] \quad \rightleftharpoons \quad [CrL_3]^-$$

$$[CrL_3]^- \quad \rightleftharpoons \quad [CrL_3]^{2-}$$

$$[CrL_3]^{2-} \quad \rightleftharpoons \quad [CrL_3]^{3-}$$

Figure 9-9. The chromium 1,2-dithiolene complex shows a number of redox states. What is the correct formulation for each of the compounds? For example, the starting complex $[CrL_3]$ could be written as a chromium(VI) complex $[Cr(\textbf{9.5})_3]$, as a chromium(V) complex $[Cr(\textbf{9.5})_2(\textbf{9.6})]$, a chromium(IV) complex $[Cr(\textbf{9.5})_2(\textbf{9.7})]$ or $[Cr(\textbf{9.5})(\textbf{9.6})_2]$, a chromium(III) complex $[Cr(\textbf{9.6})_3]$ or $[Cr(\textbf{9.5})(\textbf{9.6})(\textbf{9.7})]$, a chromium(II) complex $[Cr(\textbf{9.5})(\textbf{9.7})_2]$ or $[Cr(\textbf{9.6})_2(\textbf{9.7})]$, a chromium(I) complex $[Cr(\textbf{9.6})(\textbf{9.7})_2]$ or a chromium(0) complex $[Cr(\textbf{9.7})_3]$.

and disulfides. A typical example of the oxidation of a thiol is seen in the reaction of pyridine-2(1*H*)-thione with an excess of copper(II), which yields the corresponding disulfide. If an excess of copper(II) is not used the reaction is complicated by the formation of species resulting from the co-ordination of the copper(I) formed as a result of the reduction of copper(II) to the starting ligand.

As mentioned above, it is not *necessary* for the metal ion itself to be an oxidising agent, and co-ordinated thiolates also undergo rapid oxidation by dioxygen. Very often, in these cases, only catalytic amounts of metal ion are required for the oxidation of the thiols by dioxygen. Examples are seen in the oxidation of mercaptoacetic acid or cysteine in the presence of metal ions (Fig. 9-12). This reaction has obvious implications for the effects of transition metal ions upon proteins containing cysteine residues.

$$2RSH + 2M^{2+} \quad \rightleftharpoons \quad RSSR + 2M^+ + 2H^+$$

Figure 9-10. The oxidation of a thiol to a disulfide is often metal ions dependent. In this case the reaction involves the metal ion as the primary oxidant. In many cases the reaction is catalytic, as the reduced form of the metal may be converted to the oxidised form upon reaction with dioxygen.

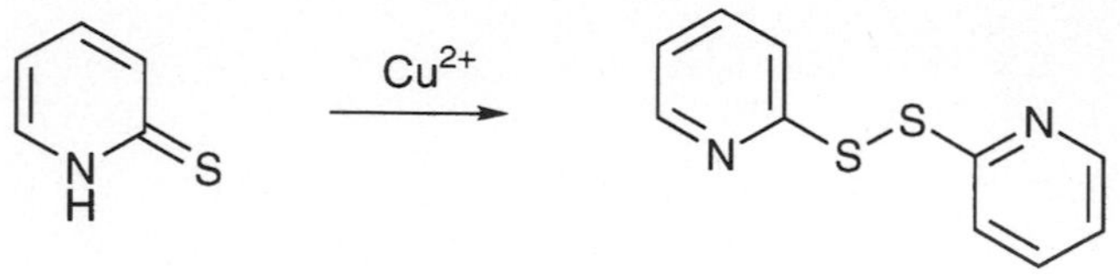

Figure 9-11. The oxidation of pyridine-2(1*H*)-thione with an excess of copper(II) gives the corresponding disulfide.

Many other sulfur compounds exhibit dimerisation reactions of this type in the presence of metal ions and oxygen. For example, dithiocarbamates are readily oxidised to the disulfides in the presence of copper salts (Fig. 9-13). Very often, attempts to form complexes with dithiocarbamate ligands are complicated by the parallel oxidation process.

All of these reactions are, in principle, reversible, and many examples are known where thiolate complexes may be prepared by the reaction of disulfides with low oxidation state metal complexes.

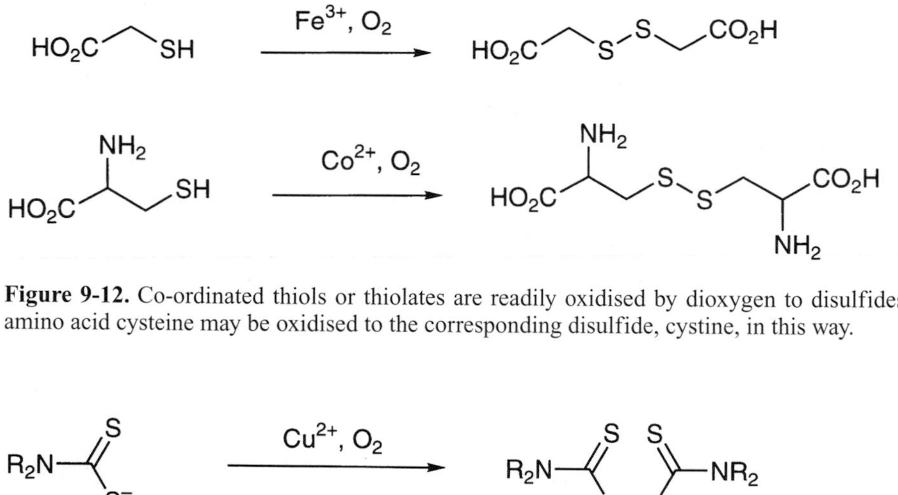

Figure 9-12. Co-ordinated thiols or thiolates are readily oxidised by dioxygen to disulfides. The amino acid cysteine may be oxidised to the corresponding disulfide, cystine, in this way.

Figure 9-13. Dithiocarbamates are readily oxidised by dioxygen in the presence of metal ions.

9.3 Oxidation by C–H Bond Cleavage

The oxidation of alcohols to aldehydes, ketones or carboxylic acids is one of the commonest reactions in organic chemistry, and is frequently achieved by transition metal complexes or salts. However, in most cases the precise mechanisms are not known, and the intermediates not fully characterised. In general, metal complexes of the alcohols are formed as transient intermediates in these reactions, but we shall not deal with these extremely important reactions in any great detail. The precise mechanisms depend upon the accessibility of the various one- and two-electron reduction products of the particular metal ion which is involved in the reaction. However, we will outline a brief indication of the mechanism. The first step involves the formation of an alcohol complex of the metal ion (Fig. 9-14). This might or might not deprotonate to the alkoxide form, depending upon the pH conditions of the reaction, the pK_a of the alcohol and the polarising ability of the metal ion.

A typical example of such a process is shown in Fig. 9-15. In the first step the alcohol ligand **9.8** is co-ordinated to ruthenium(II). The resultant ruthenium(II) complex is then oxidised to give a ruthenium(III) species. Ruthenium(III) is more polarising than ruthenium(II), and the co-ordinated alcohol is deprotonated in the ruthenium(III) complex.

These same alkoxy compounds are also the primary products in the oxidation of alcohols with high oxidation state metal oxo complexes. In a typical process, the reaction of an alcohol with the chromium(VI) compound $[HCrO_4]^-$ is shown in Fig. 9-16. The intermediate is often described as a chromate ester, but it is in all respects identical to the alkoxide complexes that we described earlier.

The precise sequence of events depends upon the combination of ligands and metal centres involved, but the key step involves a C–H bond-breaking reaction. The reaction may be viewed as a consequence of metal ion polarisation of the ligand increasing the acidity of the relevant C–H bond. Loss of a proton yields a carbanion, which undergoes an electron transfer reaction with the metal centre to yield a radical and lower oxidation state metal ion (free or co-ordinated). It must be emphasised that this is purely a formal view of the reactions.

In the case of the ruthenium-mediated oxidation of the alcohol **9.8** the overall process is as shown in Fig. 9-18. We have already noted that the deprotonation of the alco-

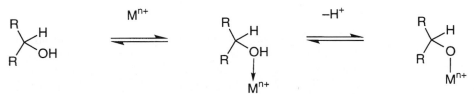

Figure 9-14. The first step in the oxidation of an alcohol by a high oxidation state metal ion is thought to be the coordination of the alcohol to the metal centre. This may or may not be followed by subsequent deprotonation, depending upon the pK_a of the alcohol and the polarising power of the metal ion.

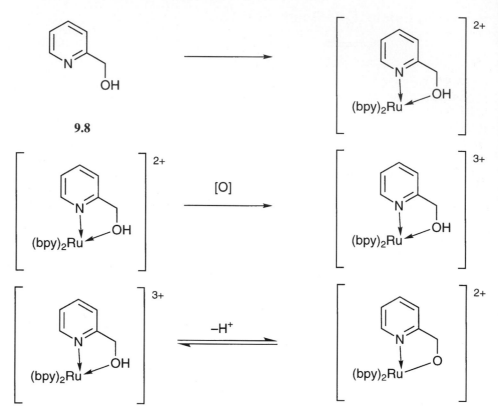

Figure 9-15. The deprotonation of the alcohol **9.8** only occurs after oxidation of the ruthenium(II) to the more polarising ruthenium(III).

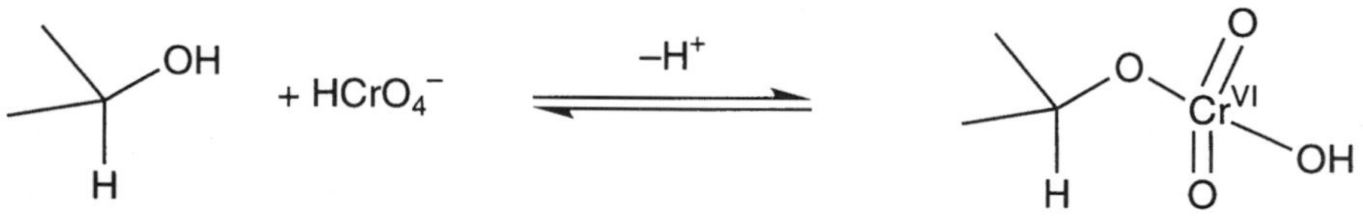

Figure 9-16. An alternative route for the formation of the alkoxy intermediate by the reaction of alcohol with a high oxidation state metal oxo complex.

hol occurs after oxidation to the ruthenium(III) state. The balance of the redox potentials of the ruthenium(III) complex of the *alcohol* and the ruthenium(III) complex of the *alkoxide* is such that a conproportionation reaction occurs, to generate a ruthenium(IV) *alkoxide* complex and the starting ruthenium(II) *alcohol* complex. The C-H bond-breaking reaction involves this ruthenium(IV) alkoxide complex. After loss of a proton, the product could be written as a ruthenium(IV) *carbanion* complex, a ruthenium(III) *radical* complex or a ruthenium(II) *carbonyl* complex. The latter description is the most appropriate.

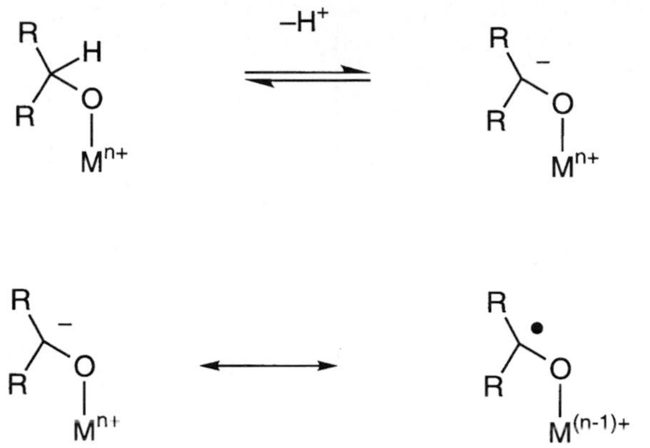

Figure 9-17. The key step in the oxidation of a co-ordinated alkoxide group is the breaking of a C-H bond.

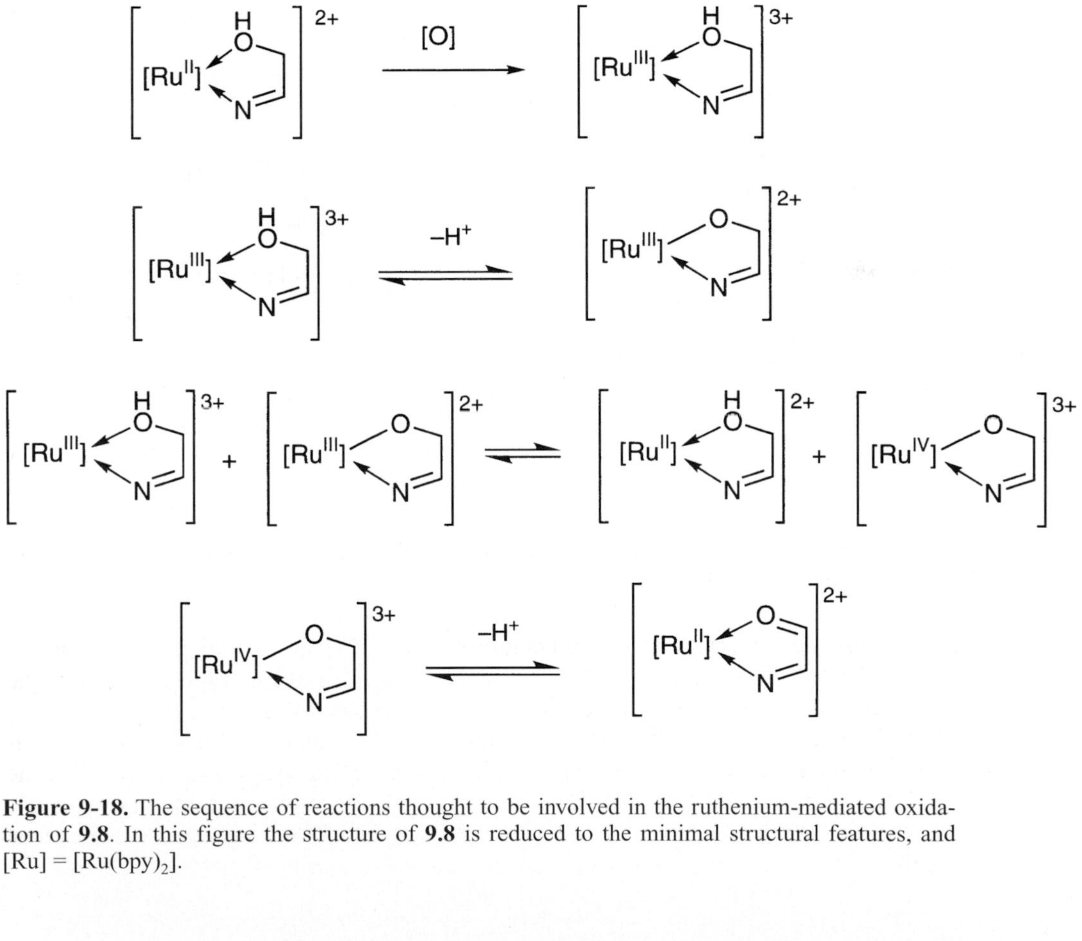

Figure 9-18. The sequence of reactions thought to be involved in the ruthenium-mediated oxidation of **9.8**. In this figure the structure of **9.8** is reduced to the minimal structural features, and [Ru] = [Ru(bpy)$_2$].

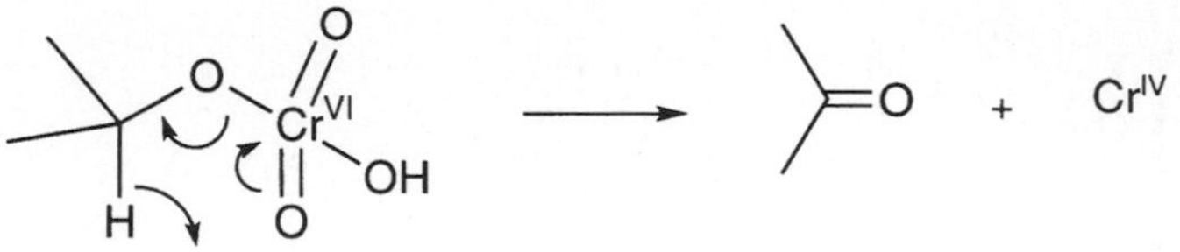

Figure 9-19. In the case of chromium(VI) and other high oxidation state alkoxy complexes, the C–H bond breaking may occur in an intramolecular *two-electron* process.

In the case of oxidation by high oxidation state oxo-complexes, there is good evidence for the involvement of a *two*-electron electrocyclic processes. In other words, the deprotonation is an intramolecular process involving the metal complex (Fig. 9-19).

It is not always necessary for the oxygen to be directly co-ordinated to the metal centre, and cases are known in which the polarisation effect of the metal ion is transmitted through a conjugated system. An example of such a metal-mediated dehydrogenation is seen in the reaction of cobalt(III) complexes of 4-pyridylmethanol with cerium(IV) compounds (Fig. 9-20). The products of the reaction are 4-pyridinecarbaldehyde and cobalt(II) salts, and it is thought that the first step of the reaction involves the formation of a cobalt(III)-radical species, which is converted to the cobalt(II) aldehyde complex by proton loss. It is interesting to note that when two-electron oxidants are used, there is no requirement for the electron transfer to cobalt(III), and no cobalt(II) is formed. This serves to illustrate the comments made at the beginning of this section regarding the relative roles of one- and two-electron processes in inorganic and organic reactions.

We have already seen that imines may be formed by the oxidative dehydrogenation of co-ordinated amines and that this is a commonly observed process, particularly in macrocyclic systems. Likely mechanisms for these dehydrogenations were suggested in Chapter 5, which emphasised the role of the variable oxidation state metal ions in the process. These reactions are quite general and many examples involving iron or ruthenium complexes have been studied in detail.

In general, the dehydrogenation of polyamine ligands co-ordinated to iron(II) results in the formation of the most conjugated products. This may be rationalised in terms of the conjugated polyimine ligands favouring the low spin d^6 configuration, with the associated high ligand field stabilisation energy. A clear example of this is seen in the aerial dehydrogenation of the ligand in tris(2,2'-bipiperidine)iron(II), Fig. 9-21.

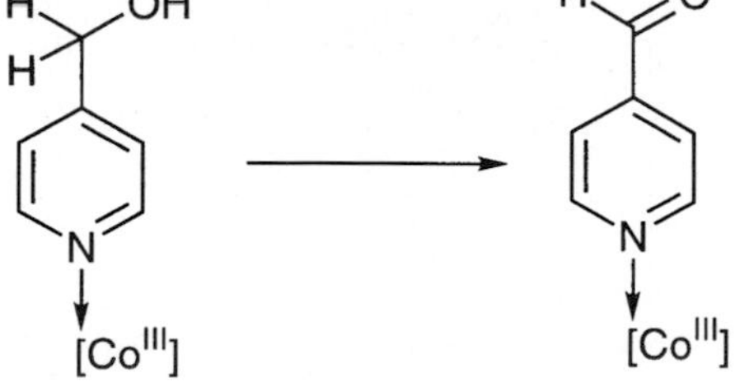

Figure 9-20. The cobalt(III) complex of 4-pyridylmethanol may be oxidised by cerium(IV) to 4-pyridinecarbaldehyde

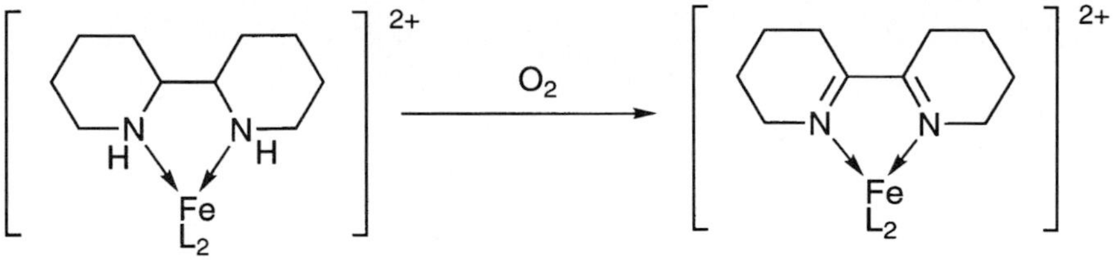

Figure 9-21. The oxidation of the iron(II) complex of 2,2'-bipiperidine by dioxygen gives the iron(II) complex of the corresponding diimine ligand.

This preference for the formation of the strong field conjugated ligand with iron(II) was seen in Chapter 5. A typical example is the formation of the conjugated diimine macrocyclic complex **9.10** upon oxidation of **9.9**. These reactions do show a significant metal-ion dependence, and the corresponding oxidation of the d^8 nickel(II) complex **9.11** gives the *non-conjugated* tetraimine complex **9.12**.

The key feature of these reactions is the interplay between metal-centred and ligand-centred radical species, as discussed in Chapter 5. To summarise, the process involves generation of a nitrogen-centred radical, which is stabilised by charge transfer from the

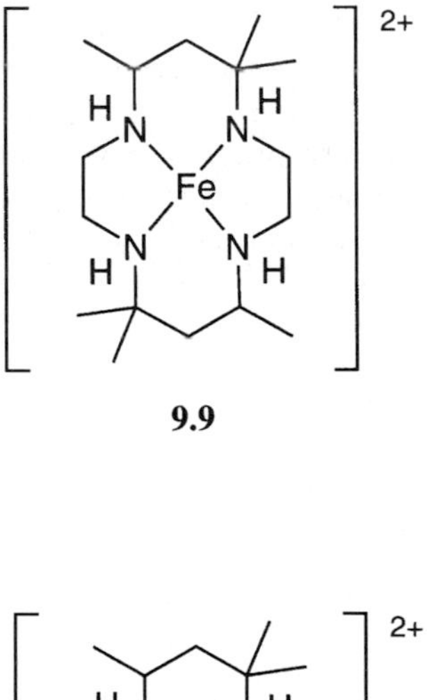

9.9

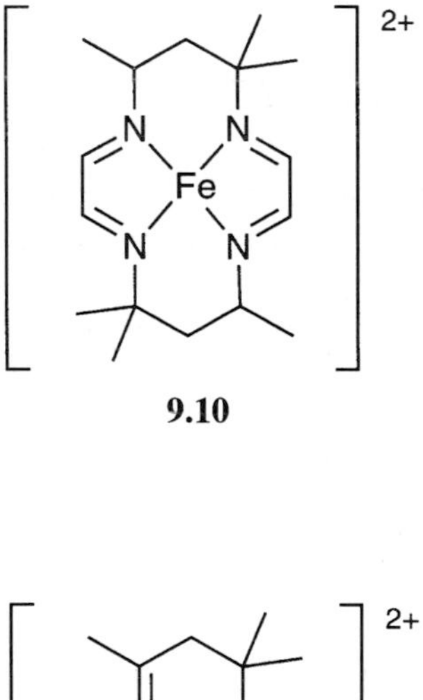

9.10

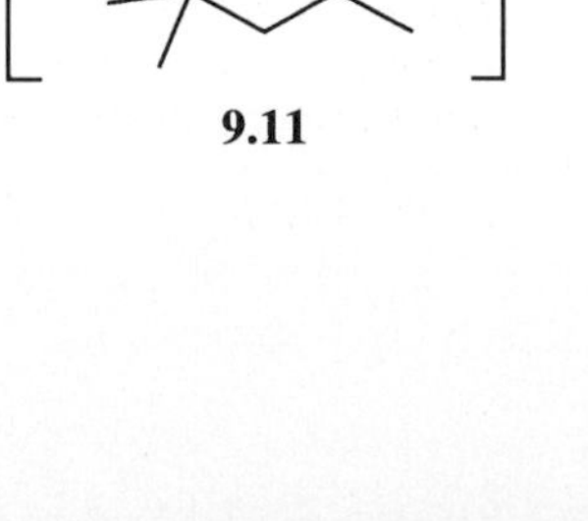

9.11

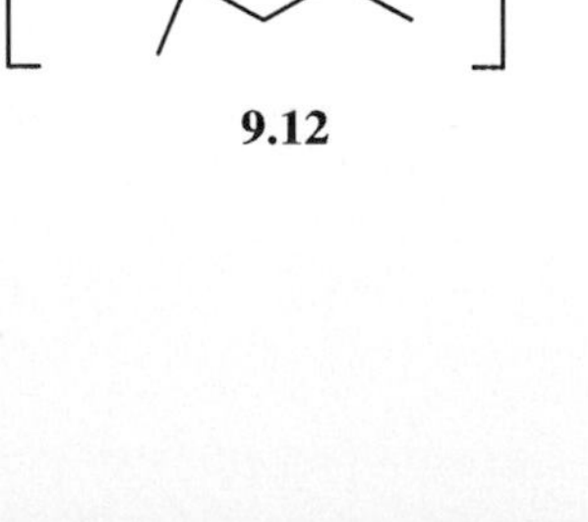

9.12

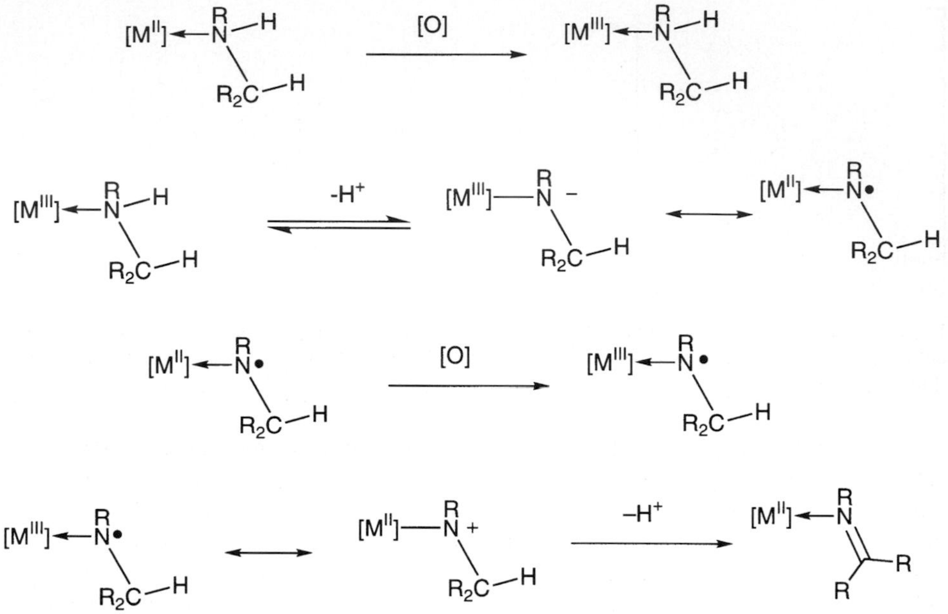

Figure 9-22. The overall sequence of reactions involved in the dehydrogenation of an amine to an imine.

metal centre. The first step probably involves oxidation of the metal centre to yield a complex in which the pK_a of the NH group is sufficiently lowered that deprotonation may occur. The metal(III)-amido species is a resonance form of a metal(II) amido *radical* and may then undergo a second oxidation to yield a metal(III) radical complex, which is tautomeric with metal(II) nitrenium ion, which in turn may lose a proton to yield the imine (Fig. 9-22).

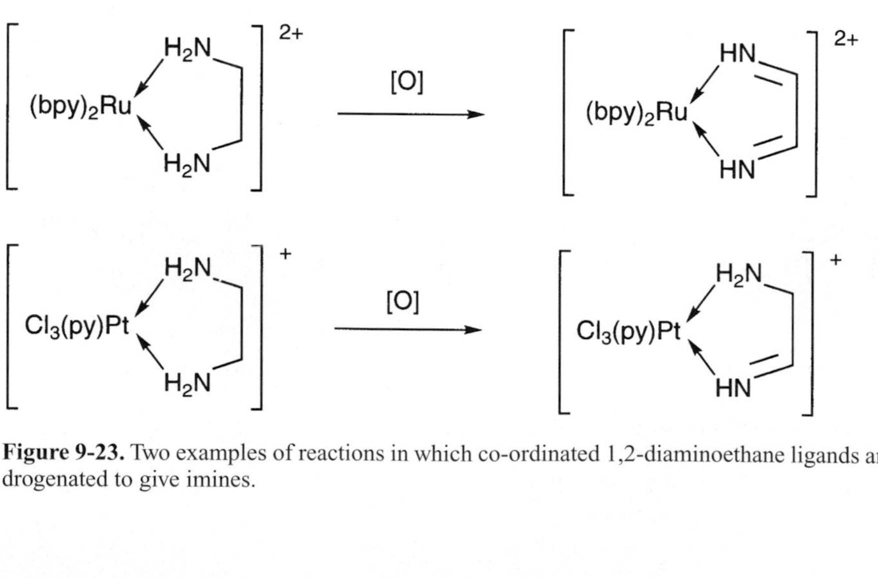

Figure 9-23. Two examples of reactions in which co-ordinated 1,2-diaminoethane ligands are dehydrogenated to give imines.

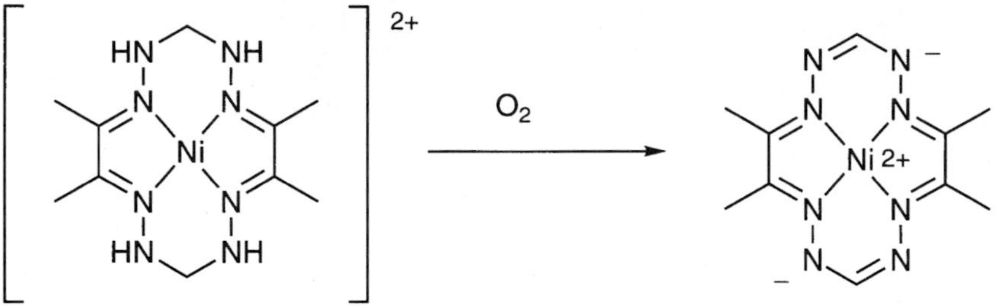

Figure 9-24. The oxidation of the nickel(II) complex formed from the template condensation of biacetyl bishydrazone with formaldehyde gives a neutral, conjugated complex, in which the ligand is doubly deprotonated.

The mechanism will vary in precise detail according to the metal. In the case of ruthenium complexes, it is quite common to observe a conproportionation and the formation of a ruthenium(IV) intermediate. In other cases, the unavailability of the metal oxidation states precludes reaction. For example, cobalt(III) complexes of cyclam cannot be oxidised to imine species because although a cobalt(II)/cobalt(III) couple is possible, the cobalt(II) oxidation state is not accessible under oxidative conditions. In the case of metal ions which can undergo two oxidation state changes, alternative mechanisms which do not involve radical species have been suggested.

Although oxidative dehydrogenation reactions are particularly well characterised with macrocyclic complexes, even very simple amine ligands such as 1,2-diaminoethane may be oxidised to the corresponding imines (Fig. 9-23).

In some cases a whole series of dehydrogenation reactions may proceed sequentially to yield aromatic or highly conjugated products. An example of this is seen in the aerial oxidation of the nickel(II) complex of the macrocycle formed by the template condensation of biacetyl bishydrazone with formaldehyde. The product of the oxidation is the fully aromatic dianionic macrocyclic complex (Fig. 9-24).

A final type of oxidation reaction associated with the loss of hydrogen atoms is seen in the oxidative dimerisation of phenols. In general, the oxidation of phenols, in either the presence or absence of metal ions, is not a clean process, and many products, derived from

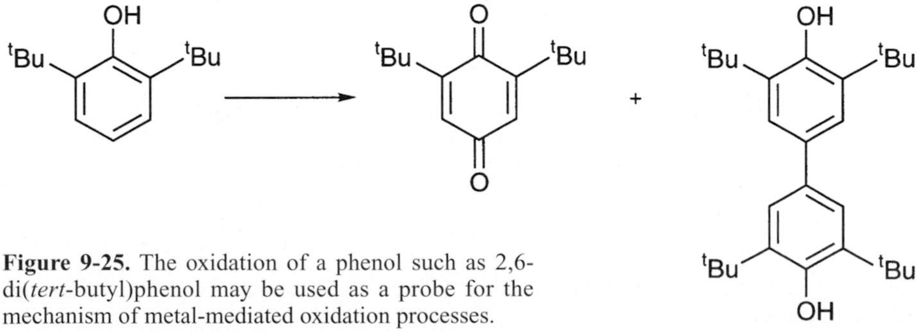

Figure 9-25. The oxidation of a phenol such as 2,6-di(*tert*-butyl)phenol may be used as a probe for the mechanism of metal-mediated oxidation processes.

various types of reaction, are formed. These side reactions may be controlled to a certain extent by the use of a highly sterically hindered phenol such as 2,6-di(*tert*-butyl)phenol. The oxidation of this ligand in the presence of metal ions gives rise to two types of organic products. The dimeric diphenol is derived by oxidative dehydrogenation, whereas the quinone is formed by an oxygen transfer reaction (Fig. 9-25). These reactions are frequently achieved by aerial oxidation in the presence of catalytic quantities of transition metal salts, particularly those of copper(II) or copper(I). The relative abundances of the two products should allow us to probe the mechanism of action of oxidation catalysts.

9.4 Oxidation by Oxygen Atom Transfer

The transfer of oxygen atom(s) from a metal centre, or from an intermediary oxygen atom donor to a ligand, is another way by which a ligand may be oxidised. The oxidation of aromatic ligands by this mechanism is quite common and is particularly noticed in copper complexes. Many such systems have been investigated as models for hydroxylase and oxygenase enzymes. A vast range of enzyme-mediated oxygen transfer reactions have been characterised, most of which involve copper- or iron-containing metalloproteins. In the typical processes shown in Fig. 9-26, the first and last reactions involve oxygen transfer, whilst the second is an electron transfer type of oxidation.

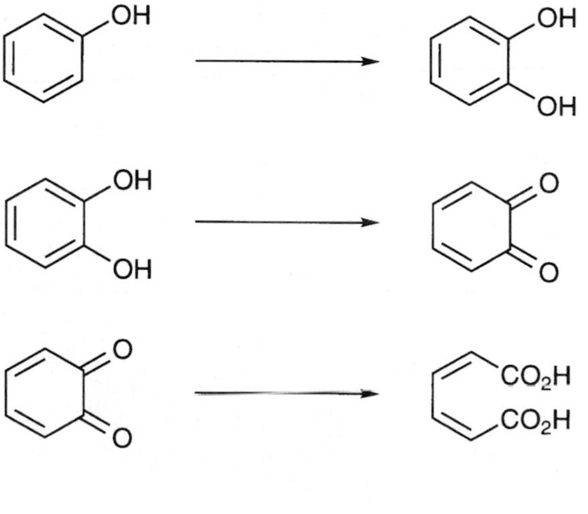

Figure 9-26. The various types of oxidation reaction catalysed by metalloenzymes. The conversion of a phenol to a 1,2-dihydroxybenzene and the ring opening oxidation both involve oxygen atom transfer to the substrate, whilst the oxidation of a 1,2-diol to a 1,2-quinone is of the type discussed earlier in this chapter.

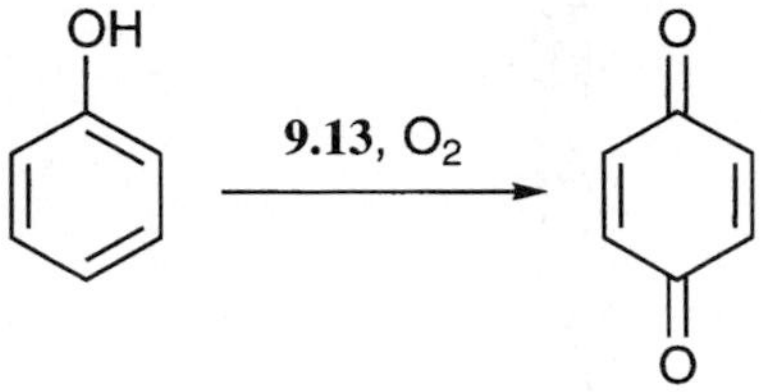

Figure 9-27. The oxidation of phenol to 1,4-benzoquinone by dioxygen is catalysed by cobalt complex **9.13**.

A large number of reactions that mimic these biological oxidations have been characterised and this is an area of active study by those attempting to understand the ways in which metal ions act in metalloproteins. A very simple model reaction is seen in the oxidation of phenol to 1,4-benzoquinone by dioxygen in the presence of cobalt(II) complexes (Fig. 9-27) A particularly useful catalyst for a variety of such oxygen transfer reactions is the complex [Co(salen)], **9.13**.

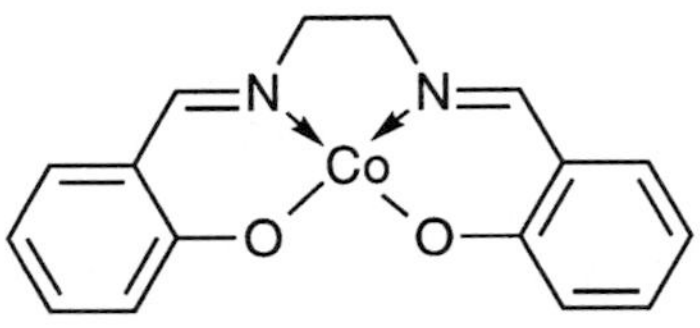

9.13

Other reactions show an even greater resemblance to those which occur in biological systems. A typical example is seen in the smooth oxidation of catechol by dioxygen in the presence of mixed pyridine/methanol solutions containing copper(I) chloride (Fig. 9-28). The cleavage products in this reaction are derived from an intermediate 1,2-quinone.

These copper-mediated reactions very often involve dinuclear intermediates, but detailed mechanistic studies on *stoichiometric* systems are relatively few. The key features are the formation of μ-peroxo or μ-superoxo complexes by electron transfer from copper(I) to dioxygen. The co-ordinated oxygen may then act as an electrophile to the aromatic ring. A possible mechanism for the *ortho*-hydroxylation of phenol by dioxygen in the presence of copper catalysts is shown in Fig. 9-29.

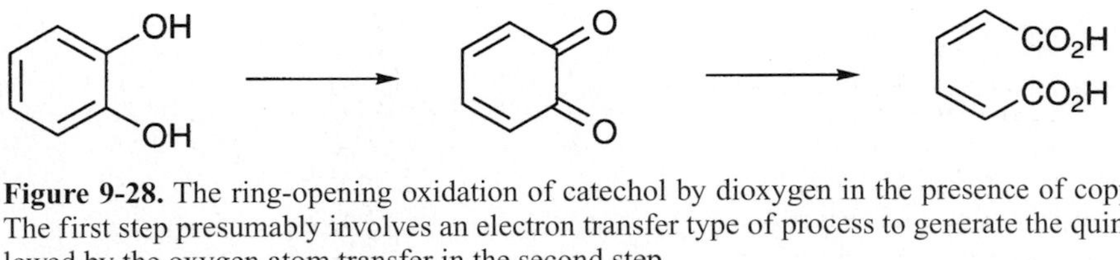

Figure 9-28. The ring-opening oxidation of catechol by dioxygen in the presence of copper salts. The first step presumably involves an electron transfer type of process to generate the quinone, followed by the oxygen atom transfer in the second step.

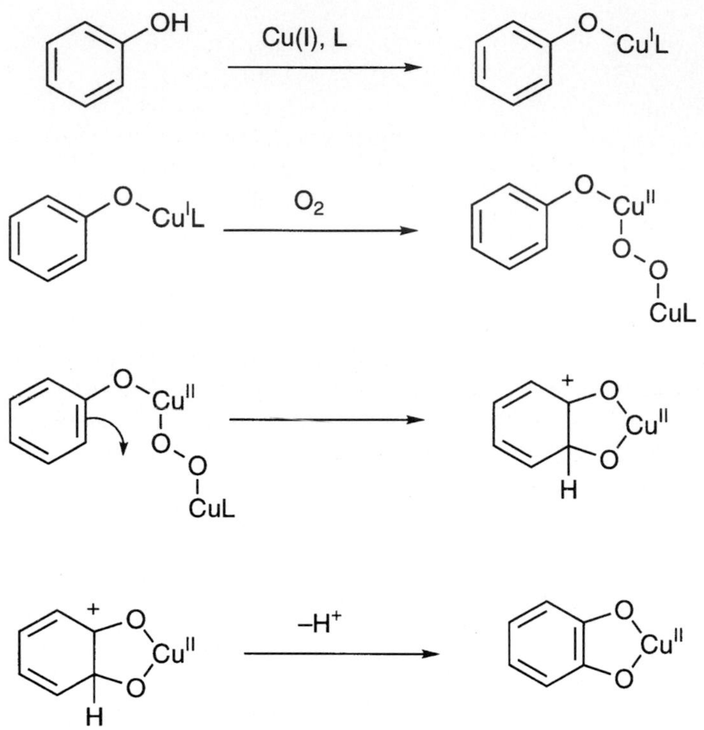

Figure 9-29. A possible mechanism for the oxidation of phenol to 1,2-benzoquinone by dioxygen in the presence of copper(I) salts. The key steps involve the formation of a peroxo or superoxo complex, followed by electrophilic attack upon the benzene ring.

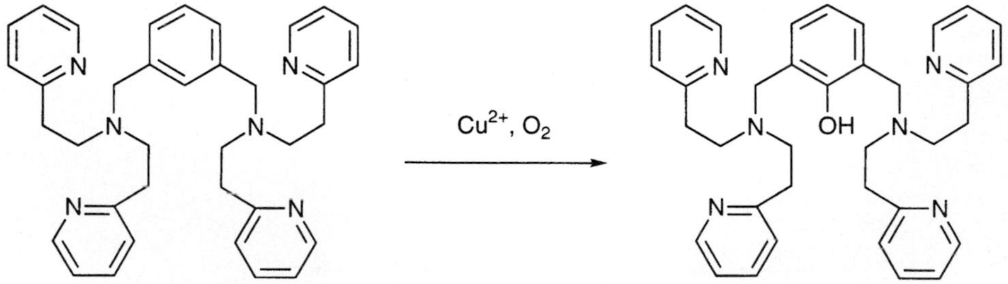

Figure 9-30. The oxidation of **9.14** by dioxygen and copper(I) generates a dinuclear copper(II) complex of the phenolic ligand **9.15**.

A typical example of the more complex reactions that may occur when aromatic compounds react with dioxygen in the presence of copper salts is seen in Fig. 9-30. When solutions of the hexadentate ligand **9.14** react with copper salts and dioxygen, a complex of a new ligand is obtained. The new phenolic compound that is formed acts as a dinucleating ligand and its dinuclear copper(II) complexes turn out to be effective oxygenation catalysts for other substrates.

Numerous other examples of reactions of this type have been described and a particularly dramatic example of a product derived from oxygen atom transfer to a co-ordinated ligand is seen in the reaction of the benzimidazole ligand **9.16** with dioxygen in the presence of a copper(I) catalysts. The isolated product of the reaction is the copper(II) complex of the benzimidazole-3-carboxylic acid, **9.17**.

A closely related reaction is seen in the copper-mediated oxygenation of bis(2-benzimidazolyl)methanes to bis(2-benzimidazolyl)ketones (Fig. 9-31). This is a general reaction shown by many bis(heteroaryl)methanes. The reaction of the iron(II) complex

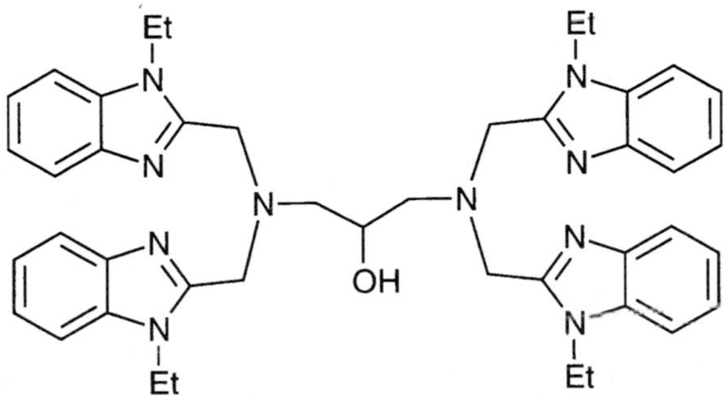

9.16

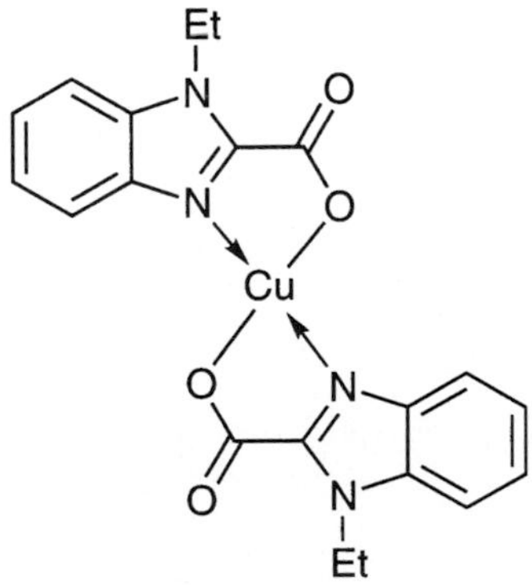

9.17

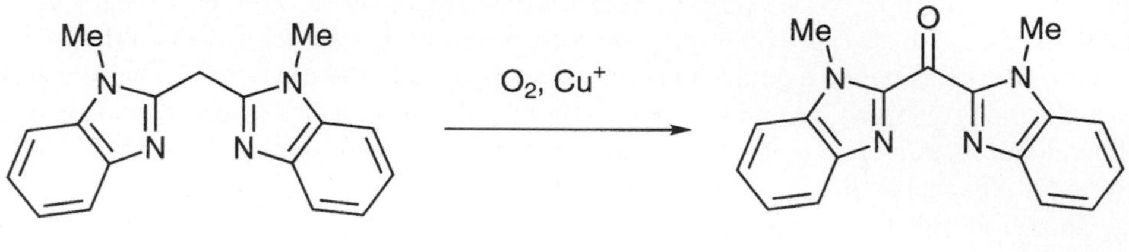

Figure 9-31. The transfer of oxygen atoms by copper/dioxygen reagents is not limited to aromatic substrates. Many bis(heteroaryl)methanes are converted to the corresponding ketones upon reaction with dioxygen in the presence of copper salts.

[Fe(**9.18**)]$_3$ with dioxygen produces complexes of **9.19**. This process has been investigated in some detail and the detailed mechanism shown in Fig. 9-32 determined. The key features, again, involve the interplay of radical and anionic species.

Superoxide is also implicated in the conversion of **9.20** to **9.21**. The first step of the reaction involves the formation of a nickel(III)–superoxide complex, which then undergoes the intramolecular oxygen atom transfer reaction. In this case, the superoxide is

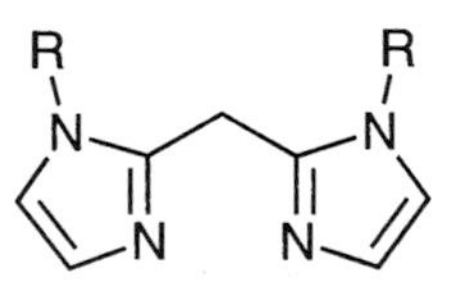

9.18

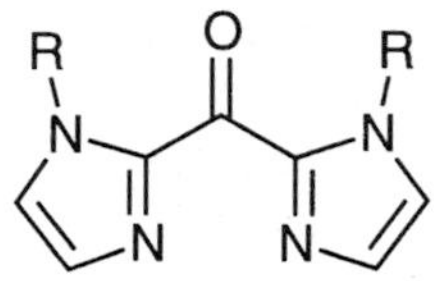

9.19

$$Fe(II) + O_2 \rightarrow Fe(III) + O_2^-$$

$$O_2^- + R_2CH_2 \rightarrow HO_2^\bullet + R_2CH^-$$

$$Fe(III) + R_2CH^- \rightarrow Fe(II) + R_2CH^\bullet$$

$$HO_2^\bullet + R_2CH^\bullet \rightarrow R_2CHO_2H$$

Figure 9-32. The oxidation of the complex [Fe(**9.18**)]$_3$ with dioxygen gives complexes of the new ligand **9.19**. The first step of the reaction involves the formation of superoxide, which is responsible for the breaking of the first C–H bond in the second step of the reaction. The overall mechanism is characterised by the interconversion of radical and anionic species by a series of one-electron transfers.

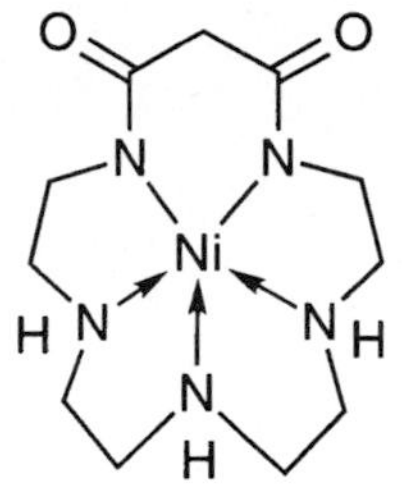

9.20

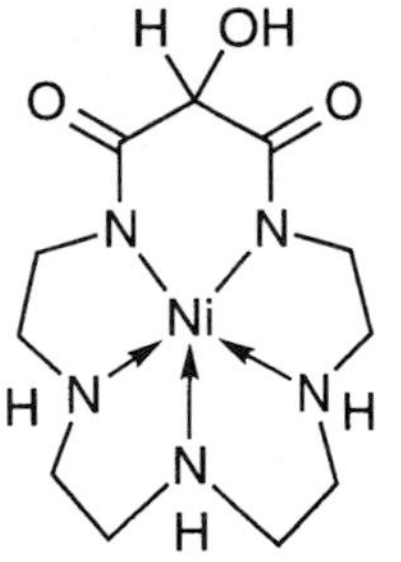

9.21

actually bound to the metal centre, whereas this is unlikely to be the case with the iron(II) complex of **9.18**.

A final example of a reaction involving the dioxygen oxidation of a co-ordinated ligand is presented in Fig. 9-33. Here, the copper(II) complex of the imine from salicylaldehyde and an amino acid reacts with oxygen to give, ultimately, a complex of salicylaldehyde oxime! The first step involves the reaction of the copper(II) complex with dioxygen to give a copper(II)–hydroperoxide complex. In the formation of this complex, we see a change in the ligand reminiscent of pyridoxal. The C=N bond moves from conjugation with the aromatic ring to conjugation with the carbonyl group. Subsequent hydrolysis of this relatively unstabilised imine gives a copper(II) complex, which undergoes an intramolecular reaction to generate the observed oxime.

Before we leave the topic of oxygen-transfer oxidations, we should at least mention the very widespread use of high oxidation state metal oxo-complexes for the oxidation of a wide variety of substrates. Although the mechanisms of these reactions are known in some detail, it is very rare for an intermediate complex to be isolated – the most common reaction conditions simply involve mixing organic substrate with the oxo-complex followed by work-up to give the organic oxidation product directly. Typical examples of such reactions involve the oxidation of alkenes with $[MnO_4]^-$ or osmium(VIII) oxide. The key steps in all of these reactions involve sequential oxidation state changes at the metal ion associated with the oxygen atom transfer. This is exemplified in Fig. 9-34 for the oxidation of alkenes to 1,2-diols by permanganate.

The accessibility of the various oxidation states is very important in these reactions. For example, the reaction of alkenes with ruthenium(VIII) oxide instead of permanganate leads to the cleavage of the C=C bond and the formation of aldehydes rather than a 1,2-diol (Fig. 9-35).

The use of metal oxo-complexes for the oxidation of aldehydes to carboxylic acids is also well-known (Fig. 9-36), although, once again, the isolation of intermediate complexes is relatively rare. In particular, high oxidation state manganese or chromium complexes are commonly used for this process.

Presumably, the oxidation of the iron(II) diimine complex derived from glyoxal and methylamine (encountered previously) by cerium(IV) occurs by a related mechanism (Fig. 9-37). This also provides an interesting example of the metal ion stabilising a particular tautomer of a ligand. The free ligand would be expected to exist as the amide tautomer **9.22**.

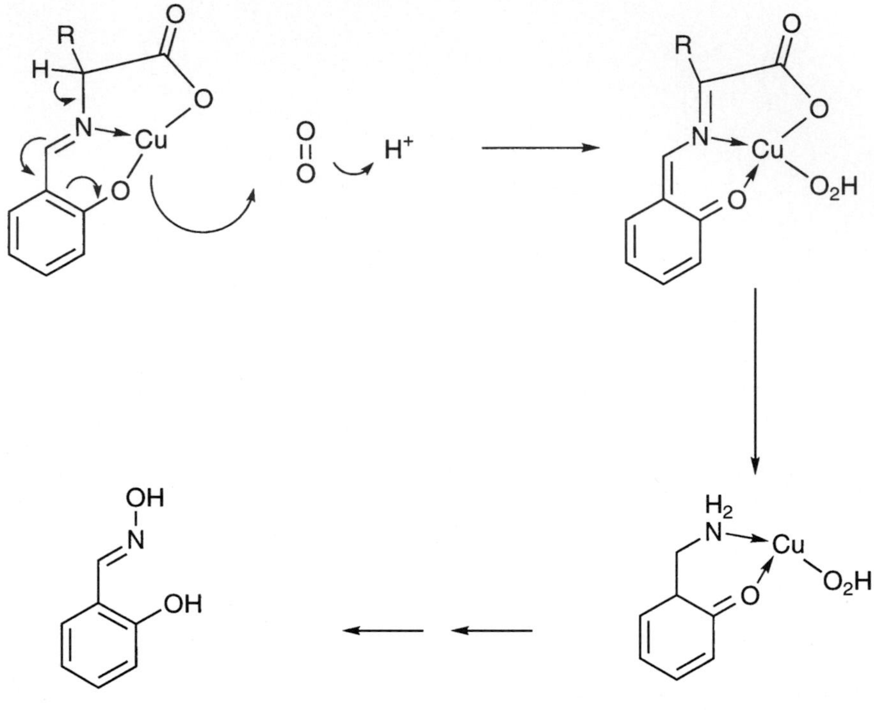

Figure 9-33. Sometimes the reaction of copper complexes with dioxygen can have dramatic consequences!

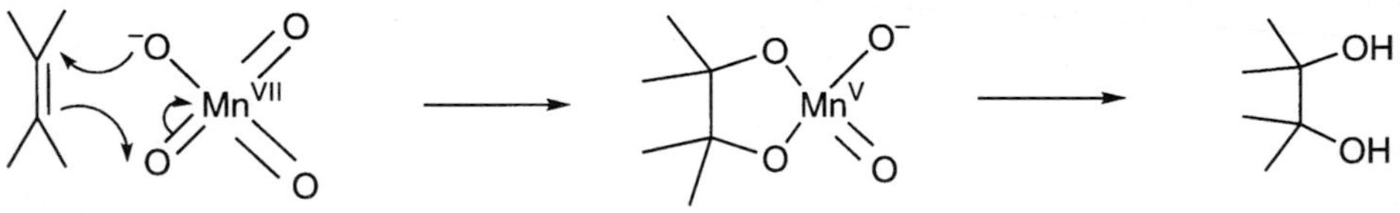

Figure 9-34. A likely mechanism for the oxidation of alkenes by high oxidation state oxo-complexes. The important feature is the formation and subsequent collapse of the cyclic ester.

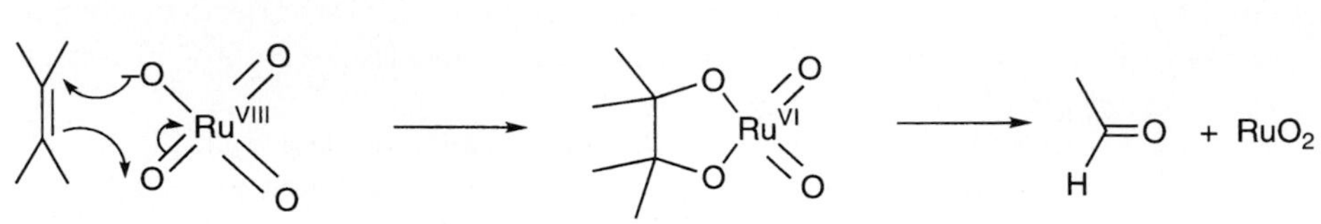

Figure 9-35. The oxidation of alkenes with ruthenium(VIII) oxide involves a cyclic ester as an intermediate.

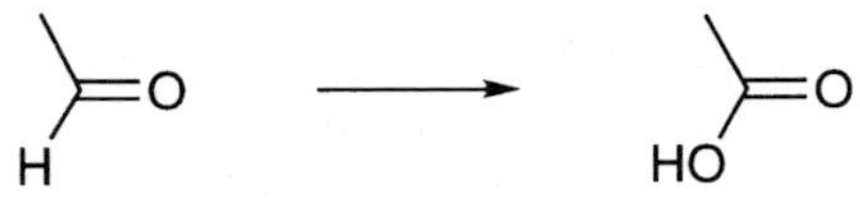

Figure 9-36. The oxidation of aldehydes to carboxylic acids is usually achieved by treatment with high oxidation state transition metal oxo complexes.

The transfer of oxygen atoms to centres other than carbon is also well-known. The commonest examples are concerned with the oxidation of co-ordinated thiolate to sulfenate or sulfinate (Fig. 9-38). Simple oxidising agents such as hydrogen peroxide are very effective in reactions of this type.

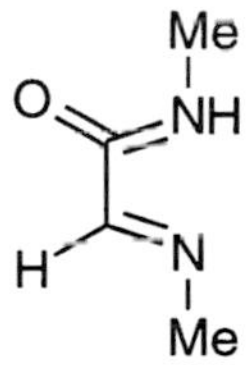

9.22

The oxidation of co-ordinated cysteine ligands may also give a variety of products. The most usually encountered reactions involve the formation of sulfenate or sulfinate as above; however, in some cases disulfide formation occurs in preference to oxygen transfer. In the example shown in Fig. 9-39, the formation of the disulfide is accompanied by decarboxylation of the amino acid!

It is very difficult to predict exactly how specific oxidising agents will react with particular co-ordinated ligands. Perhaps it is a fitting end to this section to describe the reaction of a chromium(III) mercaptoacetate complex with cerium(IV). The product of the reaction is derived by ligand oxidation, not by oxygen transfer to sulfur to give sulfoxide or sulfone, not by dimerisation to yield disulfide, but by oxidation of the methylene group to yield a thiooxalate complex (Fig. 9-40)!

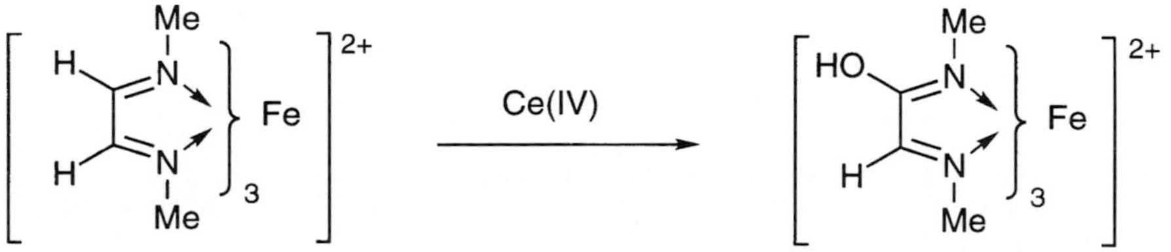

Figure 9-37. The oxidation of an iron(II) diimine complex with cerium(IV) results in an oxygen atom transfer to give a complex of the hydroxyimine tautomer of amide **9.22**.

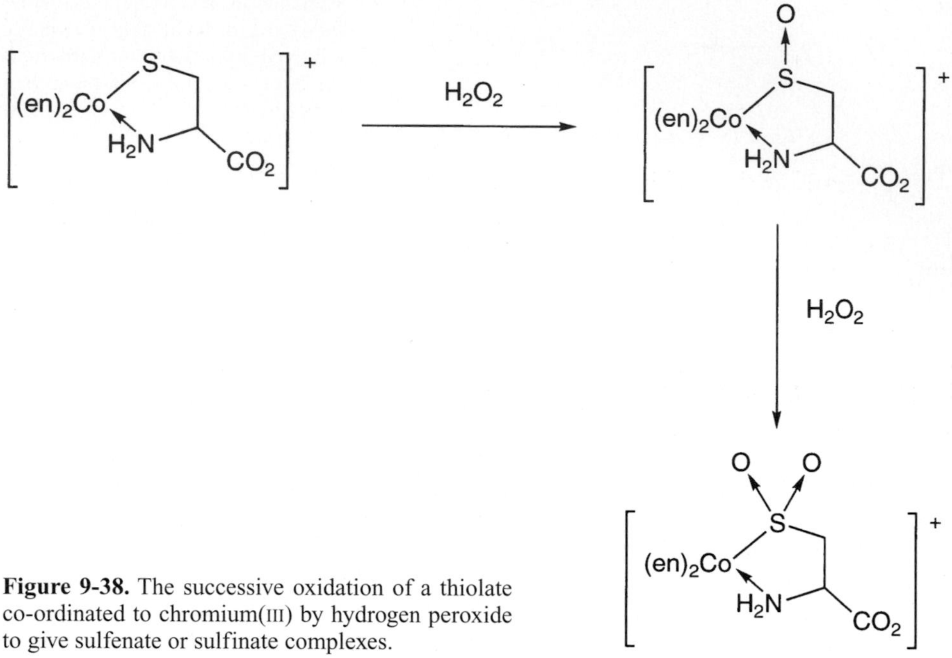

Figure 9-38. The successive oxidation of a thiolate co-ordinated to chromium(III) by hydrogen peroxide to give sulfenate or sulfinate complexes.

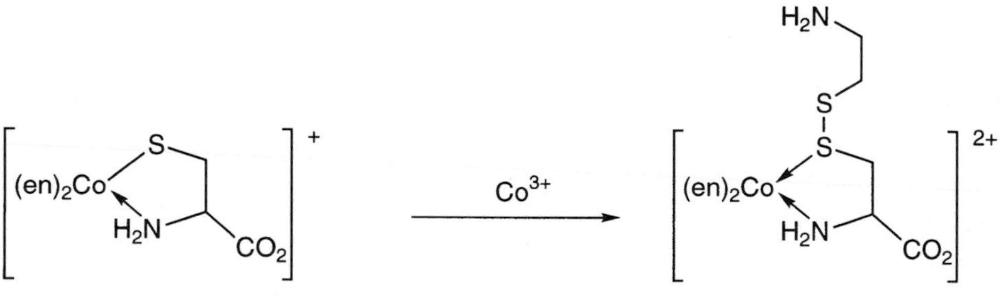

Figure 9-39. Sometimes the oxidation of thiolate complexes results in the formation of disulfide complexes rather than oxygen transfer products.

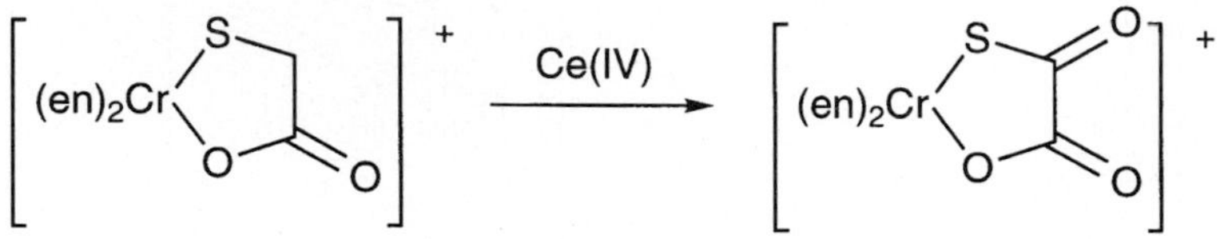

Figure 9-40. And sometimes even stranger oxidation processes occur!

9.5 Summary

In this chapter we have recognised that most of the methods available for the oxidation of organic compounds rely upon *inorganic* oxidising agents. The reasons for this lie simply in the ability of transition metal complexes to vary in oxidation state by one or more electron processes. The combination of a transition metal ion with a second oxidising agent, such as dioxygen, allows the metal to adopt a catalytic role, shuttling electrons between the ultimate oxidant (dioxygen) and the substrate.

Suggestions for further reading

1. H. Mimoun in *Comprehensive Co-ordination Chemistry, Vol. 6,* (eds. G. Wilkinson, R.D. Gillard, J.A. McCleverty), Pergamon, Oxford, **1987**.
   – A good review of the role of metal complexes in oxidation reactions.
2. *Metal Ion Activation of Dioxygen,* (ed., T.G. Spiro), Wiley, New York, **1980**.
   – A monograph on the subject of metal ion activation of dioxygen.
3. R.A. Sheldon, J.K. Kochi, *Metal-Catalysed Oxidations of Organic Compounds,* Academic, New York, **1981**.
   – A general overview.

10 Envoi

We have developed throughout this book an understanding of the many ways in which co-ordination to a metal ion may modify and control the reactivity of organic molecules. We have seen that reactivity at various sites in an organic ligand may be enhanced or diminished upon co-ordination, and that a very subtle control over the organic chemistry of the ligand may be exerted. We have also seen many examples where dramatic rate enhancements result from the co-ordination of a reaction component to a metal centre. In the majority of our examples, we were concerned with metal ions in 'normal' oxidation states of +2 or +3, and with ligands which exhibited no direct M–C interactions in their co-ordination. The choice of 'normal' oxidation state metal ions allowed the majority of the transformations discussed to occur in protic, polar solvents (often water) under aerobic conditions. Frequently, the reactions occurred at ambient or near-ambient temperature and at atmospheric pressure.

We have not considered the functions of metal ions in biological systems in any great detail, and many excellent texts exist that deal with this topic. A vast range of biological processes has been shown to be metal-ion dependent, and the functions of the metal ions in these reactions correlate very well with the basic patterns of reactivity that we have established in this book.

Biological processes usually occur, and living organisms usually exist, in aqueous aerobic conditions. High reaction temperatures or pressures are not normally accessible within living systems. The chemical processes which characterise living systems are an intricate and interlinked set of catalytic reactions. These catalytic reactions are concerned with the interconversion of organic compounds and with the transport of electrons to and from purely inorganic reactions involving dioxygen and dinitrogen (Fig. 10-1).

$$O_2 + 4H^+ + 4e^- = H_2O$$

$$N_2 + 6H^+ + 3e^- = 2NH_3$$

Figure 10-1. The "end" reactions in biology involve simple inorganic molecules such as dioxygen and dinitrogen.

The constraints under which a catalyst must work under physiological conditions preclude the *widespread* use of organometallic reagents, which are frequently air- and water-sensitive compounds. Although this is generally the case, isolated examples of metal–carbon bonded compounds are of very great biological significance. An example of such a compound is seen in coenzyme B_{12}, in which a cobalt(III) centre co-ordinated within a tetradentate macrocyclic ligand bears an axial alkyl group (Fig. 10-2).

To summarise, an efficient biological catalyst should work in water, at pH 7, at high ionic concentrations of 'simple' inorganic salts, at temperatures between 10 °C and 30 °C,

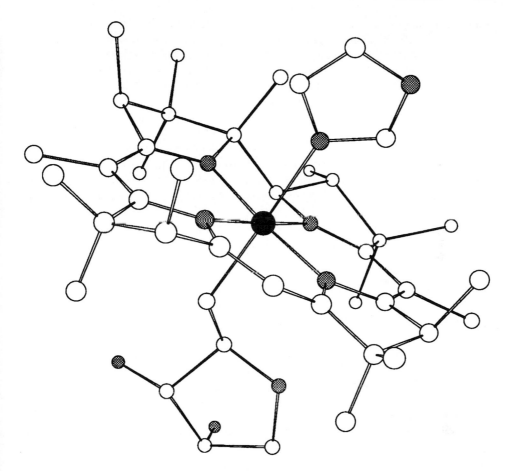

Figure 10-2. Although they are relatively rare in biology, organometallic compounds with metal–carbon bonds are important. An typical example is coenzyme B_{12}. The cobalt is co-ordinated to a tetraazamacrocyclic ligand, and has one axial nitrogen donor heterocyclic ligand and one axial alkyl group. The figure illustrates only the core of the macrocycle and the axial groups.

at atmospheric pressure and under oxidative conditions. These constraints are reminiscent of the conditions under which much of the ligand reactivity discussed in this book has occurred, and suggest that metal-directed processes might play an important role in biological catalysis.

Bio-inorganic chemistry is the study of the role of metals and other inorganic species in biological systems. This is still a relatively young subject, but it has become increasingly clear that much of the *organic* chemistry performed by living systems is aided and abetted by *inorganic* (metal) ions. In this chapter, we will very briefly consider the various applications that metal-directed reactions have found in biological systems.

The same principles of metal-directed reactivity apply to biological molecules as to smaller and more orthodox ligands. The simplest effect that a metal ion may have is in changing the conformation of the ligand, and it is convenient to consider three limiting conformational effects that a metal might display. The metal ion might maintain a large molecule in a particular conformation that is essential for its structure and function. Alternatively, the metal ion might be used to *trigger* a conformational change in a large molecule such that its function may be regulated, ameliorated or switched. Finally, the metal ion might be used to control the conformation of some other small molecule which interacts with the large biomolecule of interest.

The first conformational change is well-illustrated by those metalloproteins in which a metal ion is found maintaining the structure of a protein at a site remote from the catalytic centre. Most proteins undergo irreversible degradative changes associated with structure loss on heating above ambient temperature. This phenomenon is at odds with the observed ability of certain bacteria to live in conditions under which most proteins are denatured. Thermolysin is a hydrolytic enzyme found in a bacterium that lives in hot springs. The enzyme may be heated to 80 °C without appreciable loss of activity, whereas most proteins are denatured when maintained at 40 °C. The catalytic centre of the enzyme contains a zinc ion, but four calcium ions are also present in the structure. These calcium ions are co-ordinated to a variety of oxygen donor amino acids and to water molecules, and the remarkable thermal stability of the enzyme is associated

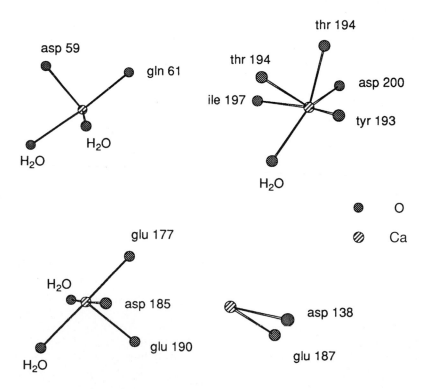

Figure 10-3. The environment of the four structural calcium ions in the enzyme thermolysin isolated from *Bacillus thermoproteolyticus*. All of the calcium ions are associated with oxygen donor ligands.

with the rigid structure imposed on the protein by co-ordination to the metal centres (Fig. 10-3).

In addition to the structure-forming function that metal ions exhibit in their interactions with proteins, they also play an important role in the chemistry of nucleic acids. Magnesium is particularly important in this role, with the structural stabilisation of nucleic acids resulting mainly from the reduction of the electrostatic repulsion between charged phosphate groups on the surface of the oligonucleotide. These examples hint at some of the problems which might be associated with the study of metal ions in biological systems. In most cases the complete three-dimensional structures of the proteins have not been determined (and even when the structure has been determined, there is no guarantee that the same structure is adopted in solution). In the absence of structural data, more 'sporting' methods have to be adopted to determine the function and environment of the metal ion. The problems of detection of a single metal ion in a molecule with a molecular mass of perhaps 800,000 are not insignificant, particularly if an ion such as Ca^{2+} or Mg^{2+}, which

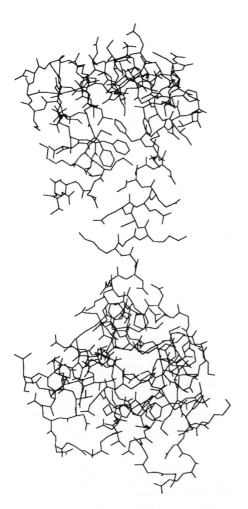

Figure 10-4. The structure of the calcium-binding protein troponin from chicken skeletal muscle. Although this is an exceptionally complicated ligand to a co-ordination chemist, the binding of calcium ions is to the hard donor sites that might be predicted. Binding of the calcium triggers a conformational change.

does not possess any convenient spectroscopic characteristics, is involved. Similarly, problems of purity of material, glassware and solvents are significant – after all, the residual concentrations of alkali metal and alkaline earth ions in 'distilled' water are usually of the same order as those needed for biological activity. Intracellular concentrations of free iron is in the micromolar region, similar to that in normal 'deionised' water. Metal ions 'within' the body of the protein, such as the calcium and zinc in thermolysin, are relatively robust and persistent, but metals that are found at the surface of the molecule, such as the magnesium in DNA, are labile and easily lost in the preparation of pure materials.

Regulatory processes such as muscle contraction are controlled by temporary conformational changes associated with metal ion co-ordination. Such regulatory processes are frequently associated with metal ions such as Mg^{2+}, Ca^{2+} and Mn^{2+} (which possesses a d^5 configuration with no crystal field preference for any particular co-ordination geometry). Muscle contraction is controlled by the binding of calcium ions to the protein troponin (Fig. 10-4), with a feed-back loop controlling the release and uptake of calcium ions from an ATP-ase, an enzyme which catalyses the hydrolysis of ATP.

When we consider the binding of small molecules at an active site, it is very difficult to separate purely conformational changes from those associated with polarisation of the ligand. In most cases the two effects are fully complementary. An example of the interplay of the two effects is seen in phosphate and polyphosphate metabolism. Almost all processes involving phosphate require magnesium or manganese ions, and these have been show to play a multiple role in co-ordinating to the phosphate or polyphosphate and controlling the configuration and also in polarising the P-O bonds to activate the phosphorus centre towards nucleophilic attack.

Polarisation of a co-ordinated substrate to a metal ion is one of the most obvious, and most widely investigated, means by which the reactivity of co-ordinated ligands may be modified. Biological systems have made widespread use of these processes, and zinc has found a particular niche for this function. Typical applications might include the polarisation of a co-ordinated ligand and its activation towards attack by a nucleophile such as water or a hydride donor such as NADH, or the polarisation of a water molecule and subsequent stabilisation of a co-ordinated hydroxide ion. The basic reactions involved in the building and transformation of proteins involve the making and breaking of amide bonds. We saw in Chapter 3 that such reactions are likely to be modified by the co-ordination of the carbonyl oxygen to a Lewis acid. We also saw that co-ordination of water to a metal ion decreased the pH and allows generation of the co-ordinated hydroxide ion at physiological pH. The addition of water to a carbonyl group could be accelerated by both of these processes. One of the fundamental problems associated with the bio-inorganic chemistry of zinc metalloproteins is the difficulty in distinguishing between mechanisms in which hydrolysis is accelerated by co-ordination of the carbonyl to the metal and those in which it is accelerated by attack of co-ordinated hydroxide. The question has not yet been fully resolved, but there is a tendency to favour metal hydroxide mechanisms for hydrolytic enzymes. Typical examples of zinc-containing hydrolytic enzymes are seen in carboxypeptidases, which hydrolyse amide groups, and carbonic anhydrase, which catalyses the hydration of carbon dioxide (Fig. 10-5).

Many of the more spectacular examples of biological processes associated with metal ions are concerned with the oxidation or reduction of organic substrates. We saw in the previous chapter the ways in which changes in the oxidation state of a metal ion

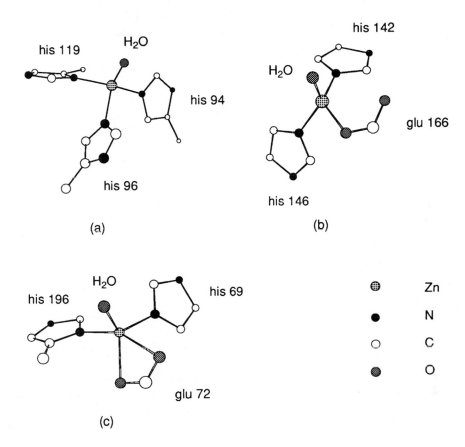

Figure 10-5. The environment of the metal in a series of zinc metalloproteins. The proteins are (a) human carbonic anhydrase II, (b) thermolysin from *Bacillus thermoproteolyticus*, and (c) bovine pancreas carboxypeptidase. Each of these enzymes is, essentially, hydrolytic.

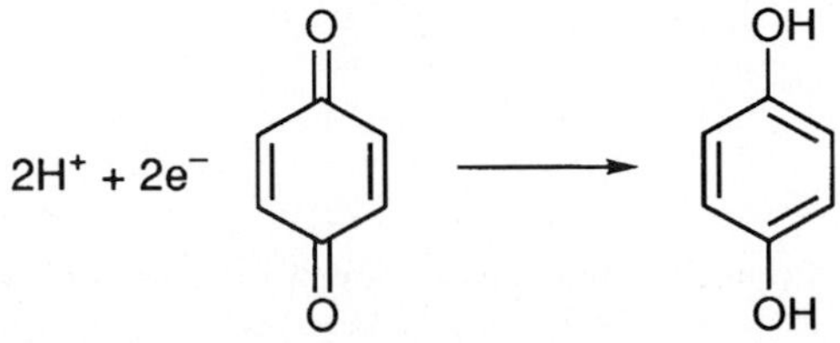

$$2H^+ + 2e^-$$

Figure 10-6. In many biological systems, quinones play an important role in the redox chains which transfer electrons to and from substrates.

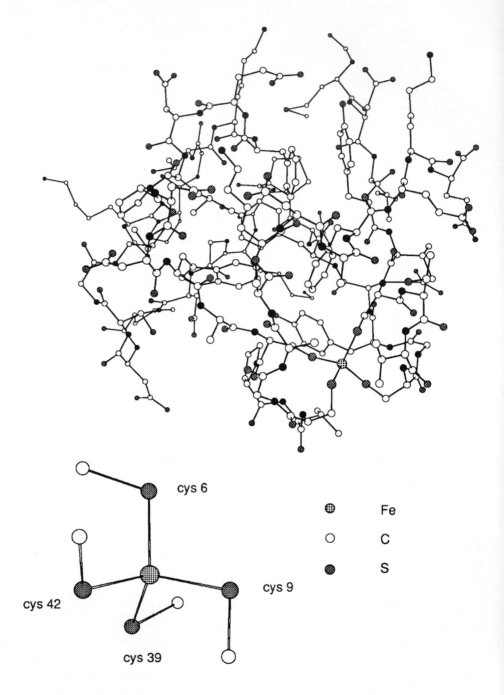

Figure 10-7. The structure about the single iron centre in rubredoxin isolated from *Desulpho-vibrio gigas*.

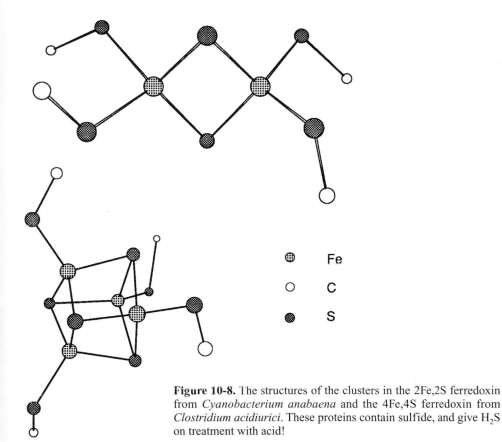

	Fe
◯	C
●	S

Figure 10-8. The structures of the clusters in the 2Fe,2S ferredoxin from *Cyanobacterium anabaena* and the 4Fe,4S ferredoxin from *Clostridium acidiurici*. These proteins contain sulfide, and give H₂S on treatment with acid!

may be coupled to redox reactions at an organic substrate. We should consider for a moment why such processes are required in biological systems. The basic source of energy for most organisms arises from the 'burning' of carbohydrates. In this process, a sugar or related molecule is converted (ultimately) to carbon dioxide and water, with the release of energy, which is stored in the cell by coupling the carbohydrate oxidation reactions to the formation of ATP. The high energy bonds within the polyphosphate provide a convenient and readily accessible form of stored chemical energy – in much the same way that human society utilises petroleum products as stored chemical energy.

The oxidation of carbohydrate involves the making and breaking of C–H bonds. This is a two electron process, and the most efficient way of achieving the oxidation will use a two electron oxidant. In many biological systems one of the key redox steps involves the two-electron conversion of a quinone to a hydroquinone (Fig. 10-6).

It is not strictly true, but we may consider that the most efficient transfer of electrons occurs when the redox potentials of the oxidant and the reductant are similar to each other. The final oxidising agent in most biological systems is dioxygen, and we find that the redox potential associated with this lies about 1 V higher than that of the quinone. We thus

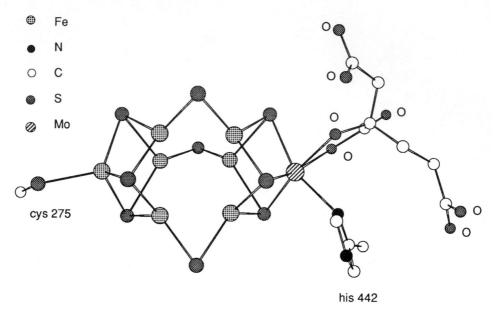

Fe

N

C

S

Mo

cys 275

his 442

Figure 10-9. The structure of the unique iron–molybdenum cluster found at the active site of nitrogenase. The molybdenum is co-ordinated to an extremely unusual homocitrate ligand. There is another iron–sulfur cluster close to this site.

need a sequence of redox active compounds with a gradation of redox potentials bridging the two ends of this potential range. Whereas it is difficult to tune the relative potentials of purely organic molecules, it is very easy to control those of metal complexes, by changing either the metal or the ligands.

Iron is the most abundant metal on earth and the commonest electron transfer agents involve iron complexes. Life is thought to have evolved in reductive conditions, in which the dominant form of iron would be as iron sulfide, not iron oxide. The simplest forms of electron transfer agents (found in plants and bacteria) involve iron with thiolate ligands. Some simple electron transfer proteins, such as rubredoxin, contain a single iron centre in an S_4 donor environment within a protein (Fig. 10-7).

Control over the redox potential may be achieved by varying the ligands and by varying the spatial arrangement of the ligands about the metal, but only relatively limited changes are possible. However, what one metal does well, two or more do better, and biology has adopted this maxim in the design of electron transfer agents. Many primitive electron transfer proteins contain clusters of iron atoms linked by sulfide ions and co-ordinated to sulfur donor amino acids of the protein (Fig. 10-8). To all intents and purposes, the protein has encapsulated a little piece of iron sulfide rock!

The proteins involved in the reduction of nitrogen to ammonia and other accessible forms contain several such clusters coupled with molybdenum centres. The structure of the central iron–molybdenum cluster at the centre of nitrogenase is shown in Fig. 10-9. Even with the detailed knowledge of this reaction site, the mode of action of nitrogenase is not understood.

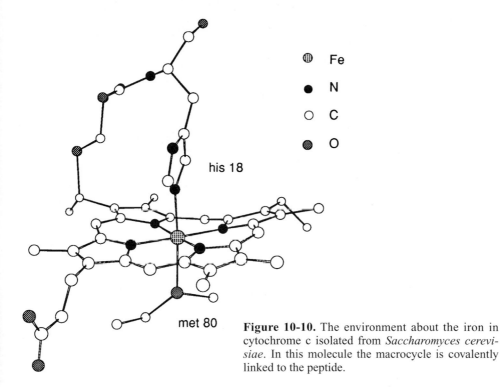

Fe
N
C
O

his 18

met 80

Figure 10-10. The environment about the iron in cytochrome c isolated from *Saccharomyces cerevisiae*. In this molecule the macrocycle is covalently linked to the peptide.

However, the changes in environment which occurred with the change from a reductive to an oxidative atmosphere rendered iron sulfide-based redox systems inconvenient, as they were very sensitive to (irreversible) oxidation. We saw in earlier chapters the facile formation of porphyrin and phthalocyanines from relatively simple precursors, and these systems were adopted for the final steps of electron transfer in oxidative conditions. The occurrence of iron centres in planar tetradentate macrocycles is ubiquitous, and metalloproteins containing such features are involved in almost every aspect of electron transfer and dioxygen metabolism. A typical example is seen in the electron transfer protein cytochrome c (Fig. 10-10).

The most common metal encountered in electron transfer systems is iron, although copper and manganese play vital functions. Merely to emphasise the complexity of the catalysts that are used in biology, the structures of the active sites of ascorbate oxidase (Fig. 10-11) and superoxide dismutase (Fig. 10-12) are presented. It is clear that we have only just begun to understand the *exact* ways in which metal ions are used to control the reactivity of small molecules in biological systems.

However, not only are the relatively abundant metal ions of biological importance. In this ante-penultimate paragraph we simply comment on some of the more surprising observations which have been made. Chromium salts are toxic in high concentrations, but small amounts are necessary for normal metabolism of insulin in humans. Zinc-containing proteases occur in some snake venoms. Molybdenum is involved in a number of biological processes which formally require large changes in oxidation state; the most

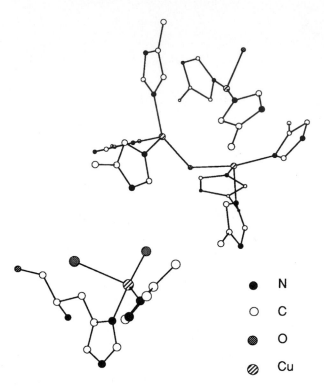

Figure 10-11. The structure of the active site in ascorbate oxidase. The enzyme contains four cop-per centres. Three of these form a triangular reaction site. Why are four copper ions needed? How does the substrate bind?

notable of these are seen in the nitrogenase enzyme complexes discussed above and the enzyme xanthine oxidase. Many plants contain bright blue copper proteins which are involved in electron transport. Vitamin B_{12} and related coenzymes are cobalt macrocyclic complexes.

Over the years to come, many more examples of metal-dependent biological processes are likely to be determined, as methods for the analysis and structural characterisation of large molecules become ever more sophisticate. One of the challenges to which the co-ordination chemist will have to rise is in understanding and explaining the function of the metal ions in these processes. It is indeed chastening to compare the efficiency of bio-logical catalysts with the best biomimetic co-ordination compounds that have been made. Zinc complexes *do* catalyse the hydration of carbon dioxide or the hydrolysis of amides, but the rates and efficiencies are many orders of magnitudes lower than those of the enzy-mic reactions. Clusters containing iron, molybdenum and sulphur have been prepared which show many of the spectroscopic properties of the active site of nitrogenase en-zymes; none of them show any ability even to bind dinitrogen, let alone to reduce it!

In conclusion, I hope that this book has served to illustrate some of the principles of reactivity in co-ordinated ligands. These principles are as applicable to chemistry in bio-logical environments as in the test tube. There is obvious scope for expansion from test

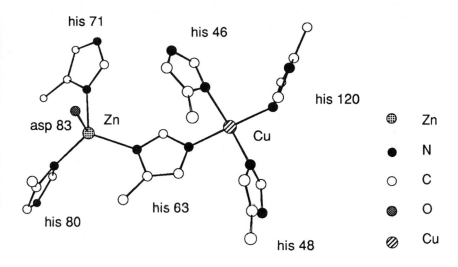

his 71

his 46

his 120

asp 83   Zn

Cu

his 63

his 80

his 48

⊕   Zn

●   N

○   C

⊛   O

⊘   Cu

Figure 10-12. The copper centre in the enzyme superoxide dismutase from spinach. Why are zinc and copper needed together? What is the importance of the bridging histidine?

tube to chemical plant. I hope that the reader has enjoyed this excursion into a slightly unfamiliar area of 'inorganic' chemistry as much as I have.

Suggestions for further reading

1. M.N. Hughes, *The Inorganic Chemistry of Biological Processes*, Wiley, London, **1981**.
   – An excellent book in its time, but a little dated now.
2. I. Bertini, H.B. Gray, S.J. Lippard, J.S. Valentine, *Bioinorganic Chemistry*, University Science Books, Mill Valley, **1994**.
   – Excellent but encyclopaedic.
3. S.J. Lippard, J.M. Berg, *Principles of Bioinorganic Chemistry*, University Science Books, Mill Valley, **1994**.
   – An excellent modern text on the topic of bioinorganic chemistry. General overview.

Index

Get the Best of Organic Synthesis!

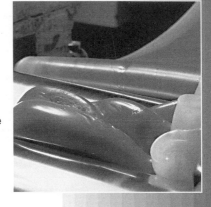
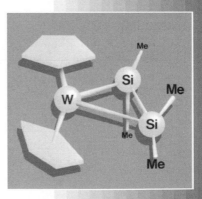